BIOPROCESSING OF AGRI-FOOD RESIDUES FOR PRODUCTION OF BIOPRODUCTS

BIOPROCESSING OF AGRI-FOOD RESIDUES FOR PRODUCTION OF BIOPRODUCTS

Edited by

Adriana Carolina Flores-Gallegos, PhD

Rosa María Rodriguez-Jasso, PhD

Cristóbal Noé Aguilar, PhD

First edition published 2022

Apple Academic Press Inc.
1265 Goldenrod Circle, NE,
Palm Bay, FL 32905 USA
4164 Lakeshore Road, Burlington,
ON, L7L 1A4 Canada

CRC Press
6000 Broken Sound Parkway NW,
Suite 300, Boca Raton, FL 33487-2742 USA
2 Park Square, Milton Park,
Abingdon, Oxon, OX14 4RN UK

Library and Archives Canada Cataloguing in Publication

Title: Bioprocessing of agri-food residues for production of bioproducts / edited by Adriana Carolina Flores-Gallegos, PhD, Rosa María Rodriguez-Jasso, PhD, Cristóbal Noé Aguilar, PhD.

Names: Flores-Gallegos, Adriana Carolina, editor. | Rodriguez-Jasso, Rosa María, editor. | Aguilar, Cristóbal Noé, editor.

Description: Includes bibliographical references and index.

Identifiers: Canadiana (print) 20200350900 | Canadiana (ebook) 2020035101X | ISBN 9781771889162 (hardcover) | ISBN 9781003048305 (ebook)

Subjects: LCSH: Agricultural wastes. | LCSH: Agriculture—Waste minimization. | LCSH: Food industry and trade—Waste minimization. | LCSH: Biological products. | LCSH: Sustainable agriculture.

Classification: LCC TD930 .B56 2021 | DDC 628/.74—dc23

Library of Congress Cataloging-in-Publication Data

Names: Flores-Gallegos, Adriana Carolina, editor. | Rodriguez-Jasso, Rosa María, editor. | Aguilar, Cristóbal Noé, editor.

Title: Bioprocessing of agri-food residues for production of bioproducts / edited by Adriana Carolina Flores-Gallegos, PhD, Rosa María Rodriguez-Jasso, PhD, Cristóbal Noé Aguilar, PhD.

Description: First edition. | Palm Bay, FL : Apple Academic Press, [2021] | Includes bibliographical references and index. | Summary: "This new volume presents original research and scientific advances in the field of the food bioprocessing, bioproducts, valorization of agricultural and food wastes, microbiology, and biotechnology. It explores the most important advances in the valorization of agri-food residues for the production of bioproducts and in the development of several bioprocessing strategies. The authors place a special emphasis on the challenges that the industry faces in the era of sustainable development and aim to facilitate the reduction of food loss and waste. This book demonstrates the potential and actual development and advances in the design and development of strategies and tools for the bioprocessing of agri-food residues for the production of bioproducts. Bioprocessing of Agri-Food Residues for Production of Bioproducts covers aspects related to biotransformation of agri-food residues such as mango seed, citrus waste, pomegranate husks, nut shells, melon peels, leaves and grains, cheese whey, among others. It discusses the high added value in bio-product recovery, such as antioxidants and prebiotics, and it evaluates their application in human health and animal nutrition. The book includes relevant information about food bioprocesses, fermentation, food microbiology, functional foods, nutraceuticals, extraction of natural products, nano- and micro technology, innovative processes/bioprocesses for utilization of by-products, alternative processes requiring less energy or water, among other topics. Key features: Describes the technological aspects involved in bioprocessing and in biotechnological management of agri-food wastes Emphasizes sustainable development and the reduction of food loss and waste Because of its qualified and innovative content, this volume is an important reference and useful tool for food technology researchers and scientists, for undergraduate and graduate students, and for professionals in the food industry"-- Provided by publisher.

Identifiers: LCCN 2020044392 (print) | LCCN 2020044393 (ebook) | ISBN 9781771889162 (hardcover) | ISBN 9781003048305 (ebook)

Subjects: LCSH: Agricultural wastes. | Food industry and trade--Waste minimization. | Agricultural processing industries--Waste minization. | Biological products. | Sustainable agriculture.

Classification: LCC TD930 .B54 2021 (print) | LCC TD930 (ebook) | DDC 660.6--dc23

LC record available at https://lccn.loc.gov/2020044392

LC ebook record available at https://lccn.loc.gov/2020044393

ISBN: 978-1-77188-916-2 (hbk)
ISBN: 978-1-77463-803-3 (pbk)
ISBN: 978-1-00304-830-5 (ebk)

About the Editors

Adriana Carolina Flores-Gallegos, PhD
Professor, School of Chemistry, Autonomous University of Coahuila, Mexico

Adriana Carolina Flores-Gallegos, PhD, is a full-time Professor at the School of Chemistry, Autonomous University of Coahuila, Mexico. She is a Biologist Chemist, specializing in microbiology; she earned her MSc and PhD in Food Science and Technology from the Autonomous University of Coahuila. She formerly worked in molecular phylogeny at Barcelona University, Spain, and in the MICALIS Institute in INRA, France. Dr. Flores-Gallegos has published 28 papers in indexed journals, several articles in Mexican journals, 15 book chapters, and one Mexican book. She has presented her contributions at scientific meetings and is a member of the Mexican Society of Biotechnology and Bioengineering.

Rosa María Rodriguez-Jasso, PhD
Professor, School of Chemistry, Autonomous University of Coahuila, Mexico

Rosa María Rodriguez-Jasso, PhD, is a full-time Professor, Co-Founder and Leader of the Biorefinery Group in the Food Research Department at the School of Chemistry, Autonomous University of Coahuila, Mexico. She obtained her PhD in Chemical and Biological Engineering from the University of Minho, Portugal. She earned her MSc degree in Food Science and Technology and a BSc in Chemical Engineering from the Autonomous University of Coahuila. Her research interests are based on the valorization of aquatic biomass and marine residues, specifically seaweed and microalgae, under the concept of a biorefinery, applying a range of key green technologies for the production and characterization of high value-added compounds and functional food ingredients. She has published 32 scientific papers in indexed journals and 22 book chapters. She has carried out several international research stays and professional visits, including at the University of North of Texas (USA), University of Georgia (USA), University of Vigo (Spain), Autonomous University of Madrid (Spain), Aix Marseille Université (France), Jacobs University (Germany), and Tokyo Institute of Technologies (Japan). She has been co-responsible for six scientific projects with national and industry funding, and she is a scientific member of the project Sargassum Sustainable

Management (CEMIE-Océano, México) as an advisor in seaweed biomass for the recovery and protection of key marine ecosystems. Currently, she is the Vice-president of the Mexican Association of Food Science (AMECA) for the period 2019–2021 and President-elect for the period 2021–2023.

Cristóbal Noé Aguilar, PhD

Professor and Director of Research and Postgraduate Programs, Autonomous University of Coahuila, Mexico

Cristóbal Noé Aguilar, PhD, is a Head of Research and Postgraduate Programs at Autonomous University of Coahuila, Mexico. His scientific work focuses on the use and development of microbial and enzymatic technologies for the design of bioprocesses for food, fine chemistry, pharmacy, biotechnology, and the environment. He is a Research Professor in the Bioprocesses and Bio-Products Research Group (DIA-FCQ/UAdeC). He was formerly the Director of the Faculty of Chemical Sciences at Autonomous University of Coahuila (2015–2018); President of the Mexican Society of Biotechnology and Bioengineering (2014–2016); and President of the Mexican Association of Food Science and Biotechnology (2008–2009). Dr. Aguilar is a member of many professional associations, including the Mexican Academy of Sciences and the National System of Investigators. He has a received many awards, including a National Food Science and Technology Award; CONACYT-Coca Cola Mexico; AgroBio National Award (2005); National Award of the Mexican Society of Biotechnology and Bioengineering (2008); National Research Award of the Mexican Academy of Sciences (2010); Coahuila State Science Technology and Innovation Award (2019); 2019 Outstanding Scientist Award from the International Bioprocess Association, presented at an event held in Miri, Sarawak, Malaysia.

Dr. Aguilar has published more than 160 papers in indexed journals and more than 40 articles in Mexican journals, and he has made more than 250 contributions in scientific meetings. He has also published many book chapters, several Mexican books, four editions of international books, and more. He has developed more than 21 research projects, including six international exchange projects.

Dr. Aguilar earned his pharmacobiology degree and master's degree in Food Science and Technology from the Autonomous University of Chihuahua, Mexico, and a PhD in Biotechnology from the Metropolitan Autonomous University, Mexico. He did his postdoctoral stay in molecular microbiology at IRD-Marseille, France).

Contents

Contributors .... *ix*
Abbreviations .... *xiii*
Preface .... *xv*

1. **Dual-Purpose Bioprocesses: Biotransformation of Agri-Food Residues and High Added-Value Bioproducts Recovery 1**
Miguel A. Medina-Morales, Ricardo Gómez-García, Marisol Cruz-Requena, and Cristóbal N. Aguilar

2. **Mango Seed Byproduct: A Sustainable Source of Bioactive Phytochemicals and Important Functional Properties 33**
Cristian Torres-León, Maria T. dos Santos Correia, Maria G. Carneiro-da-Cunha, Liliana Serna-Cock, Janeth Ventura-Sobrevilla, Juan A. Ascacio-Valdés, and Cristóbal N. Aguilar

3. **Citrus Waste: An Important Source of Bioactive Compounds 61**
Nathiely Ramírez-Guzmán, Mónica L. Chávez-González, Erick Peña-Lucio, Hugo A. Luna-García, Juan A. Ascacio Valdés, Gloria Martínez-Medina, Maria das Graças Carneiro-da -Cunha, Teresinha Gonçalves da Silva, José L. Martinez-Hernandez, and Cristóbal N. Aguilar

4. **Use of Agro-Industrial Residues to Obtain Polyphenols with Prebiotic Effect 81**
Ana Yoselyn Castro-Torres, Raúl Rodríguez-Herrera, Aidé Sáenz-Galindo, Juan Alberto Ascacio-Valdés, Jesús Antonio Morlett-Chávez, and Adriana Carolina Flores-Gallegos

5. **Valorization of Pomegranate Residues 107**
Paloma Almanza-Tovanche, Raúl Rodríguez-Herrera, Aidé Sáenz-Galindo, Juan Alberto Ascacio-Valdés, Cristóbal N. Aguilar, and Adriana Carolina Flores-Gallegos

6. **Lactic Acid Fermentation As a Tool for Obtaining Bioactive Compounds from Cruciferous 125**
Daniela Iga Buitrón, Margarita del Rosario Salazar Sánchez, Edgar Torres Maravilla, Luis Bermúdez Humaran, Juan Alberto Ascacio-Valdés, José Fernando Solanilla Duque, and Adriana Carolina Flores-Gallegos

7. **Potential of Agro-Food Residues to Produce Enzymes for Animal Nutrition 141**

Erika Nava-Reyna, Anna Ilyina, Georgina Michelena-Alvarez, and José Luis Martínez-Hernández

8. **Biotechnological Valorization of Whey: A By-Product from the Dairy Industry 159**

Hilda Karina Sáenz-Hidalgo, Alexandro Guevara-Aguilar, José Juan Buenrostro-Figueroa, Ramiro Baeza-Jiménez, Adriana C. Flores-Gallegos, and Mónica Alvarado-González

9. **Contributions of Biosurfactants in the Environment: A Green and Clean Approach 201**

Geeta Rawat and Vivek Kumar

10. **Fungal Solid-State Bioprocessing of Grapefruit Waste 225**

Ramón Larios-Cruz, Rosa M. Rodríguez-Jasso, Jose Juan Buenrostro-Figueroa, Arely Prado-Barragán, Héctor Ruiz, and Cristóbal N. Aguilar

11. **Valorization of Ataulfo Mango Seed Byproduct Based on Its Nutritional and Functional Properties 233**

Cristian Torres-León, Juan A. Ascacio-Valdés, Mónica L. Chávez-González, Liliana Serna-Cock, Nathiely Ramirez-Guzman, Alcides Cintra, Claudia López-Badillo, Romeo Rojas, Ruth Belmares-Cerda, and Cristóbal N. Aguilar

12. **Kinetic Parameters of the Carotenoids Production by *Rhodotorula glutinis* under Different Concentration of Carbon Source 253**

Ayerim Hernández-Almanza, Víctor Navarro-Macías, Oscar Aguilar, Juan C. Contreras-Esquivel, Julio C. Montañez, Guillermo Martínez Avila, and Cristóbal N. Aguilar

Index* *263

Contributors

Cristóbal N. Aguilar
Food Research Department, School of Chemistry, Autonomous University of Coahuila, Saltillo, 25280, Coahuila, Mexico, E-mail: cristobal.aguilar@uadec.edu.mx

Oscar Aguilar
Centre of Biotechnology-FEMSA. Tecnológico de Monterrey, Campus Monterrey, Nuevo León, Mexico

Paloma Almanza-Tovanche
Food Research Department, School of Chemistry, Universidad Autónoma de Coahuila, Saltillo, Coahuila, México

Mónica Alvarado-González
Centro de Investigación en Alimentación y Desarrollo A.C. Unidad Delicias, Chihuahua, México

Juan Alberto Ascacio-Valdés
Bioprocesses and Bioproducts Research Group. Food Research Department, School of Chemistry, Universidad Autónoma de Coahuila, Saltillo, Coahuila, México

Ramiro Baeza-Jiménez
Centro de Investigación en Alimentación y Desarrollo A.C. Unidad Delicias, Chihuahua, México

Ruth Belmares-Cerda
Bioprocesses and Bioproducts Research Group (BBG-DIA), Food Research Department, School of Chemistry, Universidad Autónoma de Coahuila, Saltillo, México

Luis Bermúdez Humaran
Micalis Institute, UMR INRA-AgroParisTech, Jouy en Josas, France

José Juan Buenrostro-Figueroa
Centro de Investigación en Alimentación y Desarrollo A.C. Unidad Delicias, Chihuahua, México

Ana Yoselyn Castro-Torres
Food Research Department, School of Chemistry, Universidad Autónoma de Coahuila, Saltillo, Coahuila, México

Mónica L. Chávez-González
Food Research Department, School of Chemistry, Universidad Autónoma de Coahuila Saltillo, Coahuila, México

Alcides Cintra
Departamento de Bioquímica, Centro de Ciencias Biológicas, Universidad Federal de Pernambuco, Recife, Brazil

Juan Carlos Contreras-Esquivel
Autonomous University of Coahuila, School of Chemistry, Food Research Department, Saltillo, Coahuila, Mexico

Marisol Cruz-Requena
Food Research Department, School of Chemistry, Autonomous University of Coahuila, Saltillo, 25280, Coahuila, Mexico

Maria das Graças Carneiro da Cunha
Departamento de Bioquímica, Centro de Ciencias Biológicas,
Universidade Federal de Pernambuco. Recife, Pernambuco, Brazil

Maria T. dos Santos Correia
Departamento de Bioquímica, Centro de Ciencias Biológicas,
Universidade Federal de Pernambuco. Recife, Pernambuco, Brazil

Adriana Carolina Flores-Gallegos
Research Group of Bioprocesses and Bioproducts. Department of Food Research, School of Chemistry, Universidad Autónoma de Coahuila, Saltillo, Coahuila, México

Ricardo Gómez-García
Bioprocesses and Bioproducts Research Group, Food Research Department, School of Chemistry. Autonomous University of Coahuila, Saltillo, Coahuila, Mexico

Teresinha Gonçalves da Silva
Universidad Federal de Pernambuco Recife, Brazil

Alexandro Guevara-Aguilar
Centro de Investigación en Alimentación y Desarrollo A.C. Unidad Delicias, Chihuahua, México

Ayerim Hernández-Almanza
Bioprocesses and Bioproducts Research Group (BBG-DIA), Food Research Department, School of Chemistry, Universidad Autónoma de Coahuila. Saltillo, Coahuila, México

Daniela Iga Buitrón
Food Research Department, School of Chemistry, Universidad Autónoma de Coahuila, Saltillo, Coahuila, México

Anna Ilyina
Academic staff of Nanobiosciences, School of Chemistry, Universidad Autónoma de Coahuila, Saltillo, Coahuila, Mexico

Vivek Kumar
Himalayan School of Biosciences, Swami Rama Himalayan University, Jolly Grant, Dehradun, India

Ramón Larios-Cruz
Research Group of Bioprocesses and Bioproducts, Food Research Department, School of Chemistry, Universidad Autónoma de Coahuila, Saltillo, Coahuila, México

Claudia López-Badillo
School of Chemistry, Universidad Autónoma de Coahuila, Saltillo, México

Hugo A. Luna-García
Food Research Department, School of Chemistry, Universidad Autónoma de Coahuila Saltillo, Coahuila, México

Guillermo Martínez Avila
Laboratory of Chemistry and Biochemistry, School of Agronomy,
Universidad Autónoma de Nuevo León, Monterrey, Nuevo León, Mexico

José Luis Martínez-Hernández
Food Research Department, School of Chemistry, Universidad Autónoma de Coahuila Saltillo, Coahuila, México

Gloria Martínez-Medina
Food Research Department, School of Chemistry, Universidad Autónoma de Coahuila Saltillo, Coahuila, México

Miguel A. Medina-Morales
Bioprocesses and Bioproducts Research Group. Food Research Department, School of Chemistry, Autonomous University of Coahuila, Saltillo, Coahuila, Mexico

Georgina Michelena-Alvarez
Instituto Cubano de Investigación de los Derivados de la Caña de Azúcar, Vía Blanca 804 y Central Road, San Miguel del Padrón, La Habana, Cuba

Julio C. Montañez
Bioprocesses and Bioproducts Research Group (BBG-DIA), Food Research Department, School of Chemistry, Universidad Autónoma de Coahuila. Saltillo, Coahuila, México

Jesús Antonio Morlett-Chávez
Food Research Department, School of Chemistry, Universidad Autónoma de Coahuila, Saltillo, Coahuila, México

Erika Nava-Reyna
National Centers for Disciplinary Investigation in Water, Plant, Soil and Atmosphere Relationship (CENID RASPA), Instituto Nacional de Investigaciones Forestales, Agrícolas y Pecuarias, Gomez Palacio, Durango, Mexico

Víctor Navarro-Macías
Bioprocesses and Bioproducts Research Group (BBG-DIA), Food Research Department, School of Chemistry, Universidad Autónoma de Coahuila. Saltillo, Coahuila, México

Erick Peña-Lucio
Food Research Department, School of Chemistry, Universidad Autónoma de Coahuila Saltillo, Coahuila, México

Arely Prado-Barragán
Departamento de Biotecnología, Universidad Autónoma Metropolitana Unidad Iztapalapa, Delegación Iztapalapa, Ciudad de México, México

Nathiely Ramírez-Guzmán
Food Research Department, School of Chemistry, Universidad Autónoma de Coahuila Saltillo, Coahuila, México

Geeta Rawat
Himalayan School of Biosciences, Swami Rama Himalayan University, Jolly Grant, Dehradun, India

Raúl Rodríguez-Herrera
Food Research Department, School of Chemistry, Universidad Autónoma de Coahuila, Saltillo, Coahuila, México

Rosa M. Rodríguez-Jasso
Biorefinery Group, Food Research Department, Faculty of Chemistry Sciences, Autonomous University of Coahuila, Saltillo, Coahuila, México, and Cluster of Bioalcoholes, Mexican Centre for Innovation in Bioenergy (Cemie-Bio), México

Romeo Rojas
Research Center and Development for Food Industries – CIDIA. School of Agronomy. Universidad Autónoma de Nuevo León, General Escobedo, NL, México

Héctor Ruiz
Research Group of Bioprocesses and Bioproducts, Food Research Department, School of Chemistry, Universidad Autónoma de Coahuila, Saltillo, Coahuila, México

Aidé Sáenz-Galindo
Food Research Department, School of Chemistry, Universidad Autónoma de Coahuila, Saltillo, Coahuila, México

Hilda Karina Sáenz-Hidalgo
Centro de Investigación en Alimentación y Desarrollo A.C. Unidad Delicias, Chihuahua, México

Margarita del Rosario Salazar Sánchez
Universidad del Cauca, Popayán, Cauca, Colombia

Liliana Serna-Cock
School of Engineering and Administration. Universidad Nacional de Colombia, Palmira, Valle del Cauca, Colombia

José Fernando Solanilla Duque
Universidad del Cauca, Popayán, Cauca, Colombia

Edgar Torres Maravilla
Micalis Institute, UMR INRA-AgroParisTech, Jouy en Josas, France

Cristian Torres-León
Bioprocesses and Bioproducts Research Group. Food Research Department, School of Chemistry, Universidad Autónoma de Coahuila, Saltillo, Coahuila, México

Janeth Ventura
Bioprocesses and Bioproducts Research Group. Food Research Department, School of Chemistry, Universidad Autónoma de Coahuila, Saltillo, Coahuila, México

Abbreviations

ATPS	aqueous two-phase systems
BOD	biochemical oxygen demand
BTH	butyl-hydroxy-toluene
CH4	methane
CMC	critical micelle concentration
COD	chemical oxygen demand
CONACYT	National Council of Science and Technology
CP	crude protein
CW	concentrated whey
DF	diafiltration
DIP	Dirección de Investigación y Posgrado
EAE	enzyme-assisted extraction
EBM	extractive bioconversion method
ED	electrodialysis
ETSS	enzyme thermal stability system
FCR	feed conversion ratio
FID	flame ionization detector
FRAP	ferric ion reducing antioxidant power
GA	gallic acid
GC	gas chromatography
GC-MS	gas chromatography-mass spectrometry
GLS	glucosinolates
GOS	galactooligosaccharides
GRAS	generally regards as safe
HAT	hydrogen atom transfer
HPLC	high-performance liquid chromatography
HPLC/MS	high-performance liquid chromatography/electrospray ionization mass spectrometry
ITCs	isothiocyanates
LDL	low-density lipoprotein
LF	liquid fermentation
LOI	lipid oxidation inhibition
MAE	microwave-assisted extraction
MAFRA	Ministry of Agriculture, Food, and Rural Affairs

ME	metabolizable energy
MEOR	microbial enhanced oil recovery
MF	microfiltration
MFC	microbial fuel cell
MIC	minimum inhibitory concentration
MMT	million metric tons
MS	mass spectrometry
MSET	mango seed is composed of an endocarp and testa
NDGA	nordihydroguayretic acid
NF	nanofiltration
NSP	non-starch polysaccharides
OPW	orange peel waste
PAHs	polycyclic aromatic hydrocarbons
PEF	pulsed electric field
PEG	polyethylene glycol
PGG	penta-galloyl glucose
PHA	polyhydroxyalkanoates
PLE	pressurized liquid extraction
PMFs	polymethoxylatedflavones
PSC	plant secondary compounds
RO	reverse osmosis
SCP	single-cell protein
SET	single electron transfer
SFE	supercritical fluid extraction
SmF	submerged fermentation
SSF	solid-state fermentation
Tc	total carbohydrates
TEAC	trolox equivalent antioxidant capacity
TPH	total petroleum hydrocarbons
UAE	ultrasound-assisted extraction
UF	ultrafiltration
WHO	World Health Organization
WMS	metagenome sequencing
WPC	whey protein concentrate
WPI	whey protein isolate

Preface

Bioprocessing of Agri-Food Residues for Production of Bioproducts describes original research contributions and scientific advances in the field of bioprocessing, bioproducts, valorization of agricultural and food wastes, microbiology, and biotechnology. Chapters cover broad research areas that offer original and novel highlights in research studies of bioprocessing of agricultural and food wastes and other related technologies and enhance the exchange of scientific literature.

The book was born after several working meetings of the editors, where the needs of students, teachers, and researchers were analyzed in detail. An important characteristic of the editors is that they are all professors and scientists of the Bioprocesses and Bioproducts Research Group and Biorefinery group of the Autonomous University of Coahuila, Mexico, and they have experience in teaching and research, giving them the requirements and need both for updating and for the depth. Additionally, all editors and some contributors are members of the Mexican Network for the Reduction of Food Loss and Waste (PDA Mexico), coordinated by Dr. Juliana Morales Castro (Technological Institute of Durango). The training is part of the context and the reality faced by the courses on bioprocessing and bioproducts, food biotechnology and microbiology, biorefinery, valorization of fruit wastes, agricultural residues reduction, bioactives and bioactivities, etc.

The book explores the most important advances in the valorization of agri-food residues for the production of bioproducts and the development of several bioprocessing strategies, with special emphasis on the challenges that the industry faces in the era of sustainable development. The food and agricultural industries need a frame of requirements for sustainable development goals that are related to the reduction of food loss and waste and the reuse of food residues. Detailed and up-to-date information is a useful tool in teaching and research related to sustainable agriculture and food production. This book demonstrates the potential and actual developments across the innovative advances in the design and development of strategies and tools for the bioprocessing of agri-food residues for the production of bioproducts.

The book covers aspects related to biotransformation of agri-food residues like mango seeds, citrus waste, pomegranate husks, nutshells, melon peels, leaves and grains, cheese whey, among others, for high-added-value bio-product recoveries, such as antioxidants and prebiotics, and their applications in human health and animal nutrition. The book includes relevant information about food bioprocesses, fermentation, food microbiology, functional foods, nutraceuticals, and extraction of natural products, nano- and micro technology, innovative processes/bioprocesses for utilization of by-products, alternative processes requiring less energy or water, among other topics. Due to its qualified and innovative content, the book will be an important reference for food technology research, for undergraduate and graduate students, and also for professionals in the food industry.

We are sure that the book will be of great interest and support for readers, because the book's authors and editors have been particularly focused to meet the demands and needs of students, teachers, and researchers of science and technology for the bioprocessing of agri-food residues and the production of bioproducts.

We deeply thank all the contributors who responded enthusiastically to the call issued by contributing original and novel documents. All the editors thank the contributing authors for their time and for sharing their knowledge for the benefit of students and as well as professionals. We also thank Dr. A. K. Haghi of Apple Academic Press, Inc. for his constant support and guidance from the beginning to the completion of the book. Finally, we also appreciate the facilities granted by the institutions for which we work, the Autonomous University of Coahuila.

—Adriana Carolina Flores-Gallegos, PhD
Rosa María Rodriguez-Jasso, PhD
Cristóbal Noé Aguilar, PhD

CHAPTER 1

Dual-Purpose Bioprocesses: Biotransformation of Agri-Food Residues and High Added-Value Bioproducts Recovery

MIGUEL A. MEDINA-MORALES,[1] RICARDO GÓMEZ-GARCÍA,[3] MARISOL CRUZ-REQUENA,[2] and CRISTÓBAL N. AGUILAR[2]

[1]*Department of Biotechnology, School of Chemistry, Autonomous University of Coahuila, Saltillo, 25280, Coahuila, Mexico*

[2]*Food Research Department, School of Chemistry, Autonomous University of Coahuila, Saltillo, 25280, Coahuila, Mexico, E-mail: cristobal.aguilar@uadec.edu.mx*

[3]*Universidade Católica Portuguesa, CBQF – Centro de Biotecnologia e Química Fina – Laboratório Associado, Escola Superior de Biotecnologia, Rua Diogo Botelho 1327, 4169-005 Porto, Portugal*

ABSTRACT

The exploitation of residues is a very important opportunity to add value to industrial by-products. By microbiological ways, several products from plant material via biodegradation can be obtained. In every microbial process, there is biomass production, degradation products, and high added-value compounds from the metabolism and all of these can be recovered according to the nature and purpose of the bioprocess. As previously stated, many useful outcomes from these procedures can be generated and strategies must be developed to take advantage of the microbes/substrate interaction to a great extent. Examples of current research are available to start applications and gain profits from large amounts of residues from the agri-food industry. To mention such examples are the utilization of corncobs by fungal solid-state fermentation as pretreatment for saccharification and fungal enzymes production and recovery. Husks or peels from fruits, such as orange residues can be used for fungal spore production and pecan nut powder can be fermented to release antioxidant phenolics where the fermented material can be used as fertilizer. This chapter deals with an overview of how microbial processes take place while degrading plant tissues where several bioactive compounds can be released and industrially useful enzymes are produced as well.

1.1 INTRODUCTION

It is a well-known fact that the food industry generates copious amounts of residues which often are considered as pollutants of the environment or a wasted raw material for several processes that could yield further benefits from a business point of view (Ravindran and Jaiswal, 2016). Biotechnology is playing an ever-growing role in the industrial area where bioprocesses are used to obtain enzymes, high added-value compounds, and microbial biomass for ulterior purposes (Nile et al., 2017). Food industries generate copious amounts of residues (Nazzaro et al., 2018) and management of those wastes is required where biotechnology comes handy (Nazzaro et al., 2018). The idea of having a multipurpose production line is something not properly new (Mallek-Fakhfakh and Belghith, 2015). Many microbial processes have more than one product, but the producers only focus on a single outcome most of the time (Singh et al., 2017). As it is known, microbes can produce or release a wide array of compounds such as enzymes, polymers, bioactive chemicals, among others (Mallek-Fakhfakh Belghith, 2015; Méndez-Hernández et al., 2018). In this regard, while using agro-food residues, it could be considered a renewable source of raw material for bioproducts, which can be very interesting in developing a whole-use of material in an industrial process (Pleissner et al., 2016). In light of getting the most of a microbe and a substrate, researchers are approaching the concept of consolidated bioprocessing (Kawaguchi et al., 2016). This term is mostly used in lignocellulosic exploitation for biorefinery where there is the need of having a simpler mode of operation where enzyme production, enzyme degradation, and fermentation are in a single stage. This kind of process or at least the concept can be applied to any type of bioprocess where metabolites, enzymes, and microbial biomass are produced in a single reactor for later separation. This concept, being very interesting and attractive for researchers, it could present certain advantages and limitations. In the cases of solid-state fermentation, while it is complicated to separate biomass from the substrate, enzyme and metabolite production can take place and also bioactive compound release/production as well. From an experimental point of view, many products can be obtained such as disaccharidases, glycosidases, lipases, proteases, among many others (Ravindran and Jaiswal, 2016; Melikoglu, Lin and Webb 2013 and Singh et al., 2016). High added-value compounds can also be obtained from bioprocessed agro-food residues like proanthocyanidins, flavonoids, flavones, gallic and ellagic acid (Aguilera-Carbo et al., 2008; Dai et al., 2017). Polysaccharides, oligo, and monosaccharides also could be obtained (Yang et al., 2018). At the same time, it could be possible to develop consolidated processes that can also be produced by non-enzyme microbial products (Diaz et al., 2018). The reason for the production or release of enzymes and added-value compounds is strongly correlated to the composition of the agro-food residues as it is well known that lignocellulosic fibers and the natural conformation of the value compounds in the plant cell can work as inducers of enzyme production which releases the compounds and its excretion and enzyme recovery as well (Sepúlveda et al., 2018; Lopez-Trujillo et al., 2017). While all these mechanisms of degradation are taking place, the microbial biomass is produced which is useful for spore, single-cell protein, or microbial colonies production (De La Cruz-Quiroz et al., 2017).

The main aspect of these types of processes is that to generate more than one product of interest is that either biomass, bioactive, and/or enzymes must be recovered, or at least a development of a strategy of recovery is of paramount importance. The particular goal of this chapter is to provide information about the bioproducts that can be obtained from this type of procedure. For the development of the bioprocesses of interest, fungi, yeasts, and bacteria can be used for the production of biomass, enzyme, metabolites, and high added-value compounds released from the biodegradation of plant tissue.

1.2 AGRO-FOOD CELL STRUCTURES AND BIODEGRADATION

All of the plant tissues are formed by a wide array of compounds which are essential for the organism but also have a very high added value to us (Deswal et al., 2014). In the context of biodegradation, it is practically obvious to say that if a microbial process is going to develop in plant tissue, biomass will be produced, so that aspect will be considered as given. The amount of plant tissue, which in this case is agro-food residues, and the moisture content, could define the type of microbial degradation process (Soccol, 2017). Considering that plant cell components can induce enzyme production, enzymes can be recovered and degradation products are released and can also be obtained (Lopez-Trujillo et al., 2017; Medina-Morales et al., 2017).

The main component of any plant waste is cellulose, followed by hemicellulose (Lin et al., 2013) and in some type of animal residues from food industries, chitin and chitosan are present in large amounts (Gaderer et al., 2017; Poverenov et al., 2018). Other types of polysaccharides can also be found abundantly in agro-food residues such as pectin and inulin, to name a few (Zhang et al., 2018; Singh et al., 2018). If a microbial process is carried out using a polysaccharide as carbon source, the following products can be obtained simultaneously: enzymes and/or decrease in the degree of polymerization which in turn yield oligosaccharides and bioactives which are linked to polysaccharides (Gong et al., 2015; Martins et al., 2011).

1.3 MICROBIAL DEPOLYMERIZATION OF POLYSACCHARIDES AND ITS PRODUCTS

It is important to know the structure of the previously mentioned polysaccharides because from there, the enzymes and degradation products can be estimated. Microorganisms such as fungi, bacteria, and yeast can produce enzymes that degrade polysaccharides.

1.3.1 CELLULOSE

Cellulose which is a homopolysaccharide of glucose with β-1,4 glycosidic bonds (Figure 1.1). Some microorganisms can degrade cellulose using its production of cellulase enzymes. This polysaccharide is very resistant to hydrolysis, so its biodegradation consists in the action of three enzymatic activities: Exoglucanase or cellobiohydrolase, endoglucanase, and β-glucosidase (Sánchez-Ramírez et al., 2016). Each activity can yield a type of product where both the degradation product and the enzymes are recoverable. The degradation mechanisms occur as follows: Endoglucanase and cellobiohydrolase can act simultaneously where the first break the glycosidic bonds in the inner area of cellulose and the latter releases cellobiose (glucose dimer of B-1,4 bonding) releasing shorter chains of cellulose and at each end of the short-chain, cellobiohydrolases release cellobiose. Lastly, the β-glucosidase activity holds the most important of the mentioned activities because it promotes the end product of cellulose enzymatic degradation: the release of glucose from cellobiose (Oliveira et al., 2018).

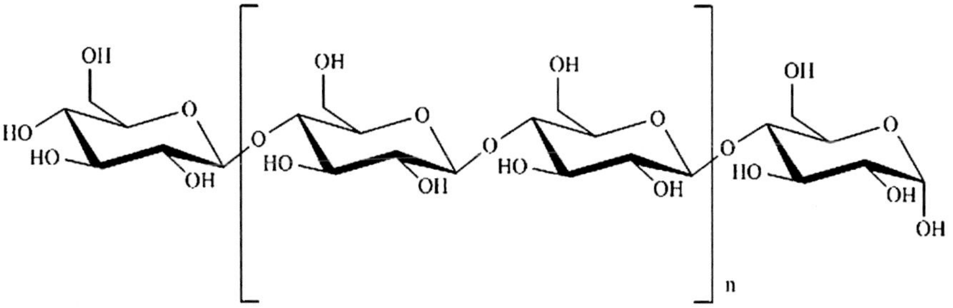

FIGURE 1.1 Structure of the cellulose polysaccharide formed by glucose linked by b-1,4 glycosidic bonds.

1.3.2 *HEMICELLULOSE*

This polymer is a heteropolysaccharide (Figure 1.2), which means that it is comprised of several types of monomeric sugars (Jiang et al., 2016). This polymer can hold in its structure several oligomers that can serve as a source of oligosaccharides or monomers that can also act as enzyme inducers. In its structure, polymers such as xylansarabinoxylans, glucuronoxylans, xyloglucans (Dondelinger et al., 2016) can be found and, while it is relatively easy to degrade hemicellulose by chemical means, enzymes are often used to degrade them and release oligomers or monomers such as xylose, arabinose, mannose, glucuronic acid, rhamnose, and galactose, to name a few (Jiang et al., 2016; Walia et al., 2017). Based on this, it is possible to also produce and recover enzymes that degrade hemicellulosic structures. For xylan hydrolysis, the microbe can produce and excrete endoxylanases and β -xylosidades (Garcia et al., 2015; Walia et al., 2017). For arabinoxylan degradation, arabinoxylanases can be produced to yield xylose and arabinose (Seiboth and Metz, 2011; Zhao, Guo, and Zhu, 2017).

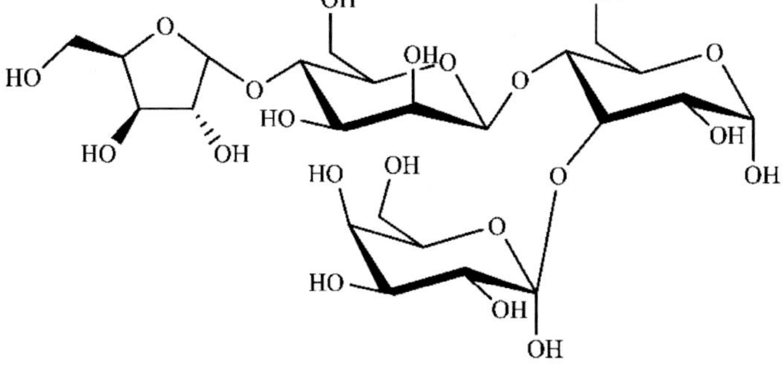

FIGURE 1.2 Common monosaccharides linked by glycosidic bonding that are found in hemicellulose (Xylan, mannose, glucose, and galactose).

1.3.3 PECTIN

This is a polysaccharide widely found in the plant's cell wall (Figure 1.3). It is mainly composed of α-linked (1,4) galacturonic acid and regions of rhamnogalacturonans. Concerning food, it is obtained primarily from citrus fruits and also from apple, apricots, guavas, and mangoes. This polysaccharide is used in food and drugs as a stabilizing agent or for juice clarification. In this case, microbes can produce pectinase, pectinesterases, polygalacturonases, and polygalacturonate lyase (Garg et al., 2016).

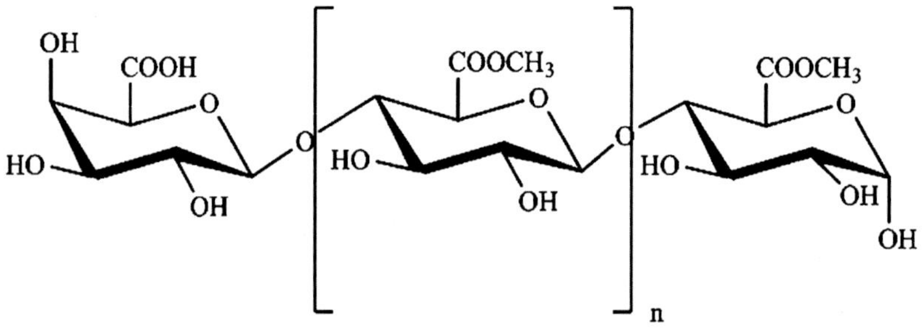

FIGURE 1.3 Methoxylated pectin found in several agro-food residues

By the action of pectin methylesterase, the methoxyl groups in pectic chains are released, leaving a carboxylic acid in the sugar structure. The mentioned change in the molecule leads to a distinct change in the gelation effect which pectin is known to produce. While this enzyme reaction can take place along with the microbial process, it is more common to isolate the enzyme and produce demethoxylated pectin or low-methoxyl pectin (Zhang et al., 2018). The pectin, in this new conformation, is able to form a gel by ionic interactions which are very useful in pharmacology for drug delivery or in the food industry because it is different from the most common use of pectin in jams or marmalades where heat and high concentrations of sugar must be added for gel formation (Yang et al., 2018). Concerning pectinase, polygalacturonase, and polygalacturonatelyase, it can be used to produce pectic oligosaccharides which can act as prebiotics and improve food digestion (Talekar et al., 2018; Zhang et al., 2018).

1.3.4 INULIN

This polymer is formed by a long chain of fructose with a glucosyl moiety termination (Figure 1.4). It is found in many food sources which leave residues such as onions and agave (Singh et al., 2018). This polysaccharide, as other examples previously addressed, can be used as a source of high added value compounds or as a source of inducers for microbial enzyme production (Nava-Cruz et al., 2014). The microbial enzymes responsible for inulin degradation are inulinases, which also can be classified as exo-inulinases and endo-inulinases (Magadum and Yadav, 2018). For the indiscriminate hydrolysis of the internal areas of inulin chains, endo-inulinase is the enzyme that releases oligosaccharides of different lengths, and exo-inulinase breaks the bonds on the ends of the chain (Huitrón et al., 2013).

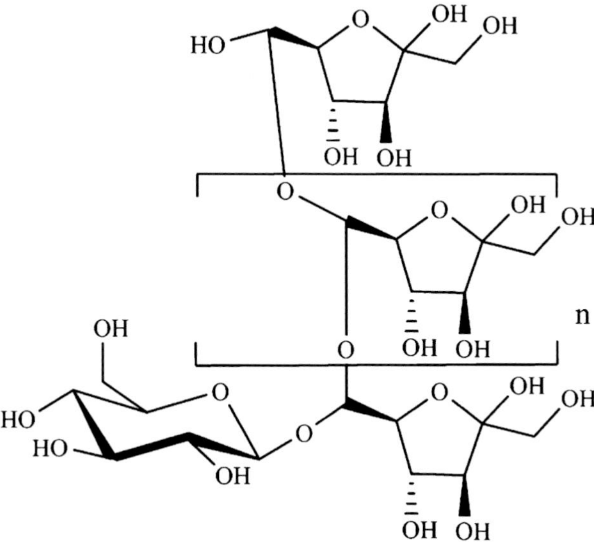

FIGURE 1.4 Inulin structure mainly formed by fructose and glucose in its end-chain.

By the hydrolytic action of inulinases, fructooligosaccharides can be released from inulin (Willems and Low, 2012). These compounds have high demand as prebiotics and as food additives (Nava-Cruz et al., 2014). There is versatility when it comes to working with inulinases. This type of enzyme, allows upon meeting certain conditions, to express transferase activity. This enzymatic activity is highly useful because it promotes the formation of small dimers or trimers related to fructooligosaccharides which also have high demand as added-value compounds (Kirchet and Morlock, 2012).

3.5 CHITIN

This polysaccharide is the second most abundant in the planet, just behind cellulose. Its structure is formed by units of N-acetyl glucosamine linked by β -(1,4) glycosidic bonds (Figure 1.5). This conformation makes it very similar to cellulose and thus very resistant (Hamdi et al., 2017). As a form of agro-food residue, it can be found in crab and shrimp wastes. It is possible to degrade chitin by using microbes. In a microbial bioprocess, chitinases and chitooligosaccharides can be obtained (Kumar et al., 2018; Liaqat and Eltem, 2018; Wang et al., 2018). Both of these products have very high added and uses in several areas. If crab and shrimp shells exploitation is desired, these residues must be treated to leave chitin free of several other compounds. Chitin, in crustacean shells, is bound to proteins and minerals, which have to be removed. In this case, chemical processes have been developed but are costly and generate pollutants, therefore, biological methods have been implemented to minimize costs and are environmental-friendly (Saravana et al., 2018). The alternative called "fermentative deproteinization" where microorganisms can degrade proteins by producing proteases and leaving chitin for ulterior purposes (Hajji et al., 2015). Related to protein and mineral removal, a lactic fermentation can be performed using *Lactobacillus* strains. In this instance, the simultaneous production of proteases and lactic acid promotes the removal of undesired compounds. An additional advantage of lactic fermentation is the low pH which prevents the growth of spoilage microorganisms. Even though it is of utmost interest the exploitation of chitinous agro-food residues, chitin must be further processed to fully evaluate the potential of its derivatives (Arbia et al., 2013).

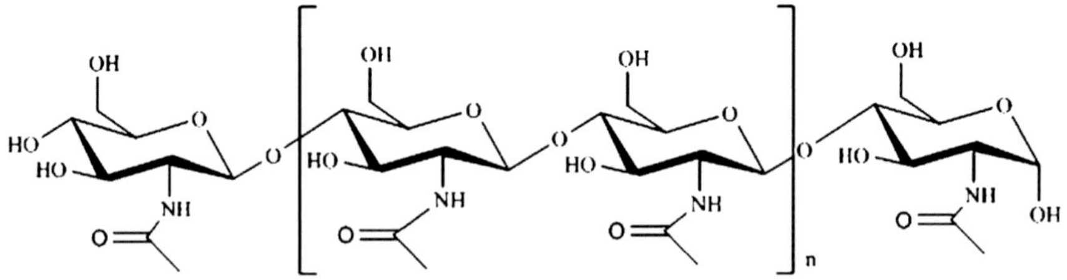

FIGURE 1.5 Chitin chain formed by n-acetyl glucosamine monomers

Although the main focus is the production of chitosan, chitin acts as an inducer for microbial production of chitinase, which in turn is a very important feature in biological control. As most of the entomological and fungal plagues, chitinase, along with lipases, cellulases, and proteases; are essential in some formulations in biological control (De La Cruz-Quiroz et al., 2017; Sosnowska et al., 2018).

1.3.6 CHITOSAN

This biopolymer is a natural polysaccharide, non-toxic, and of a linear structure consisting of β-1,4 glycosidic bonds of glucosamine (Figure 1.6). This polymer differs from chitin in the proportion of deacetylated glucosamine. Unlike chitin, chitosan is not found as an animal component and is rarely found in nature. It can be found, however, in fungi, it can be found as a free form or bound to glucans (Pechsrichuang et al., 2018; Su et al., 2017). Concerning agro-food residues, the chitin in crustacean shells must be processed to obtain chitosan. To achieve deacetylation, several methods are reported, where it stands out the biochemical pathway. Several microorganisms can produce chitin-deacetylase which releases the acetyl group from glucosamine, which increases the chitosan content in the chitin chain (Pareek et al., 2014). The main product of chitin is chitosan itself. If more products are desired, the release of chitosan oligosaccharides or chitooligosaccharides is the best choice or complete degradation of the polysaccharide to glucosamine by means of a chitosanase which can degrade the polysaccharide to glucosamine units (Nidheesh, Kumar and Suresh, 2015). The chitooligomers have a wide application array in several processes due to the biological activities and improving effects in health. Glucosamine has been used as a food supplement and also along with chondroitin, it has been used to improve joint conditions and for osteoarthritis (Li et al., 2016; Liaqat and Eltem, 2018).

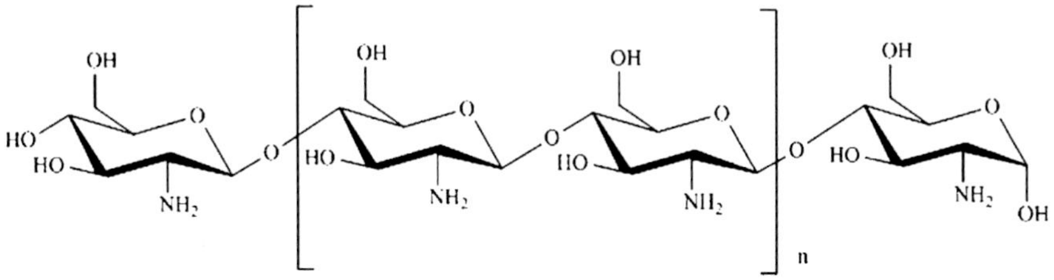

FIGURE 1.6 Chitosan polysaccharide produced by the deacetylation of chitin.

1.3.7 BIOACTIVE PHENOLICS

These molecules are located in the plant cell vacuoles. The bioactivities attributed to this type of compound such as antimicrobial, antioxidant, antiviral, and many others (Castro-López et al., 2017). Polyphenols can be found as tannins, such as condensed, hydrolyzable, and complex (Svarc-Gajic et al., 2013). Among them, flavonols, flavones, flavonoids, anthocyanins, and anthocyanidins, to name a few. These compounds are widely found in agro-industrial residues, mostly in fruit and nutshell residues (Ascacio-Valdés et al., 2016; Medina et al., 2010). There are several possibilities wherein microbial processes, both enzymes and bioactive phenolics have been or can be obtained from a single bioprocess. In this part of the chapter, some of the previously described enzymes have a significant role in the release of bioactive compounds from agro foods residues (Bei et al., 2018).

One of the enzymes that contribute to the release of antioxidant phenolics is B-glucosidase. Several compounds are bound to a glycosidic core when they are part of the plant tissue (Lee et al., 2012; Lopez-Trujillo et al., 2017). The bonds between the phenolic compounds and the sugar or sugars are glycosidic. There is evidence that phenolic glycosides can be inducers for glycoside hydrolases. In a study by Ascacio-Valdés et al. (Ascacio-Valdés et al., 2016), a solid-state fungal bioprocess was carried out using purified pomegranate polyphenols, from which most of those molecules are ellagitannins. These molecules are formed by an ellagic acid bonded with sugar, most commonly glucose. The results from this work showed that several enzymes were produced; among them, b-glucosidase, cellulase, and xylanase. This example is a basic argument, regarding this topic, for the release of bioactive compounds and enzyme production and their ulterior recovery if desired. Many, if not all, of the agro-food residues, contain bioactive compounds, which have a great number of food, pharmaceutical, and cosmetics applications (Aires, Carvalho, and Saavedra, 2016; Kawabata, Mukai and Ishisaka, 2015). There is a wide range of molecules that can be found in residues that have a very high demand at an industry level.

1.3.8 POLYPHENOLS

Several, if not all, of the applicable bioactive compounds are polyphenols (Mansour, Abdel-Shafy, and Mehaya, 2018). These molecules are present in the plant ccellvacuoles and most of them exist in a glycosylated form. As living beings, plants are composed highly by water, so these compounds possess a sugar in its structure which improves water solubility (Veitch and Grayer, 2008). There are many types of polyphenols of interest such as gallotannins, ellagitannins, anthocyanins, anthocyanidins, flavones, flavans, flavonols, among others (Figure 1.7). These compounds, as previously stated, are attributed to have bioactivities and can be found in agro-food residues. According to some sources (Lopez-Trujillo et al., 2017; Medina et al., 2010; Rani, 2014), these compounds can act as microbial enzyme inducers, and the molecules themselves can be released by biodegradation.

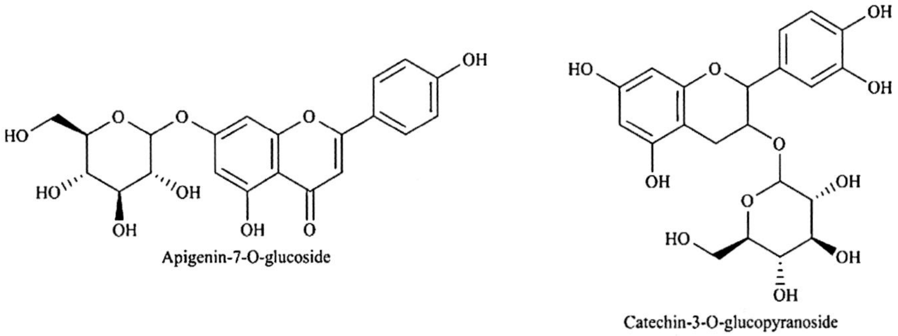

Apigenin-7-O-glucoside

Catechin-3-O-glucopyranoside

FIGURE 1.7 Phenolic acids released from gallotannins and ellagitannins.

1.3.9 GALLOTANNINS AND ELLAGITANNINS

Both of these molecules can be found in food residues. Gallotannins and gallic acid can be obtained from mango residues and ellagitannins and ellagic acid from pomegranate residues (Figure 1.8) (Agatonovic-Kustrin, Kustrin, and Morton, 2018; Ascacio-Valdés, 2014). Both of these bioactives have been obtained from microbial processes and enzymes as well. Aside from the obvious implications of released bioactive compounds released, the enzymes produced can contribute to degrade and further isolate said compounds as well as its derivates to further valorize the bioactives source.

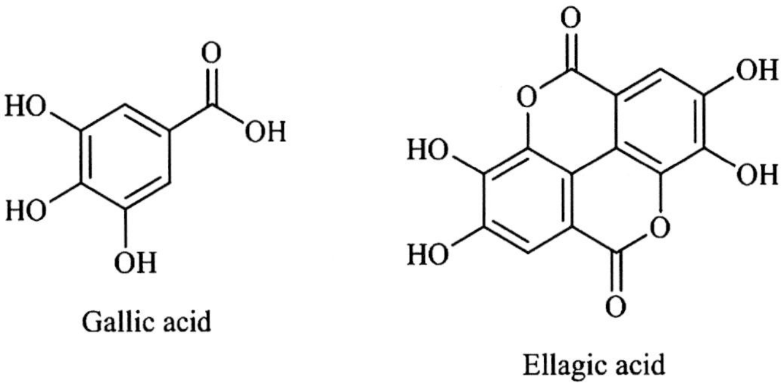

Gallic acid

Ellagic acid

FIGURE 1.8 Flavonoid structures found in agro-food residues

1.3.10 CONDENSED TANNINS

In a similar fashion as gallo and ellagitannins, condensed tannins can also be degraded to yield the respective monomers and also the enzymes responsible for its degradation (Contreras-Domínguezet al., 2006). The enzymatic depolymerization is different from the previous compounds because in this case, apart from the hydrolases needed to degrade the higher weight molecules, oxygenases must be present to cleave the C-C bonds that characterize these molecules. Fortunately, microbes can produce the enzymes required for this end (Mutabaruka, Hairiah, and Cadisch, 2007).

1.4 MICROBIAL PROCESSES

Now, for the achievement of a joint process for the production of more than one product, the microbial culture must be established (Belmessikh et al., 2013). It is of utmost importance to determine the nature of the products of interest and the manner of the interaction of the residues/substrate with microbes, whether they are fungi, yeast or bacteria. For instance, a residue that is not found in abundance but can be used for its exploitation could be adapted to a liquid or submerged microbial culture. On the other hand, if a residue is widely produced and available, a solid-state culture can be developed for production and release of metabolites (Lizardi-Jiménez and Hernández-Martínez, 2017; Ravindran and Jaiswal, 2016). In order to understand the importance of selecting the adequate microbial growth system, these must be defined and advantages and disadvantages must be addressed. There are three basic types of microbial growth which are liquid, submerged and solid cultures.

1.4.1 LIQUID CULTURES

In this case, the microbial growth can be carried out in several sizes and vessels, from culture tubes, flasks and/or in a reactor. The main difference with other cultures is that the substrate (carbon and energy source) is completely soluble in water, which composes the media along with mineral salts and the nitrogen source. Conditions such as pH, temperature, stirring and the presence or absence of O_2 differ with the microorganism which is going to degrade the substrate. Considering that this chapter is dealing with bioprocesses which are focused on producing more than one recoverable product, the most important factor is the substrate. According to literature, there are substrates which can serve to induce microbial enzymes and for degradation and release of added value compounds (Ascacio-Valdés et al., 2016; Kumar, Singh, and Singh, 2008). This case is, to our knowledge, is pivotal in the induction of glycosidases, due to the fact that even though the pomegranate extracts where used, also purified pomegranate polyphenols were used for the same goal; not only ellagitannase was produced, but also cellulase, xylanase, and b-glucosidase were detected. As stated, from the described bioprocess, ellagic acid and enzymes can be obtained from the same batch.

1.4.2 SUBMERGED CULTURES

Differing from the liquid culture, in this instance, the substrate is insoluble in the water media but the amount used is very low compared to the water content (Abd-Elhalem, 2015). The liquid fraction of the process also must contain mineral salts, a source of nitrogen and the pH, temperature and stirring levels adequate for microbial development and can be set in several size levels just like liquid cultures. In this case, enzyme production has been made by adding plant tissues such as husks, shells, peels and or skins, which all may be obtained as agroindustrial residues or by-products. From the previously mentioned factors that get into play, the substrate is the most important and the particle size must be adequate for operational and biological purposes. As an example, studies can be performed by submerging textile residues in liquid media. Using microbial strains for production of cellulases, such as fungi, can lead to the production, and if desired, recovery of cellulases (Wang et al., 2018; Yang, 2011). This principle is applicable if agro-food residues which contain enzyme inducers and recoverable high-value bioactives such as pomegranate husks, nutshells, citrus peels, among others (Sorensen et al., 2014; Oliveira et al., 2018).

1.4.3 SOLID-STATE CULTURES

This culture system is a highly important aspect in research nowadays due to the fact that saves energy, is more cost-efficient compared to submerged and liquid cultures, environmentally friendly and it gets to use high loads of substrate, which in turn, is attractive for agro-food residues usage and even consider it waste management treatment (Ravindran and Jaiswal, 2016). In this case, the operational aspects are different from the previous systems mainly because the moisture content is very low and the oxygen levels must be high, which is why this process is an aerobic culture (Soccol et al., 2017). The case of solid-state fermentation adjusts better to the context of agro-food residues in the context of the biotechnological valorization. There are several examples of how this microbial growth system is efficient for more than one product recovery. A previously mentioned process is the ellagic acid release from pomegranate residues along with high added value enzymes such as ellagitannase. In the case of the use of solid-state bioprocesses, pomegranate residues have been fermented by fungi for the same objective (Ascacio-Valdés, et al., 2016; Sepúlveda et al., 2018).

Another instance where it is possible, using fungal solid-state fermentation the production of microbial biomass, enzymes and bioactives release is published by De la Cruz et al. (De La Cruz-Quiroz et al., 2017). Even though this paper does not include bioactives, it is possible to add it to the list of products. This particular paper adds another interesting feature to the agro-food residues valorization, which is the combining of several carbon sources, from which some are considered residues. Thanks to the mixture of sugarcane bagasse, wheat bran, potato flour, olive oil and chitin, several enzymes were detected such as amylase, chitinase and endoglucanase and fungal spore production. If the residues obtained and used contained bioactive molecules, these would be released and could also be recovered, to even develop a triple-product (enzymes, microbial biomass, and bioactive compounds) bioprocess (Arora, Rani, and Ghosh, 2018; De La Cruz-Quiroz et al., 2017).

1.5 RECOVERY OF PRODUCTS BY POLYMER/SALT AQUEOUS TWO-PHASES SYSTEM

In the development of bioprocesses, the recovery strategies of the products released are of great importance. The use of aqueous two-phase systems is an attractive technique for product extraction. Currently, plant processing industries generate high quantities of organic residues which include by-products from the agri-food activities such as tomato, mango, sugar cane, pineapple, degummed fruits and vegetables, beer, milk and cellulose among others (Yusuf, 2017). As previously stated, these residues include a wide diversity of food by-products including the cores, skins, stems, pulp remnants, bagasse and seeds, which despite its nutritional value are still dismissed by industries and these matrices represent serious environmental issues (Da Silva and Jorge, 2014). Generally, agri-food by-products contain considerable amounts of biomolecules but not directly available and by this reason some studies have been addressed their aims toward the development of new and efficient methodologies for their recovery trough environmentally friendly process and sustainable chemistry which has prompted great number of research works into the handling of renewable raw materials and their by-products to obtain scale up and downstream processes of value-added products (Campos et al., 2017; Larios-Cruz et al., 2017).

Prior to purification and recovery of bioproducts, it is important to choose an optimal extraction technique. In that way, the extraction process of bioproducts from raw matrixes represents the most critical step in the downstream procedure, because of plant residues have a diverse complexity (carbohydrates, polysaccharide, proteins, polyphenolic, compounds, etc.) and some different physiochemical properties making bioproducts extraction the challenging task (Giacometti et al., 2018). Conventionally, bioproducts have been extracted by methodologies that often implies complex and time-consuming multi-step process, thus leading to low selectivity, large consumption of organic solvents, pollution and low recovery yields such as water distillation, steam distillation, cohobation, maceration, and enfleurage (Hong and Kim, 2016). In the scenery of specific bioproducts like proteins, their extraction was commonly by methods using solvents, saturated solution of salts, gel and ion-exchange chromatography these methodologies have several limitations, including low scale-up for industries, elevated costs and low yield of purification and in some cases loss of stability (Duarte et al., 2015). Recently, aqueous two-phase systems (ATPS) have arisen as a potential technique for protein extraction and purification for its advantages linking easy scale-up process, volume reduction and rapid separation (Rodrigues et al., 2010; Ruiz-Ruiz et al., 2012).

ATPS are formed by the combination of an immiscible aqueous solution of polymer-polymer, salt-salt or polymer-salt. These methods offer several advantages, compared to conventional methods as lower energy requirements, less time and costs of implementation as well as represent a non-toxic and environmentally safe alternative (Xu et al., 2017). They are categorized as methods moderately simple, present easy operation and scale-up and allow a one-step process of clarification, concentration, and purification (De Araujo-Sampaio et al., 2016). The most used polymers for biomolecules separation in ATPS are polyethylene glycol (PEG) and dextran, however, this kind of system have some disadvantages related to the elevated cost of dextran and high viscosity of the polymer phase (Ramakrishnan et al., 2016). Moreover, increasing attention has been gained to employ polymer/salt ATPS with one phase rich (top) in polymer and one rich (bottom) in salt (Prabhu, Gupta, and VenkataDasu, 2018). On the other hand, PEG has been principally employed since it presents low cost and viscosity while citrate, phosphate, and sulfate salts have been most commonly taken into consideration due to their facility to increase hydrophobic comportment between the two phases (Dong, Wan, and Cao, 2018). Moreover, in the practice of carrying out ATPS, there are three main interaction's contributions that are important to know and keep in mind before to develop the protein partitioning between the two phases, including electrostatic, hydrophobic and excluded-volume (Liu et al., 2016) as well as the physicochemical characteristics of the target molecules, such as size, shape, surface charge, molecular weight and bonding interactions (hydrogen bonding and Vander Waals) (Amaal-Alhelli et al., 2016). Also, there are some other parameters which show an critical role in the separation and purification process which comprise molecular weight and concentration of PEG, concentration and kind of salts, volume relation and temperature (Sánchez-Trasviña et al., 2017). PEG/salt systems have been extensively reported for the extraction of proteins with enzymatic properties (Table 1.1a) microorganisms and other molecules such as alkaloids (caffeine). All these reports showed that these systems (PEG + salt) based ATPS present good potential for biomolecules liquid-liquid extraction. In general, ATPS have proven to be suitable for the recovery of proteins trough aqueous extracts from different origins as composts (Mayolo-Deloisa, Trejo-Hernández, and Rito-Palomares, 2009), polluted waters (Baskaran et al., 2018), and raw plat residues (Coelho, Silveira, and Tambourgi, 2013) among others. On the other hand, concerning phenolic compounds extraction by ATPS there are a truly few reports (Table 1.1d), such as the case of Xavier et al., (Xavier et al., 2017) they employed PEG/phosphate systems for phenolics recovery from *Eucalyptus glogulus* wood wastes. Currently, great attention has been gained according to some literature reports to the applications of liquid-liquid extraction methods for the recovery of natural and safe bioproducts due to either its essential applications in pharmaceutics, cosmetic, beverage, paper, food and biofuels industries, and also with the arising concern of valorization of wastes from agri-food industries continually produced, it has been created an important research interest in order to find and establish eco-efficient processes to obtain these value-added bioproducts from these organic materials (Herculano et al., 2016; Dos-Reis et al., 2013).

In the present scenario, biotransformation of these residues regarding production and obtainment of proteins/enzymes and phenolic compounds researchers have shown a great interest in the development of these highlighted bioproducts through microbial processes. Liquid fermentation (LF) by fungi was the most commonly biotechnological process used in the production of enzymes with potential industrial applications, e.g., cellulase and xylanase are used in biofuel, textile, and paper industries (Dos-Reis et al., 2013). Presently, Table 1.1b shows some reports in the production of enzymatic extracts by LF employing agri-food residues as substrates and their partial purification using PEG/salt ATPS. Moreover, solid-state fermentation (SSF) also is a well-studied bioprocess for enzymes and phenolic compounds production (Buenrostro-Figueroa et al., 2013; Larios-Cruz et al., 2017; Medina-Morales et al., 2011), this method presents some advantages against LF involving high stability, natural extracellular production and better control parameters (temperature, pH and biomass) and less water content. There are a great number of researches concerning the production of diverse bioproducts by SSF employing agri-food residues, however extraction of them have been carried out by conventional methods, this fact opens new study avenues between researchers due to the scarcely reports exploiting PEG/salt ATPS. Recently, PEG/salt systems have been used in order to extract enzymes from SSF (Table 1.1c), such as the case of Herculano et al., (2016), they produced xylanase by *Aspergillus* strain and also they employed residues of *Ricinus communis* beans as inducer and carbon source for microbial growth, by using PEG/sodium citrate systems they obtained higher yields of enzyme recovery (268%) and purification factor (7.2) than that reported by Garai and Kumar et al., (2013) for xylanase produced in LF. Nevertheless, although there is important information considering plant residues for extraction of bioproducts, nowadays still exist unexploited sources from agri-food industries characterized by their richness in value biomolecules and their low cost of applicability to acquire high-value bioproducts.

All these results and observations reinforce the significance of PEG/salt ATPS as an effective technique for enzymes extraction from different sources. Furthermore, bioproducts from SSF coupled with ATPS could be successfully employed for the separation and partial purification of enzymes through the valorization of agri-food residues generated by industrial activities. Also, due to the great number publications available about phenolic compounds produced by SSF employing agri-food residues (Martínez-Ávila et al., 2012; Martins et al., 2011) and the scarce applications of PEG/salt systems in the recovery of these bioproducts, new scientific and industrial trends can arise to apply ATPS to improve extraction yields and purification factors.

TABLE 1.1 Spectrum of some studies focuses on the extraction of bioproducts by PEG/salt ATPS from different origins.

a) Protein/enzymes

Agri-food	Residues	Extraction conditions	Type	Yield	References
	compost	PEG MW 1000 (18.2% w/w) Phosphate (15% w/w)	Laccase from *Agaricus bisporus*	ER: 95% PF: 2.48	Cisneros-Ruiz et al., 2009
Pineapple	Steam Barks Leaves	PEG MW 4,000 (10.86% w/w) Saturated of ammonium sulfate (36.21% w/w)	Bromelin	ER: 68 % PF: 11.80	Coelho, Silveira, and Tambourgi, 2013
Potato	Peel	PEG MW 1500 (17.62% w/w) KH_2PO_4/K_2HPO_4 (15.11% w/w) NaCl (2.08 mM pH7)	Polyphenol oxidase	ER: 77.89% PF: 4.5	Niphadkar, Vetal, and Rathod, 2015
		PEG MW 3350 (11.9% w/w) Potassium phosphate (11.8% w/w)	Invertase	ER: 60.8%	Sánchez-Trasviña et al., 2015
	Aqua wastes	PEG MW 600 (50% w/w) Ammonium sulfate (40% w/w)	Crude protein	ER: 99.4%	Deswal et al., 2014
b) From LF extracts					
Corn Wheat bran Cotton	Corn cobs Bagasse Seed bagasse	PEG MW 4000 (3.5% w/w) Potassium phosphate (14% w/w)	Xylanase by *Bacillus pumilus* SBM13	ER: 70% PF: 7	Yasinok et al., 2010
Wheat bran	Bagasse	PEG MW 4000 (8.66% w/w) NaH_2PO_4 (22.4% w/w)	Xylanase by *Aspergillus candidus*	ER: 88.10% PF: 3	Garai, and Kumar, 2013
Corn and olive oil	remanents	PEG MW 4000 (20% w/w) Monobasic and dibasic sodium phosphate (20% w/w) ratio = 1.087	Lipase by *Leucosporidium scottii* L117	ER: 52.9 % PF: 10.3	Duarte et al., 2015

TABLE 1.1 *(Continued)*

Agri-food	Residues	Extraction conditions	Type	Yield	References
	Sawdust and bamboo leaves	PEG MW 600 (15% w/w) K_2HPO_4 (16% w/w)	Lignin peroxidase by *Amauroderm arugosum*	ER: 72.18% PF: 1.33	Jong et al., 2017
Passion fruit	Peel flour	PEG MW 8000 (19% w/v) Sodium citrate anhydride (23% w/v)	Polygalacturonate by *Aspergillus aculeatus*URM4953	ER: 85% PF: 1.8	Silva et al., 2018
		PEG MW 6000 (25% w/w) Potassium phosphate (12% w/w)	β galactosidase by *Pichia pastoris*	ER: 97 % PF: 4.7	Prabhu, Gupta, and VenkataDasu, 2018
c) From SSF extracts					
		PEG MW 1500 (22% w/w) Sodium citrate (14% w/w)	Phytase by *Schizophyllum commune*	ER: 367% PF: 5.43	Xavier-Salmón et al., 2014
	Ricinus communis beans	PEG MW 8000 (24% w/w) Sodium citrate (20% w/w)	Xylanase by *Aspergillus japonicus* URM5620	ER: 268% PF: 7.2	Herculano et al., 2016
Wheat bran	Bagasse	PEG MW 3000 (9.2% w/w) Phosphate (13.8% w/w) NaCl (3.3% w/w)	Lipase by *Penicillium candidum* (PCA 1/ TT031)	ER: 84% PF: 33.9	Amaal-Alhelli et al., 2016
d) Phenolic compounds					
		PEG MW 6000 (10.37% w/w) $(NH_4)_2SO4$ (9.25% w/w)	Gallic acid	EE: 89.11%	Šalić et al., 2011
		PEG MW 400 (23% w/w) Na_2SO_4 (12% w/v)	Gallic acid	EE: 96.3%	Almeida et al., 2014
		PEG MW 300 (23% w/w) Na_2SO_4 (12% w/v)	Vanillic acid	EE: 96.2%	
		PEG MW 300 (23% w/w) Na_2SO_4 (12% w/v)	Syringic acid	EE: 96.5%	

TABLE 1.1 *(Continued)*

Agri-food	Residues	Extraction conditions	Type	Yield	References
	eucalyptus wood waste	PEG MW 2000 (22% w/w) K_2HPO_4 (10.5% w/w)	Gallic acid Ellagic acid Quercetin 3-O-rhamnoside.	1.89 mg GAE/ 100 mg DW wood	Xavier et al., 2017
e) ATPS-EBM					
		PEG MW 6000 (12.99% w/w) Sodium citrate (12.09% w/w)	Xylanase	Xylose, xylobiose, and xylotriose production from xylan	Ng et al., 2013
		PEG MW 20000 (7.7% w/w) Dextran (10.3% w/w)	Cyclodextrin glycosyltransferase	Cyclodextrins bioconversion	Li, Kim, and Peeples, 2002
		PEG MW 3000 (30% w/w) Potassium phosphate (7% w/w)	Cyclodextrin glycosyltransferase	-cyclodextrin production from starch	Lian et al., 2011

MW = molecular weight (g/mol); w/w = weight/weight; ER = enzyme recovery; PF = purification factor; EE = extraction efficiencies; GAE = gallic acid equivalents; DW = dray weight; nd = non defined.

1.5.1 PEG/SALT ATPS AS AN EXTRACTIVE BIOCONVERSION METHOD (EBM)

As already mentioned, PEG/salt ATPS has been used to isolate and extract many different natural biomolecules, but still, there are many other relevant molecules very little exploited for their recoveries such as carbohydrates (cellulose and hemicellulose) and their derivatives (mono, di, oligo, and polysaccharides) through this extractive methodology. Furthermore, this green methodology has been usually explored as an extractive bioconversion method (EBM) which integrates bioconversion and purification into a single step process, e.g., extractive conversion of cellulose to ethanol and enzymatic hydrolysis of cellulose with cellulases (Li, Kim, and Peeples, 2002). In this scenario, bioconversion in ATPS is achieved with the biocatalyst present in only phase and the products either equally spread among the two-phases or they partition rather to the biocatalyst-free phase, thus, making it possible to recover products without losing the biocatalyst (Ng et al., 2013). However, in the last decade, a very low number of reports have been published to apply these joint technologies (Table 1.1e). Li et al., (2011) associated the employment of a PEG/salt ATPS for xylanase extraction as well as EBM for xylan, a natural polymer which is found in hemicellulose the second-largest polysaccharide in nature, to produce and recover a small molecule such as xylose, xylobiose, and xylotriose. The fund that ATPS composed with PEG MW 6000 (12.99% w/w) and sodium citrate (12.09% w/w) was able for the extraction of xylanase (top PEG-rich phase) and conversion of xylan into a simple mono- and disaccharides (bottom salt-rich phase) with 30.27% hydrolysis yield. Lin et al., (2016) reported production of γ-cyclodextrin from starch using ATPS-EBM by *Bacillus cereus* with optimum parameters of PEG MW 3000 (30% w/w) and potassium phosphate (7% w/w). They reported that γ-cyclodextrin was principally divided into the top PEG-rich phase with the biocatalyst in the bottom salt-rich phase. With all these facts (e.g., few information, results and unexplored fields) concerning production and recovery of polysaccharides and their derivatives by ATPS-EBM can open the possibility to obtain value-added bioproducts through fermented aqueous extracts such as oligosaccharides (prebiotics) which nowadays are strongly studied due to their health beneficial effects in human diet (Almeida-Carvalho et al., 2017) or recovery of monosaccharides for biofuels industrial purposes (Palacios et al., 2017) among others.

1.6 FINAL COMMENTS

For all mentioned above, biotechnological scientists' efforts are going in the way to understand and learn better the phenomena occurring in natural world with the purpose to find the balance between residues and the most appropriate way to use them, for future utilization and valorization to solve the worldwide issues for example, pollution and food supplies. As well as the production and extraction of value-added bioproducts from these organic sources through eco-friendly and effective methodologies without or with less environmental negative impact. Thereby, all the bioproducts generated through agro-food residues bioprocessing could increase the added value of the compounds obtained in a single process. This way, several products can be considered and obtained in a single biotechnological process, thus saving operational costs, time and resources while maintaining very low environmental risk.

KEYWORDS

- **aqueous two-phase systems**
- **bioactive compounds**
- **biomass**
- **enzymes**
- **extractive bioconversion method**
- **liquid fermentation**
- **polyethylene glycol**
- **solid-state fermentation**

REFERENCES

Abd-Elhalem, B. T., El-Sawy, M., Gamal, R. F., & Abou-Taleb, K. A., (2015). Production of amylases from *Bacillus amyloliquefaciens* under submerged fermentation using some agro-industrial by-products. *Ann. Agric. Sci., 60*, 193–202.

Agatonovic-Kustrin, S., Kustrin, E., & Morton, D. W., (2018). Phenolic acids contribution to antioxidant activities and comparative assessment of phenolic content in mango pulp and peel. *South African J. Bot., 116*, 158–163.

Aguilera-Carbo, A., Augur, C., Prado-Barragan, L. A., Favela-Torres, E., & Aguilar, C. N., (2008). Microbial production of ellagic acid and biodegradation of ellagitannins. *Appl. Microbiol. Biotechnol., 78*, 189–199.

Aires, A., Carvalho, R., & Saavedra, M. J., (2016). Valorization of solid wastes from chestnut industry processing: Extraction and optimization of polyphenols, tannins and ellagitannins and its potential for adhesives, cosmetic and pharmaceutical industry. *Waste Manag., 48*, 457–464.

Almeida, M. R., Passos, H., Pereira, M. M., Lima, Á. S., Coutinho, J. A. P., & Freire, M. G., (2014). Ionic liquids as additives to enhance the extraction of antioxidants in aqueous two-phase systems. *Sep. Purif. Technol., 128*, 1–10.

Almeida-Carvalho, E., Santos-Góes, L. M. D., Uetanabaro, A. P. T., Paranhos, D. S. E. G., Brito-Rodrigues, L., Priminho-Pirovani, C., & Miura, D. C. A., (2017). Thermo resistant xylanases from *Trichoderma stromaticum*: Application in bread making and manufacturing xylo-oligosaccharides. *Food Chem., 221*, 1499–1506.

Amaal-Alhelli, M., Abd-Manap, M. Y., Mohammed, A. S., Mirhosseini, S. H., Suliman, E., Shad, Z., Mohammed, N. K., et al., (2016). Use of response surface methodology for partitioning, one-step purification of alkaline extracellular lipase from *Penicillium candidum* (PCA 1/TT031). *J. Chromatogr. B Anal. Technol. Biomed. Life Sci., 1039*, 66–73.

Arbia, W., Adour, L., Amrane, A., & Lounici, H., (2013). Optimization of medium composition for enhanced chitin extraction from *Parapenaeus longirostris* by *Lactobacillus helveticus* using response surface methodology. *Food Hydrocoll., 31*, 392–403.

Arora, S., Rani, R., & Ghosh, S., (2018). Bioreactors in solid state fermentation technology: Design, applications and engineering aspects. *J. Biotechnol., 269*, 16–34.

Ascacio-Valdés, J. A., Aguilera-Carbó, A. F., Buenrostro, J. J., Prado-Barragán, A., Rodríguez-Herrera, R., & Aguilar, C. N., (2016). The complete biodegradation pathway of ellagitannins by *Aspergillus niger* in solid-state fermentation. *J Basic Microbiol., 56*, 329–336.

Ascacio-Valdés, J. A., Buenrostro, J. J., De La Cruz, R., Sepúlveda, L., Aguilera-Carbó, A. F., Prado-Barragán, A., Contreras-Esquivel, J. C., et al., (2014). Fungal biodegradation of pomegranate ellagitannins. *J. Basic Microbiol., 54*, 28–34.

Baskaran, D., Chinnappan, K., Manivasagan, R., & Mahadevan, D. K., (2018). Partitioning of crude protein from aqua waste using PEG 600-inorganic salt aqueous two-phase systems. *Chem. Data Collect., 15*, 143–152.

Bei, Q., Chen, G., Liu, Y., Zhang, Y., & Wu, Z., (2018). Improving phenolic compositions and bioactivity of oats by enzymatic hydrolysis and microbial fermentation. *J. Funct. Foods, 47*, 512–520.

Belmessikh, A., Boukhalfa, H., Mechakra-Maza, A., Gheribi-Aoulmi, Z., & Amrane, A., (2013). Statistical optimization of culture medium for neutral protease production by *Aspergillus oryzae*. Comparative study between solid and submerged fermentations on tomato pomace. *J. Taiwan Inst. Chem. Eng., 44*, 377–385.

Buenrostro-Figueroa, J. J., Velázquez, M., Flores-Ortega, O., Ascacio-Valdés, J. A., Huerta-Ochoa, S., Aguilar, C. N., & Prado-Barragán, L. A., (2017). Solid state fermentation of fig (*Ficuscarica* L.) by-products using fungi to obtain phenolic compounds with antioxidant activity and qualitative evaluation of phenolics obtained. *Process Biochem., 62*, 16–23.

Buenrostro-Figueroa, J., Ascacio-Valdes, A., Sepulveda, L., De La Cruz, R., Prado-Barragan, A., Aguilar-Gonzalez, M. A., Rodriguez-Herrera, R., & Aguilar, C. N., (2013). Potential use of different agro-industrial by-products as supports for fungal ellagitannase production under solid-state fermentation. *Food Bioprod. Process., 92*, 376–382.

Campos, A. D., Woitovich, N., Oliveira, A., Pastrana, L. C. M., Teixeira, J. A., & Maria, P., M., (2017). Platform design for extraction and isolation of bromelain: Complex formation and precipitation with carrageenan. *Process Biochem., 54*, 156–161.

Castro-López, C., Ventura-Sobrevilla, J. M., González-Hernández, M. D., Rojas, R., Ascacio-Valdés, J. A., Aguilar, C. N., & Martínez-Ávila, G. C. G., (2017). *Impact of Extraction Techniques on Antioxidant Capacities and Phytochemical Composition of Polyphenol-Rich Extracts., 237*, 1139–1148.

Cisneros-Ruiz, M., Mayolo-Deloisa, K., Przybycien, T. M., & Rito-Palomares, M., (2009). Separation of PEGylated from unmodified ribonuclease A using sepharose media. *Sep. Purif. Technol., 65*, 105–109.

Coelho, D. F. E., Silveira, P. J. A., & Tambourgi, E. B., (2013). Bromelain purification through unconventional aqueous two-phase system (PEG/ammonium sulfate). *Bioprocess Biosyst. Eng., 36*, 185–192.

Contreras-Domínguez, M., Guyot, S., Marnet, N., Le Petit, J., Perraud-Gaime, I., Roussos, S., & Augur, C., (2006). Degradation of procyanidins by *Aspergillus fumigatus*: Identification of a novel aromatic ring cleavage product. *Biochimie., 88*, 1899–1908.

Da Silva, A. C., & Jorge, N., (2014). Bioactive compounds of the lipid fractions of agro-industrial waste. *Food Res. Int., 493*–500.

Dai, C., Ma, H., He, R., Huang, L., Zhu, S., Ding, Q., & Luo, L., (2017). LWT-food science and technology improvement of nutritional value and bioactivity of soybean meal by solid-state fermentation with *Bacillus subtilis*. *LWT-Food Sci. Technol., 86*, 1–7.

De Araujo-Sampaio, D., Igarashi-Mafra, L., Itsuo-Yamamoto, C., De Andrade, E. F., Oberson, D. S. M., Mafra, M. R., & Castilhos, F., (2016). Aqueous two-phase (polyethylene glycol + sodium sulfate) system for caffeine extraction: Equilibrium diagrams and partitioning study. *J. Chem. Thermodyn., 98*, 86–94.

De La Cruz-Quiroz, R., Robledo-Padilla, F., Aguilar, C. N., & Roussos, R., (2017). Forced Aeration influence on the production of spores by *Trichoderma* strains. *Waste and Biomass Valorization., 8*, 2263–2270.

Deswal, D., Gupta, R., Nandal, P., & Kuhad, R. C., (2014). Fungal pretreatment improves amenability of lignocellulosic material for its saccharification to sugars. *Carbohydr. Polym., 99*, 264–269.

Diaz, A. B., Blandino, A., & Caro, I., (2018). Value added products from the fermentation of sugars derived from agro-food residues. *Trends Food Sci. Technol., 71*, 52–64.

Dondelinger, E., Aubry, N., Ben, C. F., Cohen, C., Tayeb, J., & Rémond, C., (2016). Contrasted enzymatic cocktails reveal the importance of cellulases and hemicellulases activity ratios for the hydrolysis of cellulose in presence of xylans. *AMB Express, 6*. https://doi.org/10.1186/s13568-016-0196-x (accessed on 24 July 2020).

Dong, L., Wan, J., & Cao, X., (2018). Separation of transglutaminase using aqueous two-phase systems composed of two pH-response polymers. *J. Chromatogr. A, 1555*, 106–112.

Dos-Reis, L., Roselei-Claudete, F., Da Silva-Delabona, P., Da Silva-Lima, D. J., Camassola, M., Da Cruz-Pradella, J. G., & Pinheiro-Dillon, A. J., (2013). Increased production of cellulases and xylanases by *Penicillium echinulatum* S1M29 in batch and fed-batch culture. *Bioresour. Technol., 146*, 597–603.

Duarte, A. W. F., Lopes, A. M., Molino, J. V. D., Pessoa, A., & Sette, L. D., (2015). Liquid-liquid extraction of lipase produced by psychrotrophic yeast *Leucosporidium Scottie* L117 using aqueous two-phase systems. *Sep. Purif. Technol., 156*, 215–225.

Gaderer, R., Seidl-Seiboth, V., De Vries, R. P., Seiboth, B., & Kappel, L., (2017). N-acetyl glucosamine, the building block of chitin, inhibits growth of *Neurospora crassa*. *Fungal Genet. Biol., 107*, 1–11.

Garai, D., & Kumar, V., (2013). Biocatalysis and agricultural biotechnology aqueous two phase extraction of alkaline fungal xylanase in PEG/phosphate system : Optimization by box-behnken design approach. *Biocatal. Agric. Biotechnol., 2*, 125–131.

Garcia, N. F. L., Da Silva, S. F. R., Gonçalves, F. A., Da Paz, M. F., Fonseca, G. G., & Leite, R. S. R., (2015). Production of β-glucosidase on solid-state fermentation by *Lichtheimia ramosa* in agro-industrial residues: Characterization and catalytic properties of the enzymatic extract. *Electron. J. Biotechnol., 18*, 314–319.

Garg, G., Singh, A., Kaur, A., Singh, R., Kaur, J., & Mahajan, R., (2016). Microbial pectinases: An eco-friendly tool of nature for industries. *3 Biotech, 6*. https://doi.org/10.1007/s13205-016-0371-4 (accessed on 24 July 2020).

Giacometti, J., Kovačević, D. B., Putnik, P., Gabrić, D., Bilušić, T., Krešić, G., Stulić, V., et al., (2018). Extraction of bioactive compounds and essential oils from Mediterranean herbs by conventional and green innovative techniques: A review. *Food Res., Int.*, 245–262.

Gong, W., Zhang, H., Liu, S., Zhang, L., Gao, P., Chen, G., & Wang, L., (2015). Comparative secretome analysis of *Aspergillus niger*, *Trichoderma reesei*, and *Penicillium oxalicum* during solid-state fermentation. *Appl. Biochem. Biotechnol., 177*, 1252–1271.

Hajji, S., Ghorbel-Bellaaj, O., Younes, I., Jellouli, K., & Nasri, M., (2015). Chitin extraction from crab shells by *Bacillus* bacteria. Biological activities of fermented crab supernatants. *Int. J. Biol. Macromol., 79*, 167–173.

Hamdi, M., Hammami, A., Hajji, S., Jridi, M., Nasri, M., & Nasri, R., (2017). Chitin extraction from blue crab (*Portunussegnis*) and shrimp (*Penaeuskerathurus*) shells using digestive alkaline proteases from *P. segnis* viscera. *Int. J. Biol. Macromol., 101*, 455–463.

Herculano, P. N., Moreira, K. A., Bezerra, R. P., Porto, T. S., De Souza-Motta, C. M., & Porto, A. L. F., (2016). Potential application of waste from castor bean (*Ricinus communis* L.) for production for xylanase of interest in the industry. *3 Biotech, 6*.

Hong, Y. S., & Kim, K. S., (2016). Determination of the volatile flavor components of orange and grapefruit by simultaneous distillation-extraction. *Korean J. Food Preserv, 23*, 63–73.

Huitrón, C., Pérez, R., Gutiérrez, L., Lappe, P., Petrosyan, P., Villegas, J., Aguilar, C., et al., (2013). Bioconversion of Agave tequilana fructans by exo-inulinases from indigenous *Aspergillus niger* CH-A-2010 enhances ethanol production from raw Agave tequilana juice. *J. Ind. Microbiol. Biotechnol., 40*, 123–132.

Jiang, X., Hou, Q., Liu, W., Zhang, H., & Qin, Q., (2016). Hemicelluloses removal in autohydrolysis pretreatment enhances the subsequent alkali impregnation effectiveness of poplar sapwood. *Bioresour. Technol., 222*, 361–366.

Jong, W. Y. L., Show, L. P., Ling, T. C., & Tan, Y. S., (2017). Recovery of lignin peroxidase from submerged liquid fermentation of *Amauroderma rugosum* (Blume and T. Nees) torrend using polyethylene glycol/salt aqueous two-phase system. *J. Biosci. Bioeng., 124*, 91–98.

Kawabata, K., Mukai, R., & Ishisaka, A., (2015). Quercetin and related polyphenols: New insights and implications for their bioactivity and bioavailability. *Food Funct., 6*, 1399–1417.

Kawaguchi, H., Hasunuma, T., Ogino, C., Kondo, A., (2016). Bio-processing of bio-based chemicals produced from lignocellulosic feed stocks. *Curr. Opin. Biotechnol., 42*, 30–39.

Kirchet, S., & Morlock, G., (2018). Simultaneous determination of mono-, di-, oligo- and polysaccharides via planar chromatography in 4 different prebiotic foods and 60 natural degraded inulin samples. *Journal of Chromatography A, 1569*, 212–221.

Kumar, M., Brar, A., Vivekanand, V., & Pareek, N., (2018). Process optimization, purification and characterization of a novel acidic, thermo stable chitinase from *Humicolagrisea. Int. J. Biol. Macromol., 116*, 931–938.

Kumar, R., Singh, S., & Singh, O. V., (2008). Bioconversion of lignocellulosic biomass: Biochemical and molecular perspectives. *J. Ind. Microbiol. Biotechnol., 35*, 377–391.

Larios-Cruz, R., Buenrrostro-Figueroa, J., Prado-Barragan, A., Rodríguz-Jasso, R. M., Rodriguez-Herrera, R., Julio, M. C., & Cristóbal, A. N., (2017). Valorization of grapefruit by-products as solid support for solid-state fermentation to produce antioxidant bioactive extracts. *Waste and Biomass Valorization*, https://doi.org/10.1007/s12649-017-0156-y (accessed on 24 July 2020).

Lee, Y. S., Huh, J. Y., Nam, S. H., Moon, S. K., & Lee, S. B., (2012). Enzymatic bioconversion of citrus hesperidin by *Aspergillus sojae* naringinase: Enhanced solubility of hesperetin-7-O-glucoside with in vitro inhibition of human intestinal maltase, HMG-CoA reductase, and growth of *Helicobacter pylori*. *Food Chem., 135*, 2253–2259.

Li, K., Xing, R., Liu, S., & Li, P., (2016). Advances in preparation, analysis and biological activities of single chitooligosaccharides. *Carbohydr. Polym., 139*, 178–190.

Li, M., Kim, J. W., & Peeples, T. L., (2002). Amylase partitioning and extractive bioconversion of starch using thermo separating aqueous two-phase systems. *J. Biotechnol., 93*, 15–26.

Lian, X., Li, Z., Dong, Y., Xu, B., Yong, Q., & Yu, S., (2011). Extractive bioconversion of xylan for production of xylobiose and xylotriose using a PEG6000/sodium citrate aqueous two-phase system. *Korean J. Chem. Eng., 28*, 1897–1901.

Liaqat, F., & Eltem, R., (2018). Chitooligosaccharides and their biological activities: A comprehensive review. *Carbohydr. Polym., 184*, 243–259.

Lin, C. S. K., Pfaltzgraff, L. A., Herrero-Davila, L., Mubofu, E. B., Abderrahim, S., Clark, J. H., Koutinas, A. A., et al., (2013). Food waste as a valuable resource for the production of chemicals, materials and fuels. Current situation and global perspective. *Energy Environ. Sci., 6*, 426–464.

Lin, Y. K., Show, P. L., Yap, Y. J., Arbakariya-Ariff, B., Mohammad-Annuar, M. S., Lai, O. M., Tang, T. K., et al., (2016). Production of γ-cyclodextrin by *Bacillus cereus* cyclodextrin glycosyltransferase using extractive bioconversion in polymer-salt aqueous two-phase system. *J. Biosci. Bioeng., 121*, 692–696.

Liu, Y., Zhang, Y., Wu, X., & Yan, X., (2016). Fluid phase equilibria effect of excluded-volume and hydrophobic interactions on the partition of proteins in aqueous micellar two-phase systems composed of polymer and nonionic surfactant. *Fluid Phase Equilib., 429*, 1–8.

Lizardi-Jiménez, M. A., & Hernández-Martínez, R., (2017). *Solid State Fermentation (SSF): Diversity of Applications to Valorize Waste and Biomass, 7*, 44.

Lopez-Trujillo, J., Medina-Morales, M. A., Sanchez-Flores, A., Arevalo, C., Ascacio-Valdés, J. A., Mellado, M., Aguilar, C. N., & Aguilera-Carbo, A. F., (2017). Solid bioprocess of tarbush (*Flourensia cernua*) leaves for b-glucosidase production by *Aspergillus niger*: Initial approach to fiber-glycoside interaction for enzyme induction. *3 Biotech., 7*. https://doi.org/10.1007/s13205-017-0883-6 (accessed on 24 July 2020).

Magadum, D. B., & Yadav, G. D., (2018). Fermentative production, purification of inulinase from *Aspergillus terreus* MTCC 6324 and its application for hydrolysis of sucrose. *Biocatal. Agric. Biotechnol., 14*, 293–299.

Mallek-Fakhfakh, H., & Belghith, H., (2015). Physicochemical properties of thermo tolerant extracellular β-glucosidase from *Talaromyces thermophilus* and enzymatic synthesis of cello-oligosaccharides. *Carbohydr. Res., 419*, 41–50.

Mansour, M. S. M., Abdel-Shafy, H. I., & Mehaya, F. M. S., (2018). Valorization of food solid waste by recovery of polyphenols using hybrid molecular imprinted membrane. *J. Environ. Chem. Eng., 6*, 4160–4170.

Martínez-Ávila, G. C., Aguilera-Carbó, A. F., Rodríguez-Herrera, R., & Aguilar, C. N., (2012). Fungal enhancement of the antioxidant properties of grape waste. *Ann. Microbiol, 62*, 923–930.

Martins, S., Mussatto, S. I., Martínez-Avila, G., Montañez-Saenz, J., Aguilar, C. N., & Teixeira, J. A., (2011). Bioactive phenolic compounds: Production and extraction by solid-state fermentation: A review. *Biotechnol. Adv., 29*, 365–373.

Martins, S., Mussatto, S. I., Martínez-Avila, G., Montañez-Saenz, J., Aguilar, C. N., & Teixeira, J. A., (2011). Bioactive phenolic compounds: Production and extraction by solid-state fermentation. A review. *Biotechnol. Adv., 29*, 365–373.

Mayolo-Deloisa, K., Trejo-Hernández, M. D. R., & Rito-Palomares, M., (2009). Recovery of laccase from the residual compost of *Agaricus bisporus* in aqueous two-phase systems, *Process Biochem., 44*, 435–439.

Medina, M. A., Belmáres, R. E., Aguilera-Carbo, A., Rodríguez-Herrera, R., & Aguilar, C. N., (2010). Fungal culture systems for production of antioxidant phenolics using pecan nut shells as sole carbon source. *Am. J. Agric. Biol. Sci., 5*, 397–402.

Medina-Morales, M. A., López-Trujillo, J., Gómez-Narváez, L., Mellado, M., García-Martínez, E., Ascacio-Valdés, J. A., Aguilar, C. N., & Aguilera-Carbó, A., (2017). Effect of growth conditions on β-glucosidase production using *Flourensia cernua* leaves in a solid-state fungal bioprocess. *3 Biotech., 7*. https://doi.org/10.1007/s13205-017-0990-4 (accessed on 24 July 2020).

Medina-Morales, M. A., Martínez-Hernández, J. L., De La Garza, H., & Aguilar, C. N., (2011). Cellulolytic enzymes production by solid state culture using pecan nut shell as substrate, department of food science and technology sch. *Am. Journal Agric. Biol. Sci., 6,* 196–200.

Melikoglu, M., Lin, C. S. K., & Webb, C., (2013). Stepwise optimization of enzyme production in solid-state fermentation of waste bread pieces. *Food Bioprod. Process, 91*, 638–646.

Méndez-Hernández, J. E., Loera, O., Méndez-Hernández, E. M., Herrera, E., Arce-Cervantes, O., & Soto-Cruz, N. Ó., (2018). Fungal pretreatment of corn Stover by *Fomes* sp. EUM1: Simultaneous production of readily hydrolyzable biomass and useful biocatalysts. *Waste and Biomass Valorization*. https://doi.org/10.1007/s12649-018-0290-1 (accessed on 24 July 2020).

Mutabaruka, R., Hairiah, K., & Cadisch, G., (2007). Microbial degradation of hydrolysable and condensed tannin polyphenol-protein complexes in soils from different land-use histories. *Soil Biol. Biochem., 39*, 1479–1492.

Nava-Cruz, N. Y., Medina-Morales, M. A., Martinez, J. L., Rodriguez, R., & Aguilar, C. N., (2014). Agave biotechnology: An overview. *Crit. Rev. Biotechnol., 35*, 546–559.

Nazzaro, F., Fratianni, F., Ombra, M., D'Acierno, A., & Coppola, R., (2018). Recovery of biomolecules of high benefit from food waste. *Current Opinion in Food Science, 22*, 46–54.

Ng, H. S., Ooi, C. W., Mokhtar, M. N., Show, P. L., Ariffe, A., Tan, J. S., Ng, E. P., & Ling, T. C., (2013). Extractive bioconversion of cyclodextrins by *Bacillus cereus* cyclodextrin glycosyltransferase in aqueous two-phase system. *Bioresour. Technol., 142*, 723–726.

Nidheesh, T., Kumar, P. G., & Suresh, P. V., (2015). Enzymatic degradation of chitosan and production of D-glucosamine by solid substrate fermentation of exo-b-D-glucosaminidase (exochitosanase) by *Penicillium decumbens* CFRNT15. *Int. Biodeterior. Biodegradation., 97*, 97–106.

Nile, S. H., Nile, A. S., Keum, Y. S., & Sharma, K., (2017). Utilization of quercetin and quercetin glycosides from onion (*Allium cepa* L.) solid waste as an antioxidant, urease and xanthine oxidase inhibitors. *Food Chem., 235*, 119–126.

Niphadkar, S. S., Vetal, M. D., & Rathod, V. K., (2015). Purification and characterization of polyphenol oxidase from waste potato peel by aqueous two-phase extraction. *Prep. Biochem. Biotechnol., 45*, 632–649.

Oliveira, S. D., De AraújoPadilha, C. E., Asevedo, E. A., Pimentel, V. C., De Araújo, F. R., De Macedo, G. R., & Dos, S. E. S., (2018). Utilization of agro-industrial residues for producing cellulases by *Aspergillus fumigatus* on semi-solid fermentation. *J. Environ. Chem. Eng., 6*, 937–944.

Palacios, S., Ruiz, H. A., Ramos-Gonzalez, R., Martínez, J., Segura, E., Aguilar, M., Aguilera, A., et al., (2017). Comparison of physic chemical pretreatments of banana peels for bio-ethanol production, *Food Sci. Biotechnol., 26*, 993–1001.

Pareek, N., Ghosh, S., Singh, R. P., & Vivekanand, V., (2014). Biocatalysis and agricultural biotechnology mustard oil cake as an inexpensive support for production of chitin deacetylase by *Penicillium oxalicum* SAE M-51 under solid-state fermentation. *Biocatal. Agric. Biotechnol., 3*, 212–217.

Pechsrichuang, P., Lorentzen, S. B., Aam, B. B., Tuveng, T. R., Hamre, A. G., Eijsink, V. G. H., & Yamabhai, M., (2018). Bioconversion of chitosan into chito-oligosaccharides (CHOS) using family 46 chitosanase from *Bacillus subtilis* (BsCsn46A). *Carbohydr. Polym., 186*, 420–428.

Pleissner, D., Qi, Q., Gao, C., Rivero, C. P., Webb, C., Lin, C. S. K., & Venus, J., (2016). Valorization of organic residues for the production of added-value chemicals: A contribution to the bio-based economy. *Biochem. Eng. J., 116*, 3–16.

Poverenov, E., Arnon-Rips, H., Zaitsev, Y., Bar, V., Danay, O., Horev, B., Bilbao-Sainz, C., McHugh, T., & Rodov, V., (2018). Potential of chitosan from mushroom waste to enhance quality and storability of fresh-cut melons. *Food Chem., 268*, 233–241.

Prabhu, A. A., Gupta, E., & VenkataDasu, V., (2018). Purification of β-galactosidase from recombinant *Pichia pastoris* using aqueous two-phase separation technique. *Sep. Sci. Technol.* Retrieved from: https://doi.org/10.1080/01496395.2018.1497654 (accessed on 24 July 2020).

Ramakrishnan, V., Goveas, L. C., Suralikerimath, N., Jampani, C., Halami, P. M., & Narayan, B., (2016). Extraction and purification of lipase from *Enterococcus faecium* MTCC5695 by PEG/phosphate aqueous-two phase system (ATPS) and its biochemical characterization. *Biocatal. Agric. Biotechnol., 6*, 19–27.

Rani, V., Mohanram, S., Tiwari, R., Nain, L., & Arora, A., (2014). Beta-Glucosidase: Key enzyme in determining efficiency of cellulase and biomass hydrolysis. *J. Bioprocess Biotech., 5*, 1–8.

Ravindran, R., & Jaiswal, A. K., (2016). Exploitation of food industry waste for high-value products. *Trends Biotechnol., 34*, 58–69.

Ravindran, R., & Jaiswal, A. K., (2016). Exploitation of food industry waste for high-value products. *Trends Biotechnol., 34*, 58–69.

Rodrigues, G. D., De Lemos, L. R., Da Silva, L. H. M., Da Silva, M. C. H., Minim, L. A., & Dos, R. C. J. S., (2010). A green and sensitive method to determine phenols in water and wastewater samples using an aqueous two-phase system. *Talanta*, *80*, 1139–1144.

Ruiz-Ruiz, F., Benavides, J., Aguilar, O., & Rito-Palomares, M., (2012). Aqueous two-phase affinity partitioning systems: Current applications and trends. *J. Chromatogr. A*, *1244*, 1–13.

Šalić, A., Tušek, A., Fabek, D., Rukavina, I., & Zelić, B., (2011). Aqueous two-phase extraction of polyphenols using a microchannel system-process optimization and intensification. *Food Technol. Biotechnol.*, *49*, 495–501.

Sánchez-Ramírez, J., Martínez-Hernández, J. L., Segura-Ceniceros, P., López, G., Saade, H., Medina-Morales, M. A., Ramos-González, R., et al., (2016). Cellulases immobilization on chitosan-coated magnetic nanoparticles: Application for *Agave atrovirens* lignocellulosic biomass hydrolysis. *Bioprocess Biosyst Eng.*, *40*, 9–22.

Sánchez-Trasviña, C., González-Valdez, J., Mayolo-Deloisa, K., & Rito-Palomares, M., (2015). Impact of aqueous two-phase system design parameters upon the in situ refolding and recovery of invertase. *J. Chem. Technol. Biotechnol.*, *90*, 1765–1772.

Sánchez-Trasviña, C., Mayolo-Deloisa, K., González-Valdez, J., & Rito-Palomares, M., (2017). Refolding of laccase from *Trametes versicolor* using aqueous two phase systems: Effect of different additives. *J. Chromatogr. A, 1507*, 25–31.

Saravana, P. S., Ho, T. C., Chae, S. J., Cho, Y. J., Park, J. S., Lee, H. J., & Chun, B. S., (2018). Deep eutectic solvent-based extraction and fabrication of chitin films from crustacean waste. *Carbohydr. Polym.*, *195*, 622–630.

Seiboth, B., & Metz, B., (2011). Fungal arabinan and L-arabinose metabolism. *Appl. Microbiol. Biotechnol.*, *89*, 1665–1673.

Sepúlveda, L., Wong-Paz, J. E., Buenrostro-Figueroa, J., Ascacio-Valdés, J. A., Aguilera-Carbó, A., & Aguilar, C. N., (2018). Solid-state fermentation of pomegranate husk: Recovery of ellagic acid by SEC and identification of ellagitannins by HPLC/ESI/MS. *Food Biosci.*, *22*, 99–104.

Sepúlveda, L., Wong-Paz, J. E., Buenrostro-Figueroa, J., Ascacio-Valdés, J. A., Aguilera-Carbó, A., & Aguilar, C. N., (2018). Solid state fermentation of pomegranate husk: Recovery of ellagic acid by SEC and identification of ellagitannins by HPLC/ESI/MS. *Food Biosci.*, *22*, 99–104.

Silva, C. J., De França, P. R. L., & Porto, T. S., (2018). Optimized extraction of polygalacturonase from *Aspergillus aculeatus* URM4953 by aqueous two-phase systems PEG/Citrate. *J. Mol. Liq.*, *263*, 81–88.

Singh, G., Verma, A. K., & Kumar, V., (2016). Catalytic properties, functional attributes and industrial applications of β-glucosidases. *3 Biotech.*, *6*. https://doi.org/10.1007/s13205-015-0328-z (accessed on 24 July 2020).

Singh, R. S., Chauhan, K., Pandey, A., & Larroche, C., (2018). Biocatalytic strategies for the production of high fructose syrup from inulin. *Bioresour. Technol.*, *260*, 395–403.

Singh, R. S., Chauhan, K., Pandey, A., & Larroche, C., (2018). Biocatalytic strategies for the production of high fructose syrup from inulin. *Bioresour. Technol.*, *260*, 395–403.

Singh, R., Kumar, M., Mittal, A., & Mehta, P. K., (2017). Microbial metabolites in nutrition, healthcare, and agriculture. *3 Biotech.*, *7*. https://doi.org/10.1007/s13205-016-0586-4 (accessed on 24 July 2020).

Soccol, C. R., Scopel, E., Alberto, L., Letti, J., Karp, S. G., & Woiciechowski, A. L., (2017). Recent developments and innovations in solid-state fermentation. *Biotechnol. Res. Innov.*, *1*, 25–71.

Sorensen, A., Andersen, J. J., Ahring, B. K., Teller, P. J., & Lubeck, M., (2014). Screening of carbon sources for beta-glucosidase production by *Aspergillus saccharolyticus*. *Int. Biodeterior. Biodegrad.*, *93*, 78–83.

Sosnowska, M. E., Jankiewicz, U., Kutwin, M., Chwalibog, A., & Gałązka, A., (2018). Influence of salts and metal nanoparticles on the activity and thermal stability of a recombinant chitinase from *Stenotrophomonas maltophilia* N4. *Enzyme Microb. Technol.*, *116*, 6–15.

Su, P. C., Hsueh, W. C., Chang, W. S., & Chen, P. T., (2017). Enhancement of chitosanase secretion by *Bacillus subtilis* for production of chitosan oligosaccharides. *J. Taiwan Inst. Chem. Eng.*, *79*, 49–54.

Svarc-Gajic, J., Stojanovic, Z., Segura, C. A., Arraez, R. D., Borras, I., & Vasiljevic, I., (2013). Development of a microwave-assisted extraction for the analysis of phenolic compounds from *Rosmarinus officinalis*. *J. Food Eng.*, *119*, 525–532.

Talekar, S., Patti, A. F., Vijayraghavan, R., & Arora, A., (2018). An integrated green biorefinery approach towards simultaneous recovery of pectin and polyphenols coupled with bioethanol production from waste pomegranate peels. *Bioresour. Technol.*, *266*, 322–334.

Veitch, N. C., & Grayer, R. J., (2008). Flavonoids and their glycosides, including anthocyanins. *Nat. Prod. Rep.*, *25*, 555–611.

Walia, A., Guleria, S., Mehta, P., Chauhan, A., & Parkash, J., (2017). Microbial xylanases and their industrial application in pulp and paper bio-bleaching: A review. *3 Biotech*, *7*. https://doi.org/10.1007/s13205-016-0584-6 (accessed on 24 July 2020).

Wang, D., Li, A., Han, H., Liu, T., & Yang, Q., (2018). A potent chitinase from *Bacillus subtilis* for the efficient bioconversion of chitin-containing wastes. *Int. J. Biol. Macromol.*, *116*, 863–868.

Wang, H., Kaur, G., Pensupa, N., Uisan, K., Du, C., Yang, X., & Lin, C. S. K., (2018). Textile waste valorization using submerged filamentous fungal fermentation. *Process Saf. Environ. Prot.*, *118*, 143–151,

Willems, J. L., & Low, N. H., (2012). Major carbohydrate, polyol, and oligosaccharide profiles of agave syrup. Application of this data to authenticity analysis. *J. Agric. Food Chem.*, *60*, 8745–8754.

Xavier, L., Freire, M. S., Vidal-Tato, I., & Gonzlez-Alvarez, J., (2017). Recovery of phenolic compounds from eucalyptus globulus wood wastes using PEG/phosphate Aqueous two-phase systems. *Waste and Biomass Valorization*, *8*, 443–452.

Xavier-Salmón, D. N., Walter, A., Souza-Porto, T., Aparecida-Moreira, K., De Souza-Vandenberghe, L. P., Soccol, C. R., Figueiredo-Porto, A. L., & Rigon-Spier, M., (2014). Aqueous two-phase extraction for partial purification of *Schizophyllum commune* phytase produced under solid-state fermentation. *Biocatal. Biotransformation*, *32*, 45–52.

Xu, Y. Y., Qiu, Y., Ren, H., Ju, D. H., & Jia, H. L., (2017). Optimization of ultrasound-assisted aqueous two-phase system extraction of polyphenolic compounds from *Aronia melanocarpa* pomace by response surface methodology. *Prep. Biochem. Biotechnol.*, *47*, 312–321.

Yang, P., Guo, L., Cheng, S., Lou, N., & Lin, J., (2011). Recombinant multi-functional cellulase activity in submerged fermentation of lignocellulosic wastes. *Renew. Energy*, *36*, 3268–3272.

Yang, X., Nisar, T., Liang, D., Hou, Y., & Guo, Y., (2018). Low methoxyl pectin gelation under alkaline conditions and its rheological properties: Using NaOH as a pH regulator. *Food Hydrocolloids.*, *79*, 560–571.

Yang, Y., Feng, F., Zhou, Q., Zhao, F., Du, R., Zhou, Z., & Han, Y., (2018). Isolation, purification, and characterization of exopolysaccharide produced by *Leuconostoc pseudomesenteroides* YF32 from soybean paste. *Int. J. Biol. Macromol., 114*, 529–535.

Yasinok, A. E., Biran, S., Kocabas, A., & Bakir, U., (2010). Xylanase from a soil isolate, *Bacillus pumilus*: Gene isolation, enzyme production, purification, characterization and one-step separation by aqueous-two-phase system. *World J. Microbiol. Biotechnol., 26*, 1641–1652.

Yusuf, M., (2017). *Agro-Industrial Waste Materials and their Recycled Value-Added Applications*. Review, https://dx.doi.org/10.1007/978-3-319-48281-1_48-1, 1–11. (accessed on 24 July 2020).

Zhang, S., Hu, H., Wang, L., Liu, F., & Pan, S., (2018). Preparation and prebiotic potential of pectin oligosaccharides obtained from citrus peel pectin. *Food Chem., 244*, 232–237.

Zhang, S., Hu, H., Wang, L., Liu, F., & Pan, S., (2018). Preparation and prebiotic potential of pectin oligosaccharides obtained from citrus peel pectin. *Food Chem., 244*, 232–237.

Zhang, Z., Dong, J., Zhang, D., Wang, J., Qin, X., Liu, B., Xu, X., Zhang, W., & Zhang, Y., (2018). Expression and characterization of a pectin methylesterase from *Aspergillus niger* ZJ5 and its application in fruit processing. *J. Biosci. Bioeng.* https://doi.org/10.1016/j.jbiosc.2018.05.022 (accessed on 24 July 2020).

Zhao, H. M., Guo, X. N., & Zhu, K. X., (2017). Impact of solid state fermentation on nutritional, physical and flavor properties of wheat bran. *Food Chem., 217*, 28–36.

CHAPTER 2

Mango Seed Byproduct: A Sustainable Source of Bioactive Phytochemicals and Important Functional Properties

CRISTIAN TORRES-LEÓN,[1,2] MARIA T. DOS SANTOS CORREIA,[2] MARIA G. CARNEIRO-DA-CUNHA,[2] LILIANA SERNA-COCK,[3] JANETH VENTURA-SOBREVILLA,[1] JUAN A. ASCACIO-VALDÉS,[1] and CRISTÓBAL N. AGUILAR[1]

[1]*Bioprocesses and Bioproducts Research Group.*
Food Research Department, School of Chemistry,
Universidad Autónoma de Coahuila, 25280 Saltillo, Coahuila, México,
E-mail: cristobal.aguilar@uadec.edu.mx

[2]*Departamento de Bioquímica, Centro de Ciencias Biológicas,*
Universidade Federal de Pernambuco. Recife, Pernambuco, Brazil

[3]*School of Engineering and Administration.*
Universidad Nacional de Colombia, Street 32 Chapinero, Palmira, Valle del Cauca, Colombia

ABSTRACT

Mango is a fruit with nutritional properties and also with recognized therapeutic uses. This fruit is widely grown in tropical and subtropical countries as a source of food and income for people. As mango is a seasonal fruit, about 20% of fruits are processed. Mango processing generates approximately 14,000,000 tons of bio-waste per year in the world. The mango seed is one of the main residues and represents between 10 to 25% of the fruit. Currently, this byproduct management generates high costs and are a source of environmental contamination. However, the chemical composition of mango seeds could potentiate their use as a sustainable source of high added value phytochemicals. Bioactive phytochemicals in mango seed include phenolic compounds, such as gallic acid, ellagic acid, rhamnetin, methyl gallate, and pentagalloylglucose (PGG). These compounds have a particular interest in their pharmacologic and biological activities. Various scientific research has confirmed critical functional properties in mango seed as anti-oxidant, anti-microbial, anti-cancer, anti-diabetic, anti-hemolytic, and anti-diarrheal activities. However, phytochemical profiles and biological activities can change significantly depending on provenance, the varieties of mango used, and the extraction methods used. Scientific research must continue to elucidate the effect of these factors.

Additionally, new research should be geared to evaluate activities of models that have not yet been evaluated. The mango seed offers a range of possibilities to develop value-added bioproducts. Therefore, in this chapter, we review the mango seed bioactive phytochemicals, looking in detail at their reported functional and biological activities, potential applications, and the technological aspects.

2.1 INTRODUCTION

The mango is one crucial fruit, cultivated in several tropical and subtropical regions, enjoys the status of "the king of fruits" as a result of its unique flavor, fragrance, and nutritional value (Singh et al., 2013). The mango belongs to the genus *Mangifera* and has 69 recognized species (Mukherjee & Litz, 2009) in the Anacardiaceae family (Bompard, 2009). Although in specific parts of the world, various species of *Mangifera* are cultivated as *Mangifera foetida* (Horse mango) in Asia or *Mangifera caesia* (Ataulfo mango) in Mexico, *Mangifera indica* is the species commonly cultivated in the world. Mangoes have great nutritional importance in developed countries in Asia, Africa, and America. Mango pulp contains fiber, amino acids, carbohydrates, fatty acids, minerals, organic acids, protein, vitamins and polyphenolics (Singh et al., 2013). Mango is native to South Asia, wherefrom it has spread worldwide to become one of the most cultivated fruits (Fasoli & Righetti, 2013). Mango is grown over 5,500,000 ha in 94 countries, worldwide annual production of mango is estimated at 50,000,000 tonnes. India ranks first in mango production, followed by China, Thailand, Indonesia, Mexico, Pakistan and Brazil (FAOSTAT, 2017).

Approximately, 20 % of mango fruits are processed for products such as puree, nectar, leather, pickles, minimally processed, canned slices, and chutney (Ajila et al., 2007; Dussan et al., 2014; Ravani & Joshi, 2013), this generates high amounts of waste since 35 to 60% of the fruit is discarded after processing (Torres-León et al., 2016). For instance, there are about 300,000 tonnes of dry mango seed kernels available annually in India after consumption or industrial processing of mango fruits (Soong & Barlow, 2004). The seed is not currently utilized; it is discarded as waste and becoming a source of pollution. Additionally, millions of dollars are spent to dispose of mango waste in India; the average transportation cost was found to be \$15 per tonne per trip (FICCI, 2010), which may indicate > \$20 million for the total landfilling cost. However, ethnobotanical studies indicated that the plant of mango is widely used in traditional medicine to cure vomiting, dysentery and burning (Derese & Kuete, 2017; Rajan et al., 2012), and the mango seed could be used as a source of natural compounds with high added value.

The agro-industrial byproducts have been recently posed as a source of active compounds (O'Shea, Arendt, & Gallagher, 2012), the parts discarded in the processing of fruits have more exceptional functional properties than the pulp or the finished product (Ayala et al., 2011). Therefore, the use of byproducts as mango seed can have some important applications in the food industry and human health (Torres-León et al., 2018b). Interesting biological properties such as antioxidant, antimicrobial and anticancer activity have been reported in agro-industrial waste (Serna-Cock et al., 2016). Currently, there is a concern for the search for new natural sources of antioxidants (Sindhi et al., 2013), antimicrobial agents and anti-cancer agents. Al respect, in the year 2008, the ethanol extract of mango seed was listed as one of the four food extracts with the highest biological power, compared with l-ascorbic acid (Saito et al., 2008). These biological activities can be attributed to the bioactive compounds naturally present in mango seed. Therefore, in this paper, we analyze and discuss several aspects related to the potential of mango seed as a source of bioactive phytochemicals, looking in detail at their reported functional and biological activities, potential applications and the technological aspects.

2.2 MANGO SEED

Mango fruit is a deliquescent drupe with a seed surrounded by a fleshy mesocarp (pulp) covered by an exocarp (peel) (Singh et al., 2013). As shown in Figure 2.1, the mango seed is composed of an endocarp and testa (MSET) that is thick and hard and encloses a kernel (MSK). Depending on the variety, the seed represents 10–25% of the total weight of the fruit, MSET and MSK represent approximately between 15–55% and 45–85% of the seed, respectively. Most of the research papers reported in the scientific literature regarding the extraction of active compounds have been developed in MSK. However, recent HPLC-MS studies have also reported the presence of interesting phenolic compounds in MSTE (Gómez-Caravaca et al., 2016). To facilitate the terminology, MSET and MSK will be named as mango seed in this document, with the excuse of the work of Gómez et al., (2016).

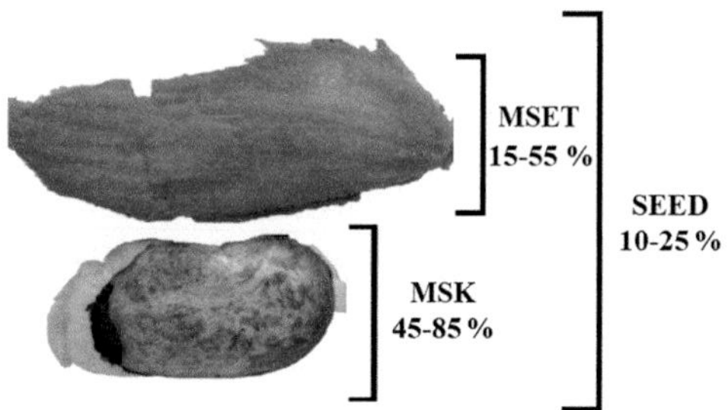

FIGURE 2.1 Structure of mango seed. MSET, mango seed endocarp and testa; MSK, mango seed kernel.

2.2.1 VALORIZATION OF MANGO SEEDS

The use of agro-industrial by-products and food wastes to obtain bioactive compounds depends on different aspects such as the type of biological waste, the compounds, and their nature. However, a general strategy should include a drying pretreatment to guarantee the microbiological and biochemical stability of the material, the separation of the compounds of the biological matrix with extraction processes, the measurement using spectrophotometric techniques (content of compounds and biological activities), and finally the identification the profile of compounds by HPLC-MS (Figure 2.2).

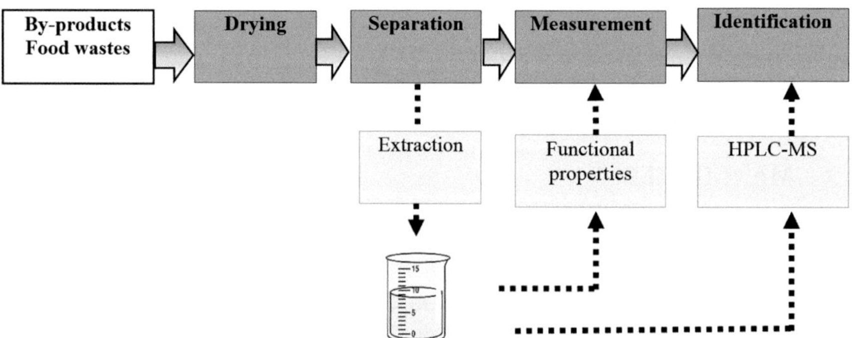

FIGURE 2.2 Methodology for the analysis of phytochemicals in agro-industrial byproducts and food waste, as a valorization strategy.

In this chapter of the book, the drying and separation will be discussed in Section 2.6 (technological aspects). In Section 2.3, we will present the reported results of total phenol content (Spectrophotometric analysis) and then the mass spectrometry (MS) results. The results of the measurement of functional properties reported in mango seeds will be discussed in detail in Section 2.4.

2.3 PHYTOCHEMICAL IN MANGO SEED

2.3.1 The Polyphenol

Mango seed contains an assortment of phytochemicals in varying concentrations, usually determined by genotypic, environmental factors, and the interaction of both factors (Dorta et al., 2014). The content of total phenols has been investigated in diversity mango varieties in countries such as Thailand, Nigeria, India, Egypt, EEUU, Brazil and Mexico. As shown in Table 2.1, the mango seed has a high total phenolic content ranging from 21.9 to 598 mg g^{-1} dry weight. Variations in chemical composition may be due to country origins, varieties used, culture conditions, drying method and extraction conditions (Figure 2.2) (Wang & Zhu, 2017). The way in which phenolic compounds are found in the biological matrix also has a great influence (in the separation operation). Phenolic compounds in mango seed may exist in free, conjugated, or bound forms. The free compounds are linked to sugars (glucose) and the bound phenols are bound to other constituents of the cell-matrix such as proteins, hemicellulose, cellulose, and lignin, so their separation is not as simple as in the case of free phenols.

The use of high performance liquid chromatography (HPLC) and gas chromatography (GC) coupled to mass spectrometry (MS) has allowed identifying phytochemical compounds in mango seed. 20 compounds were identified for Dorta et al. (2014), from the mango seed using HPLC-ESIQTOF-MS (Table 2.2). Berardini et al. (2004) reported a total of 21 phytochemicals in mango seed (Tommy Atkins) using high-performance liquid chromatography/electrospray ionization mass spectrometry (ESI-HPLC/MS). Gómez-Caravaca et al. (2016), analyzed the free and bound phenols in Keitt mango seed, using HPLC-DAD-q-TOF-MS. The authors also analyzed separately and MSET; forty-three free phenolic compounds and eight bound phenolic compounds were identified in mango MSK, and thirty-seven free polar compounds and nine bound compounds were identified in mango MSET. Finally, twelve phytochemicals were identified in the Waterlily mango seed using Gas chromatography-mass spectrometry (GC-MS) (Abdullah et al., 2014).

Dorta et al. (2014), reported that the identified compounds belong to 5 families: gallates and gallotannins; flavonoids (mainly quercetin derivatives); ellagic acid and derivatives; xanthones (principally mangiferin); and

TABLE 2.1 Total Phenol Content and Antioxidant Activity in Mango Seed

Variety	Country	Extraction method	Conditions	Total phenol content (mgGAE g^{-1})	DPPH Radical scavening activity (%)	IC_{50} μg mL^{-1})	References
Chok-a-nan	Tailandia	-	Water 1:3 (w/v)			4.1	Khammuang & Sarnthima, (2011)
Fah-lun	Tailandia	-	Water 1:3 (w/v)	399.8		5.3	Khammuang & Sarnthima, (2011)
Kaew	Tailandia	-	Water 1:3 (w/v)	293.1		16.2	Khammuang & Sarnthima, (2011)
Nam-dok-mai	Tailandia	-	Water 1:3 (w/v)	73.8		10.5	Khammuang & Sarnthima, (2011)
Ikanekpo	Nigeria	-	Methanol : Water 1:1 (v/v)	118.1		47	Prakash et al., (2011)
Ikanekpo	Nigeria	Orbital shaker	Methanol	59.6		143.3	Arogba & Omede, (2014)
kesar	India	Rotary shaker	various 1:10 (w/v)			11–120	Vaghasiya & Chanda, (2010)
Zebda	Egypt	-	Hexane 2:1 (v/w)				Abdel et al., (2012)
Zebda	Egypt	Centrifuged	Methanol: water (60:40 v/v)	21.9	95.08		Ashoush & Gadallah, (2011)
Ubá	Brazil	-	Methanol: water (60:40 v/v)	23.9	94.2		Ribeiro et al., (2008)
Tommy Atkins	USA	Water bath	2: 25 (w/v) Extraction solution	82.5			Sogi et al., (2013)
Ataulfo	México	Microwave-assisted extraction	Etanol 90% 1:60 (w/v)	598.4		78	Torres-León et al., (2017)

TABLE 2.2 Phenolic Compounds Identified in the Seed Extracts of Three Mango Varieties Produced in Spain by HPLC-ESI-QTOF-MS

Peak	Tentative identification	Variety	Rt	[M – H]–	MF
1	Galloyl glucose	Keitt	8.3	331.0671	$C_{13}H_{16}O_{10}$
2	Gallic acid	Keitt	9.2	169.0142	$C_7H_6O_5$
3	Theogallin	Keitt	10.8	343.059	$C_{14}H_{16}O_{10}$
4	Maclurin-3-C-β-D-glucoside	Keitt, Sensation, Gomera	14.3	423.0945	$C_{19}H_{20}O_{11}$
5	Methyl gallate	All the extracts	17	183.0299	$C_8H_8O_5$
6	Maclurin-3-C-(2-O-galloyl)-β-D-glucoside	Keitt, Sensation, Gomera	17.7	575.1042	$C_{26}H_{24}O_{15}$
7	Maclurin-3-C-(2,3-di-O-galloyl)-β-D-glucoside	All the extracts	21.9	727.1152	$C_{33}H_{28}O_{19}$
8	Ethyl gallate	All the extracts	24.2	197.0455	$C_9H_{10}O_5$
9	Mangiferin	Keitt, Sensation, Gomera	25	421.0776	$C_{19}H_{18}O_{11}$
10	Methyl gallate ester	All the extracts	27.7	335.0409	$C_{15}H_{12}O_9$
11	Tetra-O-galloyl-glucoside	All the extracts	35.8	787.0999	$C_{34}H_{28}O_{22}$
12	Methyl gallate ester	All the extracts	38.7	335.0409	$C_{15}H_{12}O_9$
13	Quercetin-3-O-glucoside (isoquercitrin)	All the extracts	39.7	463.0882	$C_{21}H_{20}O_{12}$
14	Ellagic acid	All the extracts	40	300.999	$C_{14}H_6O_8$
15	Quercetin-3-O-alactoside	All the extracts	40.3	463.0882	$C_{21}H_{20}O_{12}$
16	Valoneic acid dilactone	All the extracts	41.5	469.0524	$C_{24}H_{22}O_{10}$
17	Penta-O-galloyl-lucoside	All the extracts	41.5	939.1109	$C_{41}H_{32}O_{26}$
18	Quercetin-3-O-rutinoside	All the extracts	42.8	609.1813	$C_{27}H_{30}O_{16}$
19	Rhamnetin-3-[6″-2-butenoil-hexoside]	All the extracts	44.1	545.0583	$C_{26}H_{16}O_{13}$
20	Ethyl 2,4-dihydroxy-3-(3,4,5-rihydroxybenzoyl) oxybenzoate	All the extracts	45.8	349.0576	$C_{16}H_{14}O_9$

Peak, compound number; Rt, retention time (min); MF, molecular formula; [a]*Taken from Dorta et al., (2014).*

benzophenones and derivatives such as maclaurin derivatives. Methyl gallate and Penta-O-galloylglucoside (PGG) were the main compounds in the seeds of Keitt and Sensation varieties; however, a rhamnetin derivative was the principal compound in Gomera 3 seeds. Berardini et al. (2004) reported that all the compounds identified in mango seed belong to the family of gallotannins and penta-O-galloylglucoside (PGG) was the majority compound. Gómez-Caravaca et al. (2016) reported that the compounds identified belonged to the families: gallate and gallotannins, xanthones, and benzophenones and ellagic acid was the majority compound. Barreto et al. (2008), using the HPLC-ESI-MS report that the predominant compound detected in six mango cultivars of Brazil was penta-O-galloylglucoside (PGG). In general, mango seed polyphenols primarily consist of phenolic acids, tannins (gallotannins and ellagitannins), flavonoids, xanthones.

2.3.2 PHENOLIC ACIDS

Hydroxybenzoic and hydroxycinnamic acids are the two main types of phenolic acids in mango seed. Gallic acid (GA) is a hydroxybenzoic acid that possesses various functional properties (antioxidant, anti-inflammatory, antibiotic, anticancer, antiviral and cardiovascular protection) (Govea-Salas et al., 2016). In plants, gallic acid occurs in free form or the form of esters (e.g., pentagalloylglucose) (Figure 2.3). Ellagic acid is a hydroxycinnamic acid with beneficial characteristics in the human and animal physiology and health (anti-mutagenic, antimicrobial and antioxidant properties, and inhibitors of human immunodeficiency virus). This compound has generated commercial interest in recent years due to its properties, applications, and benefits to human health (Sepúlveda et al., 2011). Gómez-Caravaca et al. (2016), reported that ellagic acid was the most abundant bound compound in the mango seed from Spain (650 mg 100 g^{-1}). Soong and Barlow (2006), reported that mango seeds from Singapore are a potential source of Gallic acid and ellagic acid.

2.3.3 TANNINS

Tannins are secondary metabolites in plants and food and are generally classified into four groups: gallotannins, ellagitannins, condensed tannins and complex tannins (Ascacio et al., 2011). Gallotannis are the dominant group of phytochemical compounds in mango seed (Barreto et al., 2008; Berardini et al., 2004; Dorta et al., 2014). Berardini et al. (2004) report that mango kernels contained 15.5 mg g^{-1} of gallotanins and thus proved to be a rich source of gallotannins. The main gallotannin reported in the mango seed was penta-O-galloylglucoside (PGG) (Barreto et al., 2008; Berardini et al., 2004; Dorta et al., 2014; Torres-León et al., 2017a).

2.3.3.1 PENTA-O-GALLOYL GLUCOSE (PGG)

PGG is a hydrolyzable tannin classified as gallotannin but also participates in the formation of ellagitannins (Figure 2.3) (Torres-León et al., 2017a). This has shown important functional properties as antimicrobial, anti-inflammatory, anticarcinogenic, antidiabetic and antioxidant. However, low

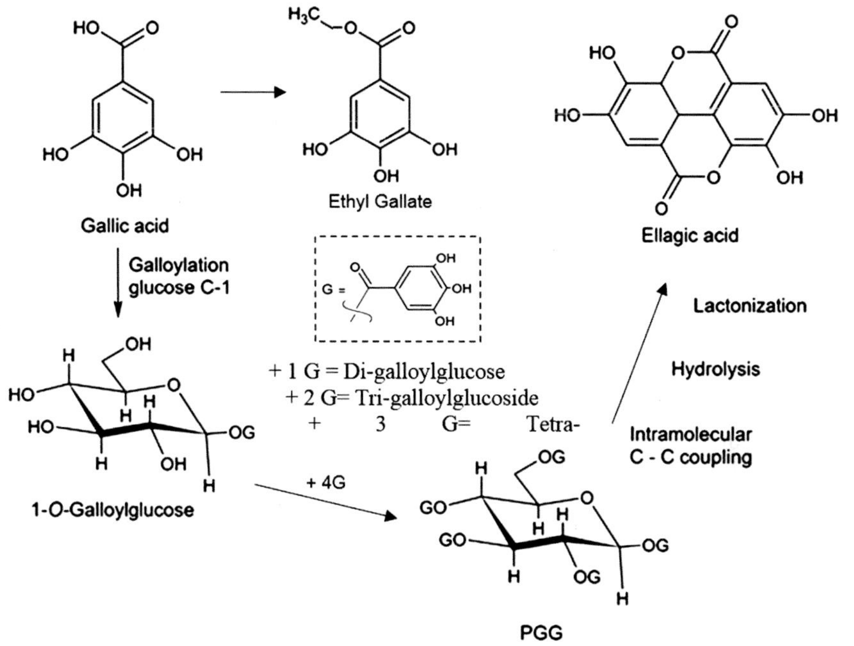

FIGURE 2.3 Main tannins in the mango seed and its participation in the route of hydrolyzable tannins.

production yields are a limitation of the evaluation of properties. PGG can be synthesized from gallic acid and glucose (Niemetz & Gross, 2005) or extracted from plants of traditional oriental medicine. In this regard, Mango seed can be a new sustainable source of PGG. In the extract of the mango seed Fahlun-Thailand has been reported a high PGG content (612.8 mg g^{-1}), this is the main polyphenol compound found in the extract (GA 4.4 mg g^{-1} and methyl gallate 6.8 mg g^{-1}) (Nithitanakool, Pithayanukul, & Bavovada, 2009). Barreto et al. (2008), reported the predominant compound detected in mango seed of Brazil was PGG, the content of PGG varied from 31.45 to 153.57 g kg^{-1} dry matter. Luo et al. (2014), reported a PGG content in MSK (China) of 34 mg/g and the MSET of 26.28 mg g^{-1}. Gómez-Caravaca et al. (2016), also reported that PGG is the main compound in Keitt mango seeds (Spain).

2.3.4 *FLAVONOIDS*

Flavonoids are low molecular weight compounds, consisting of fifteen carbon atoms, arranged in a C6–C3–C6 configuration. Flavonoids have two aromatic rings A and B, joined by a heterocyclic ring, C (Balasundram et al., 2006). The antioxidant, anti-inflammatory and anticancer capacities of these compounds are well documented (Chen & Charlie, 2013). Commercial applications of flavonoids include food additives, nutraceuticals, pharmaceuticals, and cosmetics. The global market for flavonoids was valued at over 840 million USD in 2015 and is expected to surpass 1 trillion USD beyond 2020 (Ng et al., 2019). In the mango seed has been reported the presence of flavonoids as Quercetin and Rhamnetin. Quercetin (3,30,40, 5,7-pentahydroxyflavone), is a flavonoid that has attracted significant interest because it is a potent antioxidant with proven anticancer effects (Moskaug et al., 2004) and Rhamnetin ((2-(3,4-dihydroxyfenyl)-3,5-dihydroxy-7-methoxychromen-4-one)) has also important antioxidative, antihepatotoxic, anti-carcinogenic, antiviral and anti-inflammatory properties (Ramešová et al., 2017). Dorta et al. (2014) reported that rhamnetin derivative was the principal compound in Gomera 3 seeds.

2.3.5 XANTHONES

Xanthones are heterocyclic compounds with a yellow coloration and all of them have dibenzo-γ-pyrone as the basic skeleton (Diderot et al., 2006). Mangiferin is the main Xanthona present in the mango seed (Dorta et al., 2014). Mangiferin (1,3,6,7-Tetrahydroxy-2-[3,4,5-trihydroxy-6-(hydroxymethyl) oxan-2-yl] xanthen-9-one) is a polyphenolxanthone with strong antioxidant activity; it has demonstrated effects against Alzheimer's disease (Sethiya et al., 2014).

2.4 FUNCTIONAL PROPERTIES AND HEALTH BENEFITS

2.4.1 ANTIOXIDANT

An antioxidant is any substance that, at low concentration, delays the oxidation of DNA, proteins, carbohydrates or lipids (Sindhi et al., 2013). Research in antioxidants has intensified in recent years since these compounds can reduce free radicals and prevent oxidative stress and their deleterious effects on the human body and health. Free radicals are molecules containing one or more unpaired electrons (Poprac et al., 2017). The adverse health effects reported in synthetic antioxidants used in food have also motivated the search for natural antioxidants. The habits of healthy consumption and the good approval of natural compounds in the diet have increased the search for new natural antioxidants.

Mango seed extracts have been classified with high biological activity (Saito et al., 2008), which has enhanced the determination of antioxidant activity in various varieties of mango in the world. As shown in Table 2.1, the mango seed has a high antioxidant potential (expressed in DPPH radical scavenging activity) and IC_{50} (concentration when scavenging 50% of the DPPH free radical, lower IC_{50} value represents the higher activity) with values in the range of 94–95% and 4.1–143 μg mL^{-1}, respectively. The DPPH assay is the most used spectrophotometric technique to express the antioxidant activity (*in vitro*) in natural extracts. The high antioxidant activity has been correlated with the presence of important phenolic compounds. Polyphenols inactivate free radicals by to the hydrogen atom transfer (HAT) and the single electron transfer (SET) mechanisms (Leopoldini et al., 2011).

2.4.2 ANTICANCER

Cancer is a generic term for a large group of diseases that can affect any part of the body (Torres-León et al., 2017a). Cancer is one of the major challenges facing human health care, in 2015 caused about 8.8 million deaths (WHO, 2017). As we discussed earlier, Phenolic compounds have pharmacological properties and promote protection against the damage of reactive oxygen species, which results in a beneficial activity against cancer (Farinetti et al., 2017). Scientific investigations *in vitro* have demonstrated good anti-cancer activity of mango seed in breast cancer, liver cancer and leukemia cancer cell lines (Table 2.3). It is most likely that the cytotoxic effects of the mango extract on breast cancer cells are due to the action of the polyphenolic compounds present in the extracts. The phenolic compounds found in the extract have been reported to have anticancer activity (Torres-León et al., 2017a), and these compounds can have synergistic effects (Abdullah et al., 2014). This shows that mango seed has a wide potential as a source of anti-cancer compounds. New investigations are necessary to evaluate bioavailability (biological availability) of antioxidants after ingestion, to identify the specific amount that can prevent carcinogenesis and act in advanced stages of cancer. Clinical evaluations are needed to identify the precise concentration and path of administration (Farinetti et al., 2017), the proven preventive factors suggest promising future applications for mango seed.

2.4.3 ANTIMICROBIAL

Antimicrobial resistance is a severe problem in the world. Adverse effects of available antibiotics and the constant development of bacterial resistance (against antibiotics) motivate a search for new natural antimicrobial products (Stankovic et al., 2016). The food industry also demands new antimicrobial agents. Organic acids such as malic, succinic, acetic, citric, and tartaric

TABLE 2.3 Anticancer Activity and Cytotoxicity in Mango Seed

Country	Extract type	Target	Experimental model	Mechanism-Effect-cause	IC_{50} (μg mL^{-1})	Reference
Malaysia	Ethanol in a shaker	Breast cancer cells	MCF-7 cells	Induction of oxidative stress mediated apoptosis	15	Abdullah et al. (2014)
Malaysia	Ethanol in a shaker	Breast cancer cells	MDA-MB-231 cells	Induction of oxidative stress mediated apoptosis	30	Abdullah et al. (2014)
Malaysia	Ethanol in a shaker	Normal Breast cells	MCF-10A cells	Low toxicity to normal breast cells	149	Abdullah et al. (2014)
Malaysia	Ethanol in a shaker	Breast cancer cells	The pro-apoptotic markers, Bax, cytochrome c, caspases-. -8 and −9, and anti-apoptotic markers, Bcl-2, p53 and glutathioneMDA-MB-231 cells	Extract modulates redox balance in MDA-MB-231 breast cancer cells with a tendency for apoptotic cell death	-	Abdullah et al. (2015a)
Malaysia	Ethanol in a shaker	Breast cancer cells	Bcl-2-like protein 4 (BAX), p53, cytochrome c and caspases (7, 8 and 9) in the MCF-7 cells	Extract can induce cancer cell apoptosis, likely via the activation of oxidative stress	-	Abdullah et al. (2015b)
China	Acetone in and ultrasonic	Breast cancer cells	MDA-MB-231 cells	Action of phenolic compounds such as PGG and gallic acid	8.31–11.02	Luo et al. (2014)
China	Acetone in and ultrasonic	Liver cancer cells	HepG2 cells	Action of phenolic compounds such as PGG and gallic acid	33.91–41.50	Luo et al. (2014)
China	Acetone in and ultrasonic	Leukemia cancer cells	HL-60 cells	Action of phenolic compounds such as PGG and gallic acid	14.68–28.72	Luo et al. (2014)

naturally present in fruits are used as preservative agents. However, these organic acids can be replaced by bioactive compounds of the mango seed.

The antimicrobial potential is defined as the ability of a compound or extract to inhibit the growth of microorganisms of interest. Mango seed has shown high antimicrobial activity against Gram-positive and Gram-negative bacteria (Table 2.4). Generally, the researchers have associated the antimicrobial activity with the compounds present in the mango seed; these compounds play an important role in the cell membrane of microorganisms, causing cell lysis (Asif et al., 2016). Variations in antimicrobial activity between Gram-positive and Gram-negative bacteria were associated with the difference in the cell walls of the microorganisms. Mango seed has broad potential in the development of new natural antibacterial agents in food and against resistant microorganisms. New works is also needed to evaluate the effect on new resistant bacteria that are emerging with danger to public health.

2.4.4 OTHER POTENTIAL BIOACTIVITIES AND TOXICITY

The phytochemical constituents of mango seed (described in Section 2.2) suggest that there are other potential bioactivities. For example, PGG derived from other botanical sources showed a range of bioactivities such as antidiabetic, antiviral, anti-inflammatory and improving cognitive disorders (Torres-León et al., 2017a). Gallic acid (GA) possesses anti-inflammatory, anti-biotic, anti-viral and cardiovascular protection activities (Govea-Salas et al., 2016). It may be expected that the phenolic compounds of mango seed have similar bio-functions.

2.4.4.1 ANTIHEMOLYTIC

Hemolysis is the best-studied aspect of mechanically induced erythrocytes (red blood cells) damage and is defined as the release of hemoglobin into the plasma due to a mechanical compromise of the erythrocyte membrane (Fraser, 2015). Abdel-Aty et al. (2018) determined the antihemolytic activity of mango seed on crude venom, the IC_{50} and 100% inhibition of *Cerastes cerastes* and *Eicus coloratus* hemolytic activity were observed at 21, 19 and 40 µg of mango seed, respectively. Mango seed did not induce hemolysis in red blood cells. The authors state that this activity can be explained by the ability of the phenolic compounds present in the seed to inhibit proteases and Phospholipase A2. This research demonstrates the antihemolytic property of mango seed.

TABLE 2.4 Antimicrobial Activity of Mango Seed

Country	Extract type	Microorganisms	Action/Effect	References
Gram positive bacteria				
Thailand	Aqueous 1:3 (w/v)	*Bacillus cereus ATCC 11778*	Inhibition zone diameter between 4.33–7.63 mm	Khammuang & Sarnthima (2011)
Thailand	Aqueous 1:3 (w/v)	*Bacillus cereus ATCC 7058*	Inhibition zone diameter between 2.67–6.87 mm	Khammuang & Sarnthima (2011)
India	Methanolic	*Enterococcus faecalis ATCC 29212*	MIC: 24mm, NBC: 6.25 mg/mL; Biofilm: 180CFU/mL (99.8 % eradication)	Subbiya et al. (2013)
India	Methanolic	*Enterococcus faecalis (Clinical isolate)*	MIC: 24mm, NBC: 12.5 mg/mL	Subbiya et al. (2013)
Sudan	Methanolic and ethanolic	*Staphylococcus aureus*	Inhibition zone diameter 10 mm (5–1.25 mg mL^{-1})	Kaur et al. (2010)
India	Methanolic	*Staphylococcus aureus ATCC29737*	Inhibition zone diameter between 13–15 mm (600–1200 μg)	Vaghasiya, Patel, & Chanda (2013)
India	Methanolic	*Staphylococcus aureus*	Inhibition zone diameter between 11–13 mm (600–1200 μg)	Vaghasiya et al. (2013)
Malaysia	Ethanolic	*Staphylococcus aureus*	Inhibition zone diameter between 1.23–1.5 cm (0.01–0.10 M)	Mirghani et al. (2009)
Thailand	Ethanolic	*Staphylococcus aureus*	Inhibition by ultra-structural changes in bacterial cell morphology. Inhibition zone diameter between 11–17 mm (0.625–5 mg $disc^{-1}$). MIC: 0.47 mg mL^{-1}; MBC 1.88 mg mL^{-1}. (PGG, MIC: 0.13 mg mL^{-1}; MBC 0.5 mg ml^{-1}, GA, MIC:0.19; 0.75 mg ml^{-1})	Jiamboonsri et al. (2011)
Iran	Ethanolic	*Enterococcus faecalis*	Inhibition zone diameter between 9.5–11 cm (25–50 mg L^{-1}). MIC: 25 mg L^{-1}, MBC: 50 mg L^{-1}.	Shabani & Sayadi (2014)
Iran	Ethanolic	*Staphylococcus aureus*	Inhibition zone diameter between 12.1–14 cm (25–50 mg L^{-1}). MIC: 6.25 mg L^{-1}, MBC: 12.15 mg L^{-1}	Shabani & Sayadi (2014)
Thailand	Ethanolic	*Propionibacterium acnés*	Inhibition zone diameter between 26 mm (50 mg mL^{-1}). MIC: 1.56 mg L^{-1}, MBC: 12.52 mg L^{-1}	Poomanee et al. (2018)

TABLE 2.4 *(Continued)*

Country	Extract type	Microorganisms	Action/Effect	References
Thailand	Ethanolic	*Staphylococcus aureus*	Inhibition zone diameter between 15 mm (50 mg mL^{-1}). MIC: 3.13 mg L^{-1}, MBC: 13.3 mg L^{-1}	Poomanee et al. (2018)
Malaysia	Methanolic	*Staphylococcus aureus*	Inhibition zone diameter between 21.25 mm (100 mg mL^{-1})	Kaur et al. (2010)
Gram negative bacteria				
Egypt	Methanolic 2:1 (v/w)	*Escherichia coli*	400 ppm methanol extract and 5% mango seed kernel oil showed antimicrobial activity and extended the shelf-life of pasteurized cow milk.	Abdalla, Darwish, Ayad, & El-Hamahmy (2007)
Sudan	Methanolic and ethanolic	*Escherichia coli*	Inhibition zone diameter between 5.67 - 10.33 mm	El-Gied et al. (2012)
Thailand	Aqueous 1:3 (w/v)	*Pseudomonas aeruginosa ATCC 27853*	Inhibition zone diameter between 11–7 mm (5–1.25 mg mL^{-1})	Khammuang & Sarnthima (2011)
Thailand	Aqueous 1:3 (w/v)	*Salmonella typhi DMST 5784*	Inhibition zone diameter between 3.83–5.83 mm	Khammuang & Sarnthima (2011)
Sudan	Methanolic and ethanolic	*Salmonella typhi*	Inhibition zone diameter between 7–5 mm (5–1.25 mg mL^{-1})	El-Gied et al. (2012)
Sudan	Methanolic and ethanolic	*Shigella flexnerri*	Inhibition zone diameter 5 mm (5–1.25 mg mL^{-1})	El-Gied et al. (2012)
India	Methanolic	*Escherichia coli*	Inhibition zone diameter between 8–17 mm (600–1200 μg)	Vaghasiya et al. (2013)
India	Methanolic	*Salmonella typhimurium ATCC23564*	Inhibition zone diameter between 16–17 mm (600–1200 μg)	Vaghasiya et al. (2013)
India	Methanolic	*Klebsiella aerogenes NCTC418*	Inhibition zone diameter between 13–14 mm (600–1200 μg)	Vaghasiya et al. (2013)
Malaysia	Ethanolic	*Escherichia coli*	Inhibition zone diameter between 1.3–1.5 cm (0.01–0.10 M)	Mirghani et al. (2009)

TABLE 2.4 *(Continued)*

Country	Extract type	Microorganisms	Action/Effect	References
Malaysia	Ethanolic	*pseudomonas aeruginosa*	Inhibition zone diameter between 1.3–1.4 cm	Mirghani et al. (2009)
Iran	Ethanolic	*Klebsiella pneumoniae*	Inhibition zone diameter between 11.7–13 cm (25–50 mg L^{-1}). MIC: 6.25 mg L^{-1}, MBC: 12.5	Shabani & Sayadi (2014)
Iran	Ethanolic	*Pseudomonas aeruginosa*	Inhibition zone diameter between 11.2–13.1 cm (25–50 mg L^{-1}). MIC: 12.5 mg L^{-1}, MBC: 25.	Shabani & Sayadi (2014)
India	Ethanolic	*Shigella dysenteriae*	Inhibition zone diameter between 12–16 cm (200–800 μg). MIC: 190 μg mL^{-1}	Rajan, Thirunalasundari, & Jeeva (2011)
Malaysia	Methanolic	*Escherichia coli*	Inhibition zone diameter between 17.70 mm (100 mg mL^{-1})	Kaur et al. (2010)

MIC: minimum Inhibition concentrations; MBC: minimum bactericidal concentration; CFU: Colony-forming unit.

2.4.4.2 ANTIDIABETIC

Currently, diabetes is a big problem in public health. According to the World Health Organization (WHO), about 422 million people have diabetes globally (WHO, 2016). Although insulin injection and hypoglycemic agents are effective drugs for diabetes, these compounds possess some adverse effects and have no effects on diabetes complications in the long term (Bahmani et al., 2014). A study by Irondi et al. (2016), showed that mango seed inhibits some key enzymes linked to the pathology and complications of type 2 diabetes (*in vitro*). Mango seed could, therefore, be a promising nutraceutical therapy for the management of type 2 diabetes and its associated complications.

2.4.4.3 ANTIDIARRHEAL

Diarrhea is caused by inflammation of the intestines. 526,000 deaths due to diarrhea in children younger than 5 years were estimated in 2015 (Wierzba & Muhib, 2018). Mango seed extracts significantly reduced intestinal motility and fecal score in Swiss albino mice (Rajan et al., 2012). Secondary metabolites such as polyphenolic compounds present in the seed have been implicated as having antidiarrheal activity. This study is in accordance with traditional medicine and confirms that mango seed is an effective antidiarrheal.

2.4.4.4 TOXICITY

Few studies have reported the toxicity and safety of mango seed. In cultured cells, cytotoxic effects of ethanolic seed extract of a waterlily variety grown were investigated by Abdullah et al. (2014). The extract showed significant low toxicity to normal cells (MCF-10A) (Table 2.3). Recently, Abdel-Aty et al. (2018) affirmed that because of their ability not to induce hemolysis in red blood cells, mango seed extracts are not toxic. The toxicity of mango seed extracts should be vigorously investigated using various models to provide a basis for its development as a natural source of food ingredients or therapeutic agents.

2.5 POTENTIAL INNOVATIVE USES

2.5.1 FOOD/FEED INGREDIENTS

Bioactive compounds have a wide application as a preservative in the food processing industry. The natural phenolic antioxidants inhibit oxidation reactions by itself being oxidized and also prevent the production of off-odors and tastes. Antioxidants delay the onset of oxidation and slow the reaction rate of food lipids. Mango seed retains oil stability against rancidity for up to 12 months (Abdalla et al., 2007). The mango seed can also be used to preserve the oxidative quality of meats and meat products.

Antimicrobial activity of mango seed, suggests their possible uses as antimicrobial food preservatives: 5000 ppm of mango kernel extract prevented the growth of bacteria in milk when stored at room temperature for 15 days (Abdalla et al., 2007). There are many types of food matrices to which these antioxidant compounds might be added; more studies are necessary to elucidate that substances are effective in what systems and under what condition (Brewer, 2011).

2.5.2 MEDICAL APPLICATION

The bioactive characteristics of the mango seed indicate that it has a wide potential in their development into new nutraceutical/pharmaceutical formulations. The bioactive compounds have applications as hepatoprotective agents, prevention, and treatments for Neurodegenerative diseases such as Alzheimer's disease, Parkinson's disease, and amylotrophic lateral sclerosis. They also have potential use in several pathological conditions associated with oxidative damage in cells such as rheumatoid, arthritis, cardiovascular disorders, ulcerogenic and acquired immunodeficiency diseases (Sindhi et al., 2013).

2.5.3 FORMULATION OF ACTIVE PACKAGING

Mango seed have recently aroused the attention of the scientific community in the formulation of active edible films and coatings (Belizón et al., 2018; Klangmuang & Sothornvit, 2018; Maryam et al., 2018; Maryam Adilah & Nur Hanani, 2019; Nawab et al., 2018; Nawab, Alam, & Hasnain, 2017; Torres-León et al., 2018a). The incorporation of mango seed extracts increases the antioxidant and antimicrobial potential of the formulations. Also, the extracts improve the technological properties of edible films and coatings, this effect is attributed to the properties that polyphenols have to interact with constituents of formulations (such as proteins and polysaccharides) and the surface of fruits, respectively (Torres-León et al., 2018a).

2.5.4 OTHER INGREDIENTS

Mango seed ingredients may be used for cosmetic applications. For hair care products, plant extracts may be developed in the form of natural shampoos, conditioners, hair oil, and hair tonic (Wang & Zhu, 2017). The use of mango seed in the cosmetic and therapeutic industries is a promising alternative for synthetic compounds (Sindhi et al., 2013).

2.6 TECHNOLOGICAL ASPECTS

In the processing of agro-industrial byproducts such as mango seed, the drying and extraction methods are essential unit operations that affect the activity and stability of phenolic compounds. The drying treatment may cause an enhancement of the extractability of different compounds. Freeze-drying and oven-drying with forced air led to an increase (1.6 times) in anthocyanin content compared to non-dried peel (Dorta, Lobo, & González, 2012).

Conventional drying methods rely on conductive and convective heat transfer methods, which are highly energy demanding, lead to losses the bioactive compounds (Tontul & Topuz, 2017). However, other drying techniques more efficient and reliable such as explosion puff drying (Zuo et al., 2017), combined microwave vacuum drying, infrared drying, ultrasound freeze-drying and refracting window (Kaur et al., 2017; Ochoa et al., 2012; Zotarelli, Carciofi, & Laurindo, 2015), have been proposed for the drying of mango pulp. It is important to note that these techniques have not been tested in mango seed.

In the extraction of bioactive compounds from residues of mango the drying technique, and the temperature influences to a greater extent the extraction performance, while the drying time is a less relevant factor. Sogi et al. (2013), dried mango peel and seed (Tommy Atkins variety), using different techniques (freeze-drying, hot air, vacuum and infrared). The best results were found with freeze-drying. Ekorong et al. (2015), combined the effect of drying temperature and time on total phenolic compounds and antioxidant activity of mango seed; the increase of drying temperature increased antioxidant activity while total phenolic components decreased.

On the other hand, conventional extraction techniques (soxhlet extraction, maceration, and hydrodistillation) are being used to extract bioactive components from the mango byproducts. The extraction efficiency depends on the choice of solvents. In this sense, Dorta et al. (2012), evaluate the effect of solvent (methanol, ethanol, acetone, water, methanol: water, ethanol: water, and acetone: water) on the efficiency of the extraction of antioxidants from mango peel and seed. The solvents that best obtained extracts with high antioxidant capacity were methanol, methanol: water, ethanol: water, and acetone:wáter. Although the highest content of phytochemicals was obtained with acetone, from food security (good manufacturing practices), the authors recommend using ethanol/water.

However, conventional extraction presents a series of disadvantages such as high working times, high costs, and damage to active compounds. To overcome these limitations, non-conventional extraction techniques are used (Gil-Chávez et al., 2013; Selvamuthukumaran & Shi, 2017). Some of the most promising extraction non-conventional techniques are supercritical fluid extraction (SFE), enzyme-assisted extraction (EAE), microwave-assisted extraction (MAE), ultrasound-assisted extraction (UAE), pressurized low polarity water extraction, pulsed electric field (PEF) extraction, pressurized liquid extraction (PLE), and molecular distillation. Some of these techniques comply with standards set by the Environmental Protection Agency (2015), which is why they are called green extraction techniques (Selvamuthukumaran & Shi, 2017).

Yoswathana & Eshiaghi (2013) to optimized conditions for extracts yield and total phenolic content extraction from mango seed kernel using SFE and conventional techniques such as maceration and soxhlet. The authors reported a total phenol content of 40.4 mg of tannic acid equivalent per gram of seed (six times of that from conventional techniques). SFE has also been used to obtain premium-grade cocoa butter from the waste of mango using supercritical carbon dioxide and the Soxhlet method like comparison. The total fat contents of a waste of mango varieties ranged from 64 to 135 g kg^{-1} using supercritical CO_2 extraction and from 76 to 137 g kg^{-1} using Soxhlet extraction methods (Jahurul et al., 2014). EAE has very good results in the release of phenolic compounds.

Nevertheless, the use of enzymes is limited by instability and high associated costs (Navarro-González et al., 2011). Alternative technologies such as solid-state fermentation (SSF) have been proposed for the valorization of agro-industrial waste and byproducts (Martínez et al., 2012). In the SSF, microorganisms such as fungi naturally produce enzymes that degrade the cell wall (Jamal et al., 2011), generating hydrolysis (Jamal et al.., 2011) and mobilization of compounds towards the extraction solvent. Recently, our research group (bio-uadec.com) evaluate the effect of SSF with the fungus *Aspergillus niger* GH1 in the content of phenolic compounds and the antioxidant activity of Mexican mango seed. The results showed that SSF of mango seed increased the polyphenol content in the extracts by 235%. Analysis of the free and bound fractions showed that SSF to release the bound phenols to the plant matrix, also increasing the antioxidant potential of the extracts (Torres-León et al., 2019).

Concerning MAE, our research group (bio-uadec.com) also optimized the extraction conditions of phenolic extracts of mango seed with high antioxidant activity using microwave technology (Torres-León et al., 2017b). MAE significantly increased the antioxidant activity of the extracts compared to the control. Under optimal conditions, high values of antioxidant activity (1738.2 mg Trolox g^{-1}, IC50 of 0.07 mg g^{-1}) and phenolic compounds (598.4 ± 25.80 mg AG g^{-1}) were obtained. Results superior to those reported in mango seed of other varieties and commercial antioxidants such as Trolox, Vitamin E, and BHT. The results demonstrate that MAE can be recommended as an effective non-conventional technology for the extraction of active compounds from the mango seed.

The techniques of drying and extraction are highly relevant to preserve the active compounds present in mango byproduct. Non-conventional technologies of extraction have advantages such as higher extraction yield, greater antioxidant activity, less processing time. In the same way, drying treatments influence the extraction performance of bioactive mango compounds. Methods such as explosion puff drying, refracting window, and heat pump have not been investigated in the drying of bioactive compounds, and these technologies could be a research focus. Also, the Incorporation and development of hybrid methods should be investigated.

2.7 CONCLUSION

Mango seed is a source of important phytochemicals. Mainly, polyphenolic compounds (gallotannins) such as PGG and its derivatives, it is also rich in phenolic acids such as gallic acid and ellagic and flavonoids as rhamnetin derivatives. The presence of these phytochemicals could be responsible for a range of bioactivities of mango seed as revealed by *in vitro* chemical and biological assays. Mango seed has been reported to possess antioxidative, antimicrobial, anticancer, antihemolytic, antidiabetic and antidiarrheal properties.

Research should continue to identify the profiles of the phytochemicals and the functional potential available from different seed varieties grown in the world, the effect of the extraction method on profiles of phenolic compounds and biological activities should be evaluated. New research projects are interested in the evaluation of the application. Mango seed extracts have potential as natural preservatives (antioxidant or antimicrobial) in foods. Studies of bioavailability of phenolic compounds in the body and the nutritional and sensory effects of incorporating mango seed extracts in foodstuffs are necessary. The use of mango seed can bring important social, nutritional, environmental and economic benefits.

ACKNOWLEDGMENTS

Cristian Torres-León thank the Mexican Council for Science and Technology (CONACYT) for his post-graduate scholarship.

KEYWORDS

- **agro-industrial byproducts**
- **anti-cancer**
- **antimicrobial**
- **antioxidants**
- **Mangifera indica L.**
- **mango seed**
- **mango waste**
- **penta-O-galloylglucoside**
- **phenolic compounds**

REFERENCES

Abdalla, A., Darwish, S., Ayad, E., & El-Hamahmy, R., (2007). Egyptian mango by-product 1. Compositional quality of mango seed kernel. *Food Chemistry, 103*(4), 1134–1140. https://doi.org/10.1016/j.foodchem.2006.10.017 (accessed on 24 July 2020).

Abdel-Aty, A. M., Salama, W. H., Hamed, M. B., Fahmy, A. S., & Mohamed, S. A., (2018). The phenolic-antioxidant capacity of mango seed kernels: Therapeutic effect against viper venoms. *Brazilian Journal of Pharmacognosy*, 4–11. https://doi.org/10.1016/j.bjp.2018.06.008 (accessed on 24 July 2020).

Abdullah, H., Mohammed, A., & Abdullah, R., (2014). Cytotoxic effects of *Mangifera indica* L. kernel extract on human breast cancer (MCF-7 and MDA-MB-231 cell lines) and bioactive constituents in the crude extract. *BMC Complementary and Alternative Medicine, 14*, 199. https://doi.org/10.1186/1472–6882–14–199 (accessed on 24 July 2020).

Acosta-Estrada, B., Gutiérrez-uribe, J., & Serna-saldívar, S., (2014). Bound phenolics in foods, a review. *Food Chemistry, 152*, 46–55. https://doi.org/10.1016/j.foodchem.2013.11.093 (accessed on 24 July 2020).

Ajila, C., Bhat, S., & Rao, P., (2007). Valuable components of raw and ripe peels from two Indian mango varieties. *Food Chemistry, 102*(4), 1006–1011. https://doi.org/10.1016/j.foodchem.2006.06.036 (accessed on 24 July 2020).

Ascacio, J., Buenrostro, J., Aguilera, A., Prado, A., Rodríguez, R., & Aguilar, C., (2011). Ellagitannins: Biosynthesis, biodegradation and biological properties. *Journal of Medicinal*

Plant Research, 5(19), 4696–4703. https://doi.org/10.5897/JMPR (accessed on 24 July 2020).

Asif, A., Farooq, U., Akram, K., Hayat, Z., Shafi, A., Sarfraz, F., & Aftab, S., (2016). Therapeutic potentials of bioactive compounds from mango fruit wastes. *Trends in Food Science and Technology, 53*, 102–112. https://doi.org/10.1016/j.tifs.2016.05.004 (accessed on 24 July 2020).

Ayala, J., Vega, V., Rosas, C., Palafox, H., Villa, J., Siddiqui, W., & González, G., (2011). Agro-industrial potential of exotic fruit byproducts as a source of food additives. *Food Research International, 44*(7), 1866–1874. https://doi.org/10.1016/j.foodres.2011.02.021 (accessed on 24 July 2020).

Bahmani, M., Zargaran, A., Rafieian, M., & Saki, K., (2014). Ethnobotanical study of medicinal plants used in the management of diabetes mellitus in the Urmia, Northwest Iran. *Asian Pacific Journal of Tropical Medicine, 7*(Suppl. 1), S348–S354. https://doi.org/doi:10.1016/S1995–7645 (accessed on 24 July 2020).

Balasundram, N., Sundram, K., & Samman, S., (2006). Phenolic compounds in plants and agri-industrial by-products: Antioxidant activity, occurrence, and potential uses. *Food Chemistry, 99*(1), 191–203. https://doi.org/10.1016/j.foodchem.2005.07.042 (accessed on 24 July 2020).

Barreto, J. C., Trevisan, M. T. S., Hull, W. E., Erben, G., De Brito, E. S., Pfundstein, B., & Owen, R. W., (2008). Characterization and quantitation of polyphenolic compounds in bark, kernel, leaves, and peel of mango (Mangifera indica L.). *Journal of Agricultural and Food Chemistry, 56*(14), 5599–5610. https://doi.org/10.1021/jf800738r (accessed on 24 July 2020).

Belizón, M., Fernández-Ponce, M. T., Casas, L., Mantell, C., & De La Ossa-Fernández, E. J. M., (2018). Supercritical impregnation of antioxidant mango polyphenols into a multilayer PET/PP food-grade film. *Journal of CO_2 Utilization, 25*, 56–67. https://doi.org/10.1016/j.jcou.2018.03.005 (accessed on 24 July 2020).

Berardini, N., Carle, R., & Schieber, A., (2004). Characterization of gallotannins and benzophenone derivatives from mango (Mangifera indica L. cv. 'Tommy Atkins') peels, pulp and kernels by high-performance liquid chromatography/electro spray ionization mass spectrometry. *Rapid Communications in Mass Spectrometry, 18*(19), 2208–2216. https://doi.org/10.1002/rcm.1611 (accessed on 24 July 2020).

Bompard, J. M., (2009). Taxonomy and systematics. In: Lit, R. E., (Ed.), *The Mango: Botany, Production and Uses* (2nd edn., pp. 19–41). Cambrige, Uk: CAB International Wallingford.

Brewer, M. S., (2011). Natural antioxidants: Sources, compounds, mechanisms of action, and potential applications. *Comprehensive Reviews in Food Science and Food Safety, 10*(4), 221–247. https://doi.org/10.1111/j.1541–4337.2011.00156.x (accessed on 24 July 2020).

Chen, A. Y., & Charlie, Y., (2013). A review of the dietary flavonoid, kaempferol on human health and cancer chemoprevention. *Food Chemistry, 138*(4), 2099–2107. https://doi.org/10.1016/j.foodchem.2012.11.139 (accessed on 24 July 2020).

Derese, S., & Kuete, V., (2017). *Mangifera indica* L. (Anacardiaceae). In: *Medicinal Spices and Vegetables from Africa* (pp. 451–483). https://doi.org/10.1016/B978–0-12–809286–6.00021–2 (accessed on 24 July 2020).

Dey, T., Chakraborty, S., Jain, K., Sharma, A., & Kuhad, R., (2016). Antioxidant phenolics and their microbial production by submerged and solid state fermentation process: A review. *Trends in Food Science and Technology, 53*, 60–74. https://doi.org/10.1016/j.tifs.2016.04.007 (accessed on 24 July 2020).

Diderot, T., Silvere, N., & Etienne, T., (2006). Xanthones as therapeutic agents: Chemistry and pharmacology. *Advances in Phytomedicine, 2*, 273–298. https://doi.org/10.1016/S1572–557X(05)02016–7 (accessed on 24 July 2020).

Dorta, E., González, M., Lobo, M. G., Sánchez-Moreno, C., & De Ancos, B., (2014). Screening of phenolic compounds in by-product extracts from mangoes (Mangifera indica L.) by HPLC-ESI-QTOF-MS and multivariate analysis for use as a food ingredient. *Food Research International, 57*, 51–60. https://doi.org/10.1016/j.foodres.2014.01.012 (accessed on 24 July 2020).

Dorta, E., Lobo, G., & González, M., (2012). Using drying treatments to stabilise mango peel and seed : Effect on antioxidant activity. *Food Science and Technology, 45*(2), 261–268. https://doi.org/10.1016/j.lwt.2011.08.016 (accessed on 24 July 2020).

Dorta, E., Lobo, M. G., & Gonzalez, M., (2012). Reutilization of mango byproducts: Study of the effect of extraction solvent and temperature on their antioxidant properties. *Journal of Food Science, 77*(1), C80–88. https://doi.org/10.1111/j.1750–3841.2011.02477.x (accessed on 24 July 2020).

Dussan, S. S., Torres, L. C., & Reyes, C. P. M., (2014). Effect of the edible coating on the physical-chemistry of fresh-cut "Tommy Atkins" mango and refrigerated. *Acta Agronómica, 63*(3), 212–221. https://doi.org/10.15446/acag.v63n3.40973 (accessed on 24 July 2020).

Ekorong, F., Zomegni, G., Desobgo, S., & Ndjouenkeu, R., (2015). Optimization of drying parameters for mango seed kernels using central composite design. *Bioresources and Bioprocessing, 2*(1), 8. https://doi.org/10.1186/s40643–015–0036-x (accessed on 24 July 2020).

FAOSTAT, (2017). *Food and Agriculture Organization of the United Nations*. Retrieved from Statistical Database—Agriculture website: http://www.fao.org/faostat/es/#data/QC/visualize (accessed on 24 July 2020).

Farinetti, A., Zurlo, V., Manenti, A., Coppi, F., & Mattioli, A. V., (2017). Mediterranean diet and colorectal cancer: A systematic review. *Nutrition, 43, 44*, 83–88. https://doi.org/10.1016/j.nut.2017.06.008 (accessed on 24 July 2020).

Fasoli, E., & Righetti, P. G., (2013). The peel and pulp of mango fruit: A proteomic samba. *Biochimica Et Biophysica Acta, 1834*(12), 2539–2545. https://doi.org/10.1016/j.bbapap.2013.09.004 (accessed on 24 July 2020).

FICCI, (2010). *Bottlenecks in Indian Food Processing Industry*. Retrieved from Federation of Indian Chambers of Commerce and Industry website: http://ficci.in/sedocument/20073/Food-Processing-bottlenecks-study.pdf (accessed on 24 July 2020).

Fraser, K., (2015). Mechanical stress induced blood trauma. In: Becker, S. M., & Kuznetsov, A., (eds.), *Heat Transfer and Fluid Flow in Biological Processes* (pp. 305–333). https://doi.org/10.1016/B978–0-12–408077–5.00014–6 (accessed on 24 July 2020).

Gil-Chávez, G. J., Villa, J. A., Ayala-Zavala, J. F., Basilio, H. J., Sepulveda, D., Yahia, E. M., & González-Aguilar, G. A., (2013). Technologies for extraction and production of bioactive compounds to be used as nutraceuticals and food ingredients: An overview. Comprehensive *Reviews in Food Science and Food Safety, 12*(1), 5–23. https://doi.org/10.1111/1541–4337.12005 (accessed on 24 July 2020).

Gómez-Caravaca, A. M., López-Cobo, A., Verardo, V., Segura-Carretero, A., & Fernández-Gutiérrez, A., (2016). HPLC-DAD-q-TOF-MS as a powerful platform for the determination of phenolic and other polar compounds in the edible part of mango and its by-products (peel, seed, and seed husk). *Electrophoresis, 37*(7, 8), 1072–1084. https://doi.org/10.1002/elps.201500439 (accessed on 24 July 2020).

Govea-Salas, M., Rivas-Estilla, A. M., Rodríguez-Herrera, R., Lozano-Sepúlveda, S. A., Aguilar-Gonzalez, C. N., Zugasti-Cruz, A., & Morlett-Chávez, J. A., (2016). Gallic acid decreases hepatitis C virus expression through its antioxidant capacity. *Experimental and Therapeutic Medicine, 11*(2), 619–624. https://doi.org/10.3892/etm.2015.2923 (accessed on 24 July 2020).

Irondi, E. A., Oboh, G., & Akindahunsi, A. A., (2016). Antidiabetic effects of *Mangifera indica* kernel flour-supplemented diet in streptozotocin-induced type 2 diabetes in rats. *Food Science and Nutrition, 4*(6), 828–839. https://doi.org/10.1002/fsn3.348 (accessed on 24 July 2020).

Jahurul, M., Zaidul, I., Norulaini, N., Sahena, F., Jaffri, J., & Omar, A., (2014). Supercritical carbon dioxide extraction and studies of mango seed kernel for cocoa butter analogy fats. *CyTA-Journal of Food, 12*(1), 97–103. https://doi.org/10.1080/19476337.2013.801038 (accessed on 24 July 2020).

Jamal, P., Idris, Z. M., & Alam, Z., (2011). Effects of physicochemical parameters on the production of phenolic acids from palm oil mill effluent under liquid-state fermentation by *Aspergillus niger*. *Food Chemistry, 124*(4), 1595–1602. https://doi.org/10.1016/j.foodchem.2010.08.022 (accessed on 24 July 2020).

Kaur, G., Saha, S., Kumari, K., & Datta, A. K., (2017). Mango pulp drying by refractance window method. *Agricultural Engineering International: CIGR Journal, 19*(4), 145–151.

Klangmuang, P., & Sothornvit, R., (2018). Active coating from hydroxypropyl methylcellulose-based nanocomposite incorporated with Thai essential oils on mango (cv. Namdokmai Sithong). *Food Bioscience, 23*, 9–15. https://doi.org/10.1016/j.fbio.2018.02.012 (accessed on 24 July 2020).

Leopoldini, M., Russo, N., & Toscano, M., (2011). The molecular basis of working mechanism of natural polyphenolic antioxidants. *Food Chemistry, 125*(2), 288–306. https://doi.org/10.1016/j.foodchem.2010.08.012 (accessed on 24 July 2020).

Luo, F., Fu, Y., Xiang, Y., Yan, S., Hu, G., Huang, X., & Chen, K., (2014). Identification and quantification of gallotannins in mango (Mangifera indica L.) kernel and peel and their antiproliferative activities. *Journal of Functional Foods, 8*(1), 282–291. https://doi.org/10.1016/j.jff.2014.03.030 (accessed on 24 July 2020).

Maryam, A. Z. A., & Nur, H. Z. A., (2019). Storage stability of soy protein isolate films incorporated with mango kernel extract at different temperature. *Food Hydrocolloids, 87*, 541–549. https://doi.org/10.1016/j.foodhyd.2018.08.038 (accessed on 24 July 2020).

Maryam, A. Z. A., Jamilah, B., & Nur, H. Z. A., (2018). Functional and antioxidant properties of protein-based films incorporated with mango kernel extract for active packaging. *Food Hydrocolloids, 74*, 207–218. https://doi.org/10.1016/j.foodhyd.2017.08.017 (accessed on 24 July 2020).

Moskaug, J., Carlsen, H., Myhrstad, M., & Blomhoff, R., (2004). Molecular imaging of the biological effects of quercetin and quercetin-rich foods. *Mechanisms of Ageing and Development, 125*(4), 315–324. https://doi.org/10.1016/j.mad.2004.01.007 (accessed on 24 July 2020).

Mukherjee, S., & Litz, R., (2009). Botany and importance. In: Litz, R., (ed.), *The Mango: Botany, Production and Uses* (2nd edn., pp. 1–14). Cambridge: CAB International Wallingford.

Navarro-González, I., García-Valverde, V., García-Alonso, J., & Periago, M. J., (2011). Chemical profile, functional and antioxidant properties of tomato peel fiber. *Food Research*

International, 44(5), 1528–1535. https://doi.org/10.1016/j.foodres.2011.04.005 (accessed on 24 July 2020).

Nawab, A., Alam, F., & Hasnain, A., (2017). Mango kernel starch as a novel edible coating for enhancing shelf- life of tomato (Solanum lycopersicum) fruit. *International Journal of Biological Macromolecules, 103*, 581–586. https://doi.org/10.1016/j.ijbiomac.2017.05.057 (accessed on 24 July 2020).

Nawab, A., Alam, F., Haq, M. A., Haider, M. S., Lutfi, Z., Kamaluddin, S., & Hasnain, A., (2018). Innovative edible packaging from mango kernel starch for the shelf life extension of red chili powder. *International Journal of Biological Macromolecules, 114*, 626–631. https://doi.org/10.1016/j.ijbiomac.2018.03.148 (accessed on 24 July 2020).

Ng, K. R., Lyu, X., Mark, R., & Chen, W. N., (2019). Antimicrobial and antioxidant activities of phenolic metabolites from flavonoid-producing yeast: Potential as natural food preservatives. *Food Chemistry, 270*, 123–129. https://doi.org/10.1016/j.foodchem.2018.07.077 (accessed on 24 July 2020).

Niemetz, R., & Gross, G. G., (2005). Enzymology of gallotannin and ellagitannin biosynthesis. *Phytochemistry, 66*(17 SPEC. ISS.), 2001–2011. https://doi.org/10.1016/j.phytochem.2005.01.009 (accessed on 24 July 2020).

Nithitanakool, S., Pithayanukul, P., & Bavovada, R., (2009). Antioxidant and hepatoprotective activities of Thai mango seed kernel extract. *Planta Medica, 75*(10), 1118–1123. https://doi.org/10.1055/s-0.029-1.185.507 (accessed on 24 July 2020).

O'Shea, N., Arendt, E., & Gallagher, E., (2012). Dietary fibre and phytochemical characteristics of fruit and vegetable by-products and their recent applications as novel ingredients in food products. *Innovative Food Science and Emerging Technologies, 16*, 1–10. https://doi.org/10.1016/j.ifset.2012.06.002 (accessed on 24 July 2020).

Ochoa, C., Quintero, P., Ayala, A., & Ortiz, M., (2012). Drying characteristics of mango slices using the refractance window™ technique. *Journal of Food Engineering, 109*(1), 69–75. https://doi.org/10.1016/j.jfoodeng.2011.09.032 (accessed on 24 July 2020).

Poprac, P., Jomova, K., Simunkova, M., Kollar, V., Rhodes, C. J., & Valko, M., (2017). Targeting free radicals in oxidative stress-related human diseases. *Trends in Pharmacological Sciences, 38*(7), 592–607. https://doi.org/10.1016/j.tips.2017.04.005 (accessed on 24 July 2020).

Rajan, S., Suganya, H., Thirunalasundari, T., & Jeeva, S., (2012). Antidiarrheal efficacy of *Mangifera indica* seed kernel on Swiss albino mice. *Asian Pacific Journal of Tropical Medicine, 5*(8), 630–633. https://doi.org/10.1016/S1995–7645(12)60129–1 (accessed on 24 July 2020).

Ramešová, Š., Degano, I., & Sokolová, R., (2017). The oxidative decomposition of natural bioactive compound rhamnetin. *Journal of Electroanalytical Chemistry, 788*, 125–130. https://doi.org/10.1016/j.jelechem.2017.01.054 (accessed on 24 July 2020).

Ravani, A., & Joshi, D. C., (2013). Mango and it's by product utilization: A review. *Trends in Post-Harvest Technology, 1*(1), 55–67.

Saito, K., Kohno, M., Yoshizaki, F., & Niwano, Y., (2008). Extensive screening for edible herbal extracts with potent scavenging activity against superoxide anions. *Plant Foods for Human Nutrition, 63*(2), 65–70. https://doi.org/10.1007/s11130–008–0071–2 (accessed on 24 July 2020).

Selvamuthukumaran, M., & Shi, J., (2017). Recent advances in extraction of antioxidants from plant by-products processing industries. *Food Quality and Safety, 1*(1), 61–81. https://doi.org/10.1093/fqs/fyx004 (accessed on 24 July 2020).

Sepúlveda, L., Ascacio, A., Rodríguez, R., Aguilera, A., & Aguilar, C., (2011). Ellagic acid: Biological properties and biotechnological development for production processes. *African Journal of Biotechnology, 10*(22), 4518–4523. https://doi.org/10.5897/AJB10.2201 (accessed on 24 July 2020).

Serna-Cock, L., García-Gonzales, E., & Torres-León, C., (2016). Agro-industrial potential of the mango peel based on its nutritional and functional properties. *Food Reviews International, 32*(4), 364–376. https://doi.org/10.1080/87559129.2015.1094815 (accessed on 24 July 2020).

Sethiya, N., Trivedi, A., & Mishra, S., (2014). The total antioxidant content and radical scavenging investigation on 17 phytochemical from dietary plant sources used globally as functional food. *Biomedicine and Preventive Nutrition, 4*(3), 439–444. https://doi.org/10.1016/j.bionut.2014.03.007 (accessed on 24 July 2020).

Sindhi, V., Gupta, V., Sharma, K., Bhatnagar, S., Kumari, R., & Dhaka, N., (2013). Potential applications of antioxidants: A review. *Journal of Pharmacy Research, 7*(9), 828–835. https://doi.org/10.1016/j.jopr.2013.10.001 (accessed on 24 July 2020).

Singh, Z., Singh, R., Sane, V., & Nath, P., (2013). Mango-postharvest biology and biotechnology. *Critical Reviews in Plant Sciences, 32*(4), 217–236. https://doi.org/10.1080/07352689.2012.743399 (accessed on 24 July 2020).

Sogi, D., Siddiq, M., Greiby, I., & Dolan, K., (2013). Total phenolics, antioxidant activity, and functional properties of "Tommy Atkins" mango peel and kernel as affected by drying methods. *Food Chemistry, 141*(3), 2649–2655. https://doi.org/10.1016/j.foodchem.2013.05.053 (accessed on 24 July 2020).

Soong, Y. Y., & Barlow, P. J., (2004). Antioxidant activity and phenolic content of selected fruit seeds. *Food Chemistry, 88*(3), 411–417. https://doi.org/10.1016/j.foodchem.2004.02.003 (accessed on 24 July 2020).

Soong, Y., & Barlow, P., (2006). Quantification of gallic acid and ellagic acid from longan (*Dimocarpus longan* Lour.) seed and mango (*Mangifera indica* L.) kernel and their effects on antioxidant activity. *Food Chemistry, 97*, 524–530.

Stankovic, N., Mihajilov, T., Zlatkovic, B., Stankov, V., Mitic, V., Jovic, J., & Bernstein, N., (2016). Antibacterial and antioxidant activity of traditional medicinal plants from the Balkan peninsula. *NJAS-Wageningen Journal of Life Sciences, 78*, 21–28. https://doi.org/10.1016/j.njas.2015.12.006 (accessed on 24 July 2020).

Tontul, I., & Topuz, A., (2017). Effects of different drying methods on the physicochemical properties of pomegranate leather (pestil). *LWT-Food Science and Technology, 80*, 294–303. https://doi.org/10.1016/j.lwt.2017.02.035 (accessed on 24 July 2020).

Torres-León, C., Ramirez, N., Londoño, L., Martinez, G., Diaz, R., Navarro, V., & Aguilar, C. N., (2018b). Food waste and byproducts: An opportunity to minimize malnutrition and hunger in developing countries. *Front. Sustain. Food Syst., 2*, 52. https://doi.org/10.3389/fsufs.2018.00052 (accessed on 24 July 2020).

Torres-León, C., Ramirez-Guzman, N., Ascacio-Valdes, J., Serna-Cock, L., Dos, S. C. M. T., Contreras-Esquivel, J. C., & Aguilar, C. N., (2019). Solid-state fermentation with Aspergillus niger to enhance the phenolic contents and antioxidative activity of Mexican mango seed: A promising source of natural antioxidants. *LWT-Food Science and Technology, 112*, 108236. https://doi.org/10.1016/j.lwt.2019.06.003 (accessed on 24 July 2020).

Torres-León, C., Rojas, R., Contreras-Esquivel, J. C., Serna-Cock, L., Belmares-Cerda, R. E., & Aguilar, C. N., (2016). Mango seed: Functional and nutritional properties. *Trends in Food Science and Technology, 55*, 109–117. https://doi.org/10.1016/j.tifs.2016.06.009 (accessed on 24 July 2020).

Torres-León, C., Rojas, R., Serna-Cock, L., Belmares-Cerda, R., & Aguilar, C. N., (2017b). Extraction of antioxidants from mango seed kernel: Optimization assisted by microwave. *Food and Bioproducts Processing, 105*, 188–196. https://doi.org/10.1016/j.fbp.2017.07.005 (accessed on 24 July 2020).

Torres-León, C., Ventura-Sobrevilla, J., Serna-Cock, L., Ascacio-Valdés, J. A., Contreras-Esquivel, J., & Aguilar, C. N., (2017a). Pentagalloylglucose (PGG): A valuable phenolic compound with functional properties. *Journal of Functional Foods, 37*, 176–189. https://doi.org/10.1016/j.jff.2017.07.045 (accessed on 24 July 2020).

Torres-León, C., Vicente, A. A., Flores-lópez, M. L., Rojas, R., Serna-cock, L., Alvarez-pérez, O. B., & Aguilar, C. N., (2018a). Edible films and coatings based on mango (var. Ataulfo) by-products to improve gas transfer rate of peach. *LWT-Food Science and Technology, 97*, 624–631. https://doi.org/10.1016/j.lwt.2018.07.057 (accessed on 24 July 2020).

Wang, S., & Zhu, F., (2017). Chemical composition and biological activity of staghorn sumac (*Rhus typhina*). *Food Chemistry, 237*, 431–443. https://doi.org/10.1016/j.foodchem.2017.05.111 (accessed on 24 July 2020).

WHO, (2016). *Global Report on Diabetes*. Geneva.

WHO, (2017). *Cancer*. Retrieved from World Health Organization website: http://www.who.int/mediacentre/factsheets/fs297/en/ (accessed on 24 July 2020).

Wierzba, T. F., & Muhib, F., (2018). Exploring the broader consequences of diarrheal diseases on child health. *The Lancet Global Health, 6*(3), e230–e231. https://doi.org/10.1016/S2214–109X(18)30047–0 (accessed on 24 July 2020).

Yoswathana, N., & Eshiaghi, M., (2013). Subcritical water extraction of phenolic compounds from mango seed kernel using response surface methodology. *Asian Journal of Chemistry, 25*(3), 1741–1744.

Zotarelli, M. F., Carciofi, B. A. M., & Laurindo, J. B., (2015). Effect of process variables on the drying rate of mango pulp by Refractance Window. *Food Research International, 69*, 410–417. https://doi.org/10.1016/j.foodres.2015.01.013 (accessed on 24 July 2020).

Zuo, H., Gaoa, J., Yuan, J., Deng, H., Yang, L., Weng, S., & Xu, X., (2017). Fatty acid synthase plays a positive role in shrimp immune responses against *vibrio parahaemolyticus* infection. *Fish and Shellfish Immunology, 60*, 282–288.

CHAPTER 3

Citrus Waste: An Important Source of Bioactive Compounds

NATHIELY RAMÍREZ-GUZMÁN,[1] MÓNICA L. CHÁVEZ-GONZÁLEZ,[1] ERICK PEÑA-LUCIO,[1] HUGO A. LUNA-GARCÍA,[1] JUAN A. ASCACIO VALDÉS,[1] GLORIA MARTÍNEZ-MEDINA,[1] MARIA DAS GRAÇAS CARNEIRO-DA -CUNHA,[2] TERESINHA GONÇALVES DA SILVA,[2] JOSÉ L. MARTINEZ-HERNANDEZ,[1] and CRISTÓBAL N. AGUILAR[1]

[1]Food Research Department, School of Chemistry, Universidad Autónoma de Coahuila Saltillo, Coahuila, México, E-mail: cristobal.aguilar@uadec.edu.mx

[2]Universidad Federal de Pernambuco Recife, Brazil

ABSTRACT

This chapter analyzes and describes the importance of the valorization of citrus residues since it is one of the most widely consumed fruit families worldwide due to its high nutritive contribution to the body, which are the main sources of vitamins and minerals. The citrus processing industries, as juices producers, generates wastes as husks, pomaces, and seeds, which becomes unsustainable for the environment since they are generally not used for something different as consumption or animal feeding. For this reason, the reuse of these leftovers is a critical opportunity area, as these wastes represent an attractive source of bioactive compounds such as sugars, essential oils, pectins, fiber, etc. which can be used in industrial sectors such as food production, agriculture, cosmetics, even pharmaceutical due to its high percentage of polyphenols with potential antioxidant, anti-inflammatory, anti-carcinogenic activities, etc.

3.1 INTRODUCTION

Citrus fruits compose a large group of universally edible fruits, from genus *Citrus*, subfamily *Aurontioideae*, family *Rutaceae* and *Geraniales* order; growing in more than 80 countries in 30–35 degrees north and south of equatorial line, in tropical and subtropical environments (Berk, 2016; Ladaniya, 2008b), where generous members are sweet orange (*Citrus sinensis*), tangerine (*Citrus reticulata*), grapefruit (*Citrus paradisi*), lemon (*Citrus limon*), and lime (*Citrus aurantifolia*) (Zheng, Zhang, Quan, Zheng and Xi, 2016). They are recognized for characteristics as their refreshing fragrance, thirst-quenching ability, for being vitamin C supply (Ladaniya, 2008a) as well as plenty substances with biological and industrial importance as carbohydrates (Glucose, trehalose, fructose,) essential oils, other vitamins (riboflavin, niacin) carotenoids (lutein B-carotene, lycopene), lipids (palmitic, stearic, oleic) acids, enzymes, flavonoids (Hesperidin, naringin, diosmetin, anthocyanins), limonoids (Limonin, limonic acid), polyphenols (phloroglucionol, phenolic acids, coumarins) polysaccharides (pectin, xylan, araban, galactan) and organic acids (citric acid, malic acid, succinic acid) (Boukroufa, Boutekedjiret and Chemat, 2017; Sharma, Mahato, Cho and Lee, 2017a). Due to the nutritional value and health benefits after their consumption, represent an important food group and performs the most consumed fruits at global level (Satari and Karimi, 2018).

Preliminary studies from FAO, estimate in 2016 a production of 124 246 thousand tonnes of citrus fruits at a worldwide level (FAO, 2016). The dominant produced variety with 61% are orange, followed by tangerine with 22%, then lemon and limes with 11% and finally grapefruit with 6% (Berk, 2016), where countries as Brazil, China, India, Mexico, Spain, and the USA produce over two thirds of world citrus fruits (Satari and Karimi, 2018), for example, for 2016 in the northern hemisphere the main oranges producers constitutes USA (5,371 t), China (7,000 t) , India (6,850 t), Spain (3 641 t) and Mexico (3,535 t) while in south hemisphere the main producer is Brazil (14,350 t) (FAO, 2016); despite the fresh consume is growing specially in developing countries (Sharma et al., 2017a), at least ~50% of citrus is processed in products as juice or marmalade where between 50–60% of fruit is converted in waste (Negro, Mancini, Ruggeri and Fino, 2016)and as a consequence large quantities of residues are produced, in the citrus fruits instance, the main wastes generated are the non-edible constitutes as peel flavedo, albedo, and seeds (Sharma, Mahato, Cho and Lee, 2017b). Citrus wastes are voluminous, heterogeneous, complex and with high biodegradation ratio; also present a low pH (3–4), elevated organic and aqueous content (Negro et al., 2016; Sharma et al., 2017a). This class of food-derived wastes are generally reduced to a low protein cattle feed or disposed in landfills (Lohrasbi et al., 2010), bringing economic and environmental consequences, due to their facility for decomposition, and problematics as high transporting costs or lack of disposal sites (Pourbafrani et al., 2010). In response to this challenging situation, food science is searching for possible choices for this rich compound wastes appraisement.

Due to their complex and valuable composition of this residues, emerging approaches for their valorization include bioethanol production (John et al., 2017), or isolation of high added-value components with importance in food, pharmaceuticals and cosmetics industries (Boukroufa et al., 2017; Bustamante et al., 2016; Céliz, Díaz, and Daz, 2018). This chapter aims to highlight and collect information about the potential of abundant food waste with chemical and biotechnological potential.

3.2 BIOACTIVE COMPOUNDS PRESENT IN CITRUS WASTES

The citrus industry is one of the most relevant agricultural activities in Mexico and in the world. The leading production is branched through ten countries, heading China and Brazil with 42% of production, Mexico ranks fourth in the list (Ruiz Barreda, 2016). The main product of the citrus industry is found in fruit juices, where the main consumers are European countries and E.E.U.U. In Mexico, the main producers of citrus fruits are the states of Veracruz, San Luis Potosí, Nuevo León, Michoacán, and Tamaulipas. At national level, the product generated is marketed primarily in the domestic market, and the rest is exported to E.E.U.U and some countries in South America, mainly in juices and segments (Ruiz Barreda, 2016). The main citrus fruits produced in Mexico are lemon, orange, and grapefruit; it is predicted that by 2024, citrus production will be 10.17 million tons in Mexico (SAGARPA, 2016).

It has been proved that citrus peel is rich in a plenty number of bioactive substances which exhibits benefits for health with antioxidant, anti-proliferative, anti-inflammatory and cardiovascular system effects (Lachos-Perez et al., 2018). Bioactive compounds can be defined as essential and non-essential compounds that are found in nature and they have some beneficial impact on health (Biesalski et al., 2009), among the bioactive compounds, the most studied are the antioxidant compounds.

Antioxidants are compounds that delay oxidation by inhibiting the formation of free radicals or interrupting their propagation. The effectiveness of antioxidants is related to their activation energy, redox potential, among other factors. The most effective antioxidants are those that can interrupt the chain of formation of free radicals; they are usually formed by aromatic or phenolic compounds. Antioxidants act providing a proton to free radicals, resulting in a intermediate radical; these intermediate radicals are stabilized by electron resonance delocalization within the aromatic ring and the formation of quinone structures(Brewer, 2011). Free radicals attack molecules with greater stability, taking electrons from them, and becoming free radicals by starting a chain reaction. Normally in cells, there exist a balance between oxidizing and antioxidant species, which can be altered by the concentration of both oxygen and antioxidants. When the concentration of oxygen increases and the concentration of antioxidants decreases, a process of oxidative stress occurs, which causes damage to the cell's molecules (Shalaby and Shanab, 2013).

The human body uses mechanisms to avoid increasing the concentration of reactive oxygen species using antioxidants, which can be trough enzymatic and non-enzymatic mechanisms. Among the enzymatic antioxidants are superoxide dismutases, catalases, and glutathione peroxidases plus a large variety of small molecules distributed throughout the body. Among the non-enzymatic antioxidants are tocopherol, glutathione, Vitamin C, β-carotene, antioxidants can be natural or synthetically formed. The polyhydroxyflavones, flavones, isoflavones, flavanols, and chalcones, are compounds that have high antioxidant activity and are found in plants (Carocho and Ferreira, 2013).

3.2.1 POLYPHENOLS

Polyphenols are the widest range of non-energetic compounds found in foods of plant origin. Its structure is integrated mainly by one or more phenolic rings. Its origin comes from the secondary metabolism of plants participating as part of their metabolism or as defense mechanisms in the face of stress situations (Quiñones and Aleixandre, 2012). There is a variety of polyphenols classes and subclasses, which are defined by the number of rings and the structural elements that make them up. Due to the wide variety of existing polyphenolic compounds, it has been classified in different ways as their source of origin, their biological function, natural distribution, and chemical structure. (Belščak-Cvitanović, Durgo, Huđek, Bačun-Družina and Komes, 2018). Chemically polyphenols can be classified by the number of rings they contain, by the components attached to them that differentiate them from flavonoids, flavones, etc. (Cutrim and Cortez, 2018).

3.2.2 FLAVONOIDS

Flavonoids are natural pigments found in vegetables, which have an antioxidant role, protecting cells against several factors that affect it, such as

UV radiation and pollution among other factors. Humans obtain it through diet and supplements. Its structure is principally based on the occurrence of polyphenols, with low molecular weight and their skeleton are very similar to diphenylpyranes. These compounds have a great capacity to chelate the Fe^{+} ion (Martínez-Castillo et al., 2018). Flavonoids represent the mainly polyphenols in the diet. They can be divided by their degree of oxidation: in flavones, isoflavones, flavanols, anthocyanins, and proanthocyanins (Scalbert and Williamson, 2000).

During the growth of citrus fruits, different phytochemicals are produced, including low molecular weight phenolic compounds like hydroxybenzoic acid and hydroxycinnamic acid (Sharma et al., 2017). Using UPLC and based on the retention times of different citrus fruit, sixteen types of polyphenolic molecules were identified as predominant on citrus fruits, of which seven are flavanones while the rest are polymethoxylated polyphenols. Among which stand out the eriocitrin, narginine, hesperidin, like flavanones and sinensetine, nobiletin like flavonoids polymethoxylated (Zhao et al., 2017).

3.2.3 ESSENTIAL OILS

Essential oils are volatile and complex compounds, characterized by a strong smell formed by aromatic plants as secondary metabolites. They are obtained from steam or hydrodistillation. They are known for their antimicrobial, analgesic, anti-inflammatory properties. They are very complex compounds because they are mixtures of between 20–60 different compounds of which two compounds are those that appear in higher concentrations. Generally, these predominant compounds are those that confer their biological properties.

The most prominent groups in the essential oils are terpenes or terpenoids attached to aromatic rings and other aliphatic compounds characterized by low molecular weight (Bakkali and Idaomar, 2008). Essential oils are extracted from different parts of the plant such as seeds, flowers, roots, leaves, etc. And they are obtained by different methods such as the use of solvents, the use of high pressures and more modern methods such as the use of microwaves (Chellappandian et al., 2018). The essential oils are formed by a volatile and a nonvolatile part which can be more than 200 compounds. The volatile fraction is mostly composed of monoterpenes, sesquiterpenes, and oxygenated compounds. The non-volatile fraction may contain waxes, fatty acids, carotenoids, flavonoids, etc. (Vieira et al., 2018). Among the principal substances of citrus essential oils are terpenes, terpenoids and phenylpropanoids, of which the most are: limonene, terpinene, pinene, and others (Pateiro et al., 2018). Which, due to their antimicrobial and aromatic capacities, which are used in different production areas such food, cosmetics and pharmaceutical products (Dhital et al., 2018).

3.2.4 PECTINS

Pectins are polysaccharides with a complex structure, mainly in the cell wall of plants. They are obtained mainly from the citrus peels and are used as gelling agents in the food industries (Axelos, Thibault and Lefebvre, 1989). Pectins are made up of at least eight different types of polysaccharides, of which the most common are rhamnogalacturonan I, II and xylogalacturonan. (Pasandide et al., 2017) Structurally, pectins are formed mainly by a linear fraction or also known as homogalacturonan, which is the smooth part of the pectin chain and consists of galacturonic acid polymers linked in α1–4 glycosidic bonds (Fracasso et al., 2018).

Pectin also has beneficial effects on health such as reducing cholesterol and glucose levels, anti-proliferative and anti-cancer effects, as well as stimulating the immune response. Currently, the principal sources of pectin are apple pulp and citrus peel, but due to the high demand for this product research is done to find new sources of pectin (Fracasso et al., 2018; Pasandide et al., 2017). In citrus fruits they are found mainly in the dietary fiber, forming between 65–70% of the total fiber, being the majority compound in the dietary fiber (Gutiérrez and Pascual, 2016).

3.2.5 HEMICELLULOSE

Hemicellulose is a heterogeneous polymer, which can be formed by pentoses, hexoses, and/or uronic acids, among other sugars (Rivas Siota, 2014). In citrus fruits, hemicellulose is part of the insoluble part of dietary fiber, which has been shown to have health benefits (Wang et al., 2015). The high content of hemicellulose and similar compounds in citrus allows the consideration of these products as raw material other products generation such paper and bioethanol as fuel (González-Velandia, Daza-Rey, Caballero-Amado and Martínez-González, 2016; Sánchez Riaño, Gutiérrez Morales, Muñoz Hernández and Rivera Barrero, 2010). Around 80% of the compounds in the citrus fiber are carbohydrates, which pectin (42.25%) and cellulose (15.95%) are the most common polysaccharides. Hemicellulose is also one of the components of citrus fiber, which has a very high viscosity due to its water retention capacity (Lundberg, Pan, White, Chau and Hotchkiss, 2014). The main source of cellulose comes from the shell and the residues of the medulla from the extraction of the juice (Ho and Lin, 2008).

3.3 CONVENTIONAL, EMERGING METHODS (ULTRASOUND-ASSISTED EXTRACTION, MICROWAVE-ASSISTED EXTRACTION)

For an efficient extraction of bioactive compounds is crucial examine that the recovery yields are high-dependent on extraction solvents, and the extraction method applied (as the most relevant point), this because the bioactive compounds (considered as secondary vegetable metabolites) are generally in low concentrations or it is difficult to extract them and recover them from the plant matter. An efficient extraction method is necessary two points of view including not only the proper recovery of the bioactive components of the plant matrix, but also facilitates their identification and characterization (Ribeiro et al., 2009). In this section, the most important conventional and emerging extraction methods are shown.

Conventional extraction methods for bioactive vegetal molecules include maceration, infusion, and reflux (Wong et al., 2015) as well as the use of Soxhlet, steam trawling, among others. In table 1, some of the characteristics of these methods are shown.

In summary, it can be said that conventional extraction methods of bioactive compounds are difficult to work with, require long working periods, and require a large number of organic solvents that can cause environmental damage and generate low extraction yields (Olaya and Mendez, 2003). Also, these methods involve the degradation of thermo-labile bioactive compounds (Wong et al., 2015). Due to it is necessary to find a strategy for extraction and recovery of biologically active substances of phytochemical origin from an efficient, sustainable, and resource-efficient point of view. In this sense, emerging methods or technologies can be used.

3.3.1 EMERGING METHODS OF EXTRACTION

The concept emerging or alternative method alludes to techniques different from conventional ones, whose characteristics are the reduction in energy consumption and the use of solvents that ensure the quality of extracts and bioactive compounds (Saini et al., 2014). Furthermore, using these emerging methods is also intended to reduce chemical risks, reduce adverse effects, and reduce as far as possible the consumption of non-renewable resources (Manaha, 2007). According to the above, extraction techniques have been developed using ultrasound and microwaves (Patil and Akamanchi, 2017). Techniques as ultrasound provide plenty of improvements inside this field, such as easy and quick handling, low levels of energy consumption and better performance in the extraction of biological molecules (Brar et al., 2013), as

TABLE 3.1 Conventional Extraction Methods and Their Characteristics

Methodology	Characteristics	Advantages	Disadvantages
Steamtrawling	Essentialoilsextraction	High quality and purity	– Unpractical equipment – Slow droplet collection – Volatile and water-insoluble extract – Sometimes requires a second distillation
CO_2 extraction	Similar to steam trawling	Good quality oils	– Costly equipment and infrastructure(Gennaro and Remington, 2003)
Tinctures	A hydro-alcoholic inert solvent, maceration or percolation are required	No special equipment required	– Plant material should be left to stand for 2–14 days with sporadic agitation (Patil and Akamanchi, 2017)
Pressing	The plant material is pressed against a sponge which is then squeezed out.	Good quality oils	– Recommended only for citrus fruits
Effleurage	The plant is spread over fat and distilled with alcohol.	– Considered the only option for plants that cannot be distilled	– Good yields are not generated
Soxhlet	The plant material is placed in a cartridge inside a chamber of the equipment, where it is continuously extracted.	– Control over solvent consumption – Good yields – Good quality of the extracts[1]	– Slow process – Not recommended for thermolabile compounds – Strictly necessary the final stage of evaporation
Infusion	The plant matter is immersed in cold or boiling water[4]	– It can be done at home	– A limited concentration of compounds is obtained
Reflux	Continuous extraction, the matrix is placed in a cellulose cartridge inside an extraction chamber	– Allows to exhaust the compounds in the matrix	– Not suitable for thermolabile substances – Requires a heating and cooling system for the condenser

well as good yields in short periods using low temperatures (Tomas et al., 2001), selectivity, and improvement in the quality of the extracts (Tiwari, 2015). This technique is particularly recommended for the extraction of volatile and semi-volatile organic compounds.

The beginning step of this procedure is the application of ultrasonic waves through the solvent generating acoustic cavitation; then air bubbles are generated in the solvent (these bubbles increase or decrease in size during pressure fluctuations, where they then collapse and release energy, reflected as an increase in temperature and medium pressure). Subsequently, a phenomenon of rupture in the cells is carried out with an intense mass transfer, allowing the penetration of the solvent (Tomas et al., 2001), and therefore the release of the phytocompounds of interest is produced (Tchabo et al., 2015). This cavitation phenomenon causes the growth of bubbles inside cell fluids or near solid surfaces (Dobias et al., 2010).

During the ultrasound-assisted extraction process, a process of swelling of the plant material also occurs. The amounts of solvent absorbed may vary according to the species and parts of the plant. However, they always absorb an extra volume, indicating that ultrasonic application into dry plants with aqueous media generates a progressive hydration process (Wu et al., 2014). Toma et al. (2001) showed that this hydration process favors the extraction of certain substances, since ultrasound hydrates the lamellae (present in plant cell membranes), making the material more malleable. This effect is stronger when using a frequency of 20 kHz than 500 kHz. Once the lamella has disintegrated, the plant cells are exposed to the solvent extraction process. In their work they demonstrated that there is a transformation of the EtOH solvent (40%), causing a slight decomposition to acetic aldehyde when irradiation is prolonged for more than 24 hours due to the oxidation of dissolved air. Consequently, a short exposition time in ultrasound extraction is decisive to avoid damage in the quality of the extract. However, the fundamental point out that a disadvantage of ultrasound-assisted extraction is the unavoidable use of solvent (Perino et al., 2016).Generally, ultrasounds applicability for phytochemicals extraction generates an extensive reproducibility in short times, reducing solvent consumption, increasing yields, and the use of low energy consumption. It has been reported that this technique is more efficient than traditional extraction methods mainly when used to obtain bioactive compounds from plant sources and agro-industrial residues such as citrus peels. (Trejo et al., 2015).

Extraction assisted with microwave represent a profitable and innovative technique in the hydrodistillation of essential oils fields due to simplifying the recovery of bioactive molecules and allows maintenance of their quality. It has a fast heating system that allows temperature control to avoid thermal gradients and degradations. It is also a method focused on carrying out ecological protocols, as it saves solvent consumption, time, and energy (Perino et al., 2016).

The first stage of the microwave-assisted extraction process is the homogeneous, efficient and rapid heating of the solvent and the plant matrix. Subsequently, the plant matrix absorbs the energy, causing cell rupture due to internal overheating (which facilitates the recovery of bioactive compounds) (Zhang et al., 2014) so that the solvent penetrates directly into the sample, transferring the heat of the solvent through the pores of the plant matrix. Finally, a minimum amount of CO_2 will be generated and because the amount of energy used is almost entirely consumed, the amount of time used in the process decreases. Microwave-assisted extraction can also be used without the need for solvents, a technique known as "hydro diffusion and microwave gravity," which is a process that reduces solvent consumption by using water (solvent) *in situ* from the matrix to extract the bioactive hydrophilic compounds. For this reason, it is usually applied in fresh plant samples (Melgar et al., 2016).

These methods are suitable for the extraction of bioactive compounds. For example, Rojas et al. (2014) evaluated the effect of solvents such as water, acetone, ethanol, and methanol at different concentrations using ultrasound-assisted extraction for 2 min to extract the total content of bioactive compounds and the antioxidant capacity of some fruits such as blackberry and citrus fruits from Castilla. On the other hand, Valadez et al. (2017) studied the extraction of bioactive compounds and the antioxidant capacity exhibited by extracts from diverse fruit waste, as well as citrus fruits, using mixtures of 100% methanol, 75% methanol-water, 50% methanol-water and 25% methanol-water. The extracts were obtained by microwave-assisted extraction and ultrasound in extraction times of 30, 60 and 120 minutes.

In conclusion, it is very important to generate information about new methods for the extraction and recovery of bioactive compounds, since conventional methods offer some disadvantages that affect the performance, stability, and quality of the compounds obtained. All this is also important because these methods are known as emerging or alternative methods have great application in the use of fruit residues including citrus fruits.

3.4 CITRUS WASTES A SOURCE OF FUNCTIONAL INGREDIENTS

Citrus fruits represent a plenty demanded product by worldwide consumers as fresh produce, nevertheless seed and peel are discarded as waste containing a wide collection of other compounds with substantial antioxidant activity. Worldwide cultivation of citrus fruits has an important increment in the course of the past few years attaining 82 million tons (Rafiq et al., 2016). Citrus processing instigates enormous volumes of citrus by-product such represent about 50% of their weigh, in consequence, it generates a large number of waste citrus fruits causing damage to the environment, however, these by-products may be revalued, because its composition can be applied to different areas of the industry and also offers several benefits: economic gains, environmental protection and an increase in the rate of life.

3.5 BIOTECHNOLOGICAL APPLICATIONS OF CITRUS WASTES

Citrus by-product consists principally in skins, pulp, and seeds that contain a source of essential oils, sugars, pigments, fat, acids, insoluble carbohydrates, enzymes, flavonoids, pectin (Boukroufa et al., 2017). Citrus essential oils extracted from citrus by-products can be used in several kinds of food, such as flavoring ingredients in food or pharmaceuticals, essentially for its anti-inflammatory and antibacterial effects, these properties are generally characterized by its key component: D-limonene, corresponding to more than 95% (Boukroufa et al., 2017).

Citrus peel represents the main pectinases source in several industries but also apple pomace, the higher quantity of pectin in citrus peel can be utilized as a substrate from microorganism to produce pectinolytic enzymes; Orange bagasse composition is rich in soluble carbohydrates, like fructose, glucose, sucrose and pectin due to can be used as fermentation products generation including enzymes (Ahmed et al., 2015).

Citrus waste performs as an attractive lignocellulosic substrate for bioethanol production due to their carbohydrates richness and small lignin content, Orange peel waste (OPW), represent the solid waste generated after juice extraction, is meaningful lignocellulosic feedstock for the production of bioethanol, composed by: peel, juice sacs, rag (cores and segment membranes), and seeds, that amounts to 50–70% of the fresh fruit weight (John et al., 2017).

The production of bioethanol can be achieved for the presence of pectinolytic, xylanolytic, and cellulolytic enzymes that promote the breakdown of complex carbohydrates. Genus as *Aspergillus* and *Trichoderma* represent the main utilized microorganisms, due to their abundant xylanolytic, cellulolytic and pectinolytic enzymes (John et al., 2017). There are different kinds of citrus waste that can be used to its application to produce bioethanol; for example; Zhou et al., 2008 reported an ethanol yield of 4% from orange peels and John et al. (2017) reported a yield of ethanol of 18% using sweet lime peel, this indicates that sweet lime peel can produce bioethanol more effectively than orange peels.

Carotenoids are pigments with two or several units of hydrocarbons in their structure, more than 600 carotenoids have been identified in nature and about in citrus fruit peels such as orange since 1937 by Zeichsmeter and Tuzson (Boukroufa et al., 2017). ß-carotene, with α-carotene, β-cryptoxantin, lycopene, lutein and zeaxanthin are the main carotenoids found in citrus fruit and they constitute a main source of vitamin A and may protect from development of degenerative diseases such as macular degeneration, cancer, and heart diseases (Granado et al., 1992). Extracting these compounds has been developing different techniques to improve the yields and quality of the extracts, such as microwave-assisted extraction, ultrasound extraction, supercritical fluid extraction and electronic-technologies (Boukroufa et al., 2017).

Phenolic and flavonoids molecules can be found in citrus peel, playing important roles against diverse physiological threats and also studies have revealed its pharmacological and biological viewpoints such as hypolipidemic, hypoglycemic and anti-inflammatory perspectives (Ashraf, Butt and Iqbal, 2017). Due to its ease in accessibility of fruits by-products, citrus peel is one of the cheap sources of polyphenols for value-added and designer food products. There are more than 4,000 structural variants of flavonoids that have been identified and characterized for their prophylactic potential (Ashraf et al., 2017).

3.6 BIOACTIVE COMPOUNDS PRESENT IN CITRUS WASTES

The principal biological active molecules known for health improvers came from the phytochemical origin, especially phenolics in fruits and vegetables (Rafiq et al., 2016).

Citrus fruits possess a nutrient richness, with molecules such as vitamins A and C, folic acid and dietary fiber, and are a source of bioactive compounds, like flavonoids, coumarins, limonoids and carotenoids (Ding et al., 2012; Turner & Burri, 2013), presenting considerable amounts of flavanones, flavones, flavonols, and anthocyanins, where the main flavonoids are the flavanone (Nakajima, Macedo, and Macedo, 2014). Peels extract of some citrus fruits was measured by the Folin–Ciocalteu assay to make a comparison on the total phenol content (Figure 3.1).

Ramful, et al. (2010) that evaluated orange, clementine, mandarine, tangor, tangelo and Pamplemousse peels detected poncirin (2.49–18.85 mg/g FW), rhoifolin (4.54–10.39 mg/g FW), didymin (3.22–13.94 mg/g FW), rutin (8.16–42.13 mg/g FW), diosmin (4.01–18.06 mg/g FW), isorhoifolin (1.72–14.14 mg/g FW), neohesperidin (3.20–11.67 mg/g FW), hesperidin (83.4–234.1 mg/g FW), neoeriocitrin (8.8–34.65 mg/g FW) and narirutin (5.05–21.23 mg/g FW).

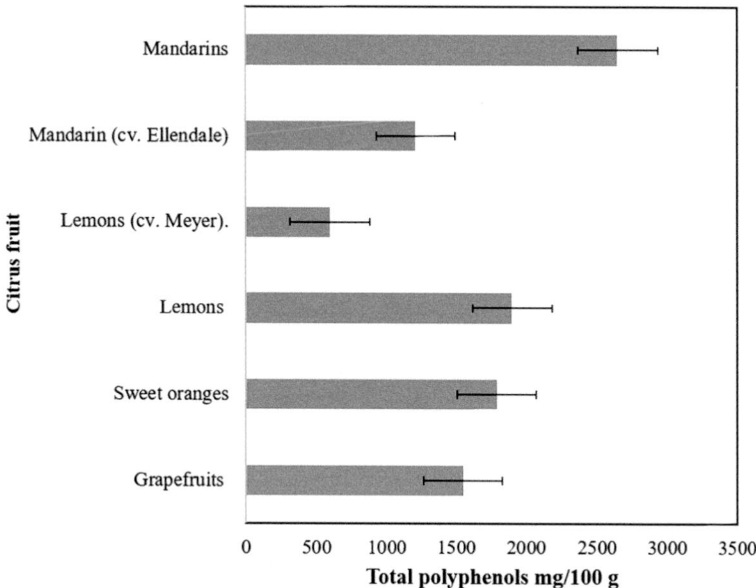

FIGURE 3.1 Total polyphenols content from different citrus peels by the Folin-Ciocalteu assay (Modified from Rafiq et al. 2016).

Flavonoids are polyphenolic molecules with a phenyl benzopyrone structure, constitute for two benzene rings (C6) joined by a linear three-carbon chain (C3), with a carbonyl group at the C position.

Exists different citrus flavonoids including a class of glycosides, like hesperidin and naringin and another class of O-methylatedaglycones of flavones like nobiletin and tangeretin, which are relatively two common polymethoxylatedflavones (PMFs) (Li et al., 2014).In citrus fruits, peels are reported as the main source of polymethoxylated flavones compared to other edible parts of the fruit.

The citrus flavonoids are reported as health-improvers, with diverse properties, including anticancer, antiviral and antiinflammatory activities, also as reducing capillary fragility, and restricting human platelet aggregation (Boukroufa et al., 2017). Different by products in high quantities, are generated from cereal, fruity or vegetable sources and they can be used as a source of add-value products. They supply of dietary fiber as well as biologically active molecules such as polyphenols or essential oils, provides an economic and wellness benefit to the producer and consumer respectively (Rafiq et al., 2016); Some kinds of peel citrus can contain different levels of dietary fiber that are shown in Figure 3.2.

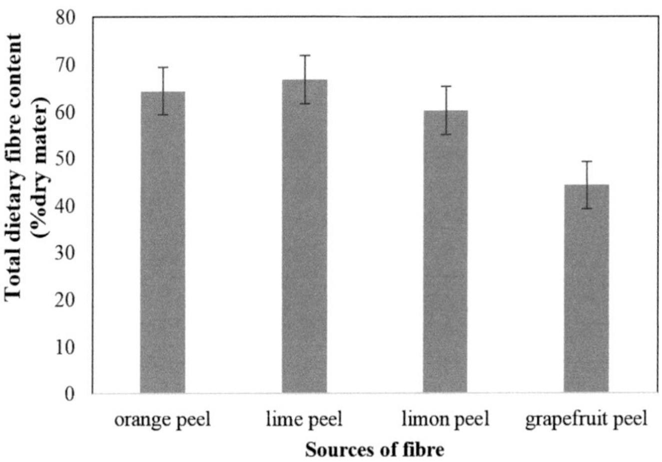

FIGURE 3.2 Different kinds of dietary fiber from citrus peels (Modified from Rafiq et al., 2016).

Citrus peel could be considered to be a potential source of pectin which is composed of white, spongy and cellulosic tissue (Rafiq et al., 2016). The consumption of dietary fiber can also avoid some diseases, obesity, cáncer, diabetes and cardiovascular disease. It promotes a reduction in blood cholesterol level, serum triglyceride level, and liver total lipids (Terpstra et al., 2002).

3.7 PHARMACEUTICAL AND BIOLOGICAL EFFECTS OF CITRUS WASTE ON HUMAN HEALTH

Phytochemical rich foods are considered an excellent alternative to maintain good health. Some studies have demonstrated its positive properties over human health; for example, its bioactive compounds can function as hypolipidemic, hypoglycemic and anti-inflammatory. It has been reported that overall diets containing citrus peel powder and extract delineated percent reduction in cholesterol level by 1.15% and 2.05%, respectively, and also it was observed that the nutraceutical diet elicited more percent decrement in cholesterol level (8.55%) in contrast to functional diet (5.80%) (Ashraf et al., 2017). Menichini et al. (2011) reported that citrus peel alleviates metabolic syndrome like coronary heart diseases, decreasing free fatty acids, hepatic and plasma triglycerides and increasing fecal excretion of triglycerides. The nutraceutical diet including citrus peels actives can promote an increase in the insulin secretion, for example, Ashraf et al., (2017) reported that functional and nutraceutical diet holds insulinotropic properties that improved insulin secretion by 2.07% and 3.21%, correspondingly.

Nobiletin is a flavonoid (O-methylated flavone) isolated from citrus peels. It has been reported to have anti-diabetic and hypocholesterolemic effects (Tsutsumi et al., 2014). Fayek et al. (2017) reported that citrus waste from Mandarin, Sweet orange, white grapefruit, and Lime decrease the levels of plasma cholesterol, triacylglyceride, and glucose (mg/dl) and also that hypocholesterolemic and hypotriglyceridemic effect of the tested samples is directly proportional to their content of nobiletin (Fayek et al., 2017).

Citrus fruits are well supplied of nutrients, such as vitamins A and C, folic acid and dietary fiber. Furthermore, these fruits are a source of biologically active substances, like flavonoids, coumarins, limonoids and carotenoids (Ding et al., 2012; Turner & Burri, 2013). Citrus phenolics has been evaluated to know its effect on the obesity, specifically in the adipocytes apoptosis, Nakayima et al. (2014) observed that the addition of polymethoxyflavones of citrus (100 mM) caused an increase in intracellular calcium, which induced significant calpain quantities and caspase-12, two proteins associated with programmed cell death, therefore, the reduction in the number of adipose cells due to apoptosis could assist in maintaining weight loss (Nakajima et al., 2014).

3.7.1 ANTIOXIDANT AND ANTIMICROBIAL ACTIVITY OF CITRUS PEELS WASTE

Antioxidants and phenolics are molecules that allows the enhancement of viability for bioprocessing of food-waste streams; it has been reported that antioxidants can inhibit the carcinogenesis (Redcorn, Fatemi, and Engelberth, 2018). Citrus peel waste presented phenolic compounds that can increase the antioxidant activity. Otang and Ofalayan (2016), reported that *C. limón* exhibit comparable antioxidant capacity to the synthetic antioxidant, it consists of 68 percent d-limonene, a prominent antioxidant with skin benefits due to its cleansing and purifying properties. Lou et al., (2017) studied the antioxidant activity of the *Kumquat*, during the testing of DPPH (2,2-diphenyl-1-picrylhydrazyl free radical scavenging method) and hydroxyl free radical scavenging activities of kumquat extracts, ethyl acetate fraction showed better antioxidant activities than both dichloromethane and butanol fractions. Shyi et al. (2014), investigated calamondin and the results indicated that is rich in a great amount of biological active molecules, such as flavonoids including polymethoxyflavones, phenolic acids, and limonoids, the DPPH scavenging potency of extracts from calamondin peel was significantly higher than calamondin pulp and the highest DPPH scavenging potency was observed in the 80°C water extract (Lou, Hsu and Ho, 2014).

A great number of studies have reported the inhibitory activities of Citrus peel; Otang et al. (2016) reported that the highest antibacterial activity was obtained with the acetone extract of *C. limon* against *E. faecalis* and *B. subtilis* with inhibition zone diameters of 23 and 20mm. It can be attributed to the content of the main active molecules in lemon; monoterpenes such as limonene, γ-terpinene, β-pinene, and the aldehydes geranial and neral (Otang and Afolayan, 2016). The antimicrobial capacity of extracts with hot water immature kumquat peel shows an improved inhibitory activity against gram-positive bacteria than gram-negative bacteria; the most affected bacteria was *Bacillus cereus* with a minimum inhibitory concentration (MIC) of 25 mg/mL (Lou and Ho, 2016). Wang et al. (2012) evaluated the essential oil of kumquat peel and it showed a profit antimicrobial effect at microgram range concentrations.

Some bacteria and fungus have developed antibiotic resistance that is a continual problem, the growth of *P. aeruginosa*, *S. aureus*, and *E. coli*, which previously known being resistant to multiple antibiotics, was weakly inhibited by both extracts of *C. limon*, while *S. pyogenes* and *B. cereus* and the fungus *C. krusei* was not inhibited by the ethanolic extracts of *C. limon*. In another study, the antibacterial effect of the essential oil from *Fortunella crassifolia was* evaluated on viable counts of different microorganism as *E. coli, B. subtilis* and *B. cereus* using a sterilized beef extract, BE, 3% in deionized water), exhibit a complete inhibition for those bacteria tested. In Wang et al., 2012 studies is possible to observe how bacteria, in contact with essential oil, result in a reduction in CFU count during the first hour, exerting the maximum effect using the lowest concentration of essential oil, against *B. cereus.*

Anti-hepatitis B virus activity of the extracts in dichloromethane, ethyl acetate, n-butanol, acetone, and methanol solvents of peel and pulp from calamondin fruits was tested; and shows ethyl acetate and acetone extracts of the peel lows the virus surface antigen expression of hepatitis B virus by 41.6% and 71.4%, respectively, at a dose of 50 mg/mL (Lou and Ho, 2016).

KEYWORDS

- **bioactivities**
- **citrus waste**
- **flavonoids**
- **pectin**
- **polyphenols**

REFERENCES

Ahmed, I., Anjum, M., Azhar, M., Akram, Z., & Tahir, M., (2015). Science direct bioprocessing of citrus waste peel for induced pectinase production by *Aspergillus niger* ; its purification and characterization. *Journal of Radiation Research and Applied Sciences, 9*(2), 148–154.

Ashraf, H., Butt, M. S., & Iqbal, M. J., (2017). *Asian Pacific Journal of Tropical Biomedicine, 7*(10), 870–880.

Axelos, M. A. V., Thibault, J. F., & Lefebvre, J., (1989). Structure of citrus pectins and viscometric study of their solution properties. *International Journal of Biological Macromolecules, 11*(3), 186–191.

Bakkali, F., & Idaomar, M., (2008). *Biological Effects of Essential Oils: A Review, 46*, 446–475.

Belščak-Cvitanović, A., Durgo, K., Huđek, A., Bačun-Družina, V., & Komes, D., (2018). Overview of polyphenols and their properties. In: *Polyphenols: Properties, Recovery, and Applications* (pp. 3–44). Elsevier.

Berk, Z., (2016). Chapter 1-Introduction: History, production, trade, and utilization. In: Berk, Z., (ed.), *Citrus Fruit Processing* (pp. 1–8). San Diego: Academic Press.

Biesalski, H. K., Dragsted, L. O., Elmadfa, I., Grossklaus, R., Müller, M., Schrenk, D., & Weber, P., (2009). Bioactive compounds: Definition and assessment of activity. *Nutrition, 25*(11–12), 1202–1205.

Boukroufa, M., Boutekedjiret, C., & Chemat, F., (2017). *Resource-Efficient Technologies Development of a Green Procedure of Citrus Fruits Waste Processing to Recover Carotenoids*, *3*, 252–262.

Brar, S. K., Dhillon, G. S., & Soccol, C. R., (2013). *Biotransformation of Waste Biomass into High-Value Biochemicals*. Springer Science & Business Media.

Brewer, M. S., (2011). Natural antioxidants: Sources, compounds, mechanisms of action, and potential applications. *Comprehensive Reviews in Food Science and Food Safety, 10*(4), 221–247.

Bustamante, J., Stempvoort, S. V., García-gallarreta, M., Houghton, J. A., Briers, H. K., Budarin, V. L., & Clark, J. H., (2016). Microwave assisted hydro-distillation of essential oils from wet citrus peel waste. *Journal of Cleaner Production, 137*, 598–605.

Carocho, M., & Ferreira, I. C. F. R., (2013). A review on antioxidants, pro-oxidants and related controversy: Natural and synthetic compounds, screening and analysis methodologies and future perspectives. *Food and Chemical Toxicology*, *51*, 15–25.

Céliz, G., Díaz, R., & Daz, M., (2018). Biocatalysis and agricultural biotechnology obtaining hesperetin 7-O-glucosyl 6 ″ -O-laurate , a high lipophilic flavonoid ester, from citrus waste. *Biocatalysis and Agricultural Biotechnology, 13*, 25–30.

Chellappandian, M., Vasantha-Srinivasan, P., Senthil-Nathan, S., Karthi, S., Thanigaivel, A., Ponsankar, A., & Hunter, W. B., (2018). Botanical essential oils and uses as mosquitocides and repellents against dengue. *Environment International*, *113*, 214–230.

Cutrim, C. S., & Cortez, M. A. S., (2018). A review on polyphenols: Classification, beneficial effects and their application in dairy products. *International Journal of Dairy Technology, 71*(3), 564–578.

Dhital, R., Mora, N. B., Watson, D. G., Kohli, P., & Choudhary, R., (2018). Efficacy of limonene nano coatings on post-harvest shelf life of strawberries. *LWT*, *97*, 124–134.

Dobiáš, P., Pavlíková, P., Adam, M., Eisner, A., Bevnová, B., & Ventura, K., (2010). Comparison of pressurized fluid and ultrasonic extraction methods for analysis of plant antioxidants and their antioxidant capacity. *Cent. Eur. J. Chem., 8*(1), 87–95.

FAO, (2016). Citrus fruit fresh and processed statistical bulletin 2016. *Statistical Bulletin*, 77. Retrieved from http://www.fao.org/3/a-i8092e.pdf (accessed on 24 July 2020).

Fayek, N. M., El-shazly, A. H., Abdel-monem, A. R., Moussa, M. Y., Abd-elwahab, S. M., & El-tanbouly, N. D., (2017). Comparative study of the hypocholesterolemic, antidiabetic effects of four agro-waste citrus peels cultivars and their HPLC standardization. *Revista Brasileira de Farmacognosia, 27*(4), 488–494.

Fracasso, A. F., Perussello, C. A., Carpiné, D., Petkowicz, C. L. D. O., & Haminiuk, C. W. I., (2018). Chemical modification of citrus pectin: Structural, physical and rheologial implications. *International Journal of Biological Macromolecules*, *109*, 784–792.

Gennaro, A. R., (2003). *Remington: Farmacia*. Editorial médica panamericana. (ed) Buenos Aires.

González-Velandia, K. D., Daza-Rey, D., Caballero-Amado, P. A., & Martínez-González, C., (2016). Evaluación de las propiedades físicas y químicas de residuos sólidos orgánicos a emplearse en la elaboración de papel. *Luna Azul, 43*(43), 499–517.

Gutiérrez, E., & Pascual, G., (2016). Caracterización de cáscara de mandarina (Citrusreticulata) en polvo e inclusión en una formulación panaria. *Agronomía Colombiana, 34*, 776–778.

Ho, S. C., & Lin, C. C., (2008). Investigation of heat treating conditions for enhancing the anti-inflammatory activity of citrus fruit (*Citrusreticulata*) peels. *Journal of Agricultural and Food Chemistry, 56*(17), 7976–7982.

John, I., Yaragarla, P., Muthaiah, P., Ponnusamy, K., & Appusamy, A., (2017). *Resource-Efficient Technologies Statistical Optimization of Acid Catalyzed Steam Pretreatment of Citrus Peel Waste for Bioethanol Production, 3*, 429–433.

Lachos-Perez, D., Baseggio, A. M., Mayanga-Torres, P. C., Maróstica, M. R., Rostagno, M. A., Martínez, J., & Forster-Carneiro, T., (2018). Subcritical water extraction of flavanones from defatted orange peel. *The Journal of Supercritical Fluids*, *138*, 7–16.

Ladaniya, M. S., (2008a). *Citrus Fruit*.

Ladaniya, M. S., (2008b). Introduction. In: *Citrus Fruit: Biology, Technology and Evaluation* (pp. 1–11).

Lohrasbi, M., Pourbafrani, M., Niklasson, C., & Taherzadeh, M. J., (2010). Bioresource Technology Process design and economic analysis of a citrus waste biorefinery with biofuels and limonene as products. *Bioresource Technology, 101*(19), 7382–7388.

Lou, S., & Ho, C., (2016). Science direct phenolic compounds and biological activities of small-size citrus : Kumquat and calamondin. *Journal of Food and Drug Analysis, 25*(1), 162–175.

Lou, S., Hsu, Y., & Ho, C., (2014). Science direct flavonoid compositions and antioxidant activity of calamondin extracts prepared using different solvents. *Journal of Food and Drug Analysis, 22*(3), 290–295.

Lundberg, B., Pan, X., White, A., Chau, H., & Hotchkiss, A., (2014). Rheology and composition of citrus fiber. *Journal of Food Engineering*, *125*, 97–104.

Manahan, S. E., (2007). Introducción a la química ambiental. Editorial Reverté S.A. Primera edición. México.

Martínez-Castillo, M., Pacheco-Yepez, J., Flores-Huerta, N., Guzmán-Téllez, P., Jarillo-Luna, R. A., Cárdenas-Jaramillo, L. M., & Shibayama, M., (2018). Flavonoids as a Natural treatment against *Entamoeba histolytica*. *Frontiers in Cellular and Infection Microbiology*, *8*, 209.

Melgar-Lalanne, G., Hernández-Álvarez, A. J., Jiménez-Fernández, M., & Azuara, E., (2016). Oleoresins from *Capsicum* Spp.: Extraction methods and bioactivity. *Food Bioprocess Technol.*, 1–26.

Nakajima, V. M., Macedo, G. A., & Macedo, J. A., (2014). LWT-Food Science and Technology Citrus bioactive phenolics : Role in the obesity treatment. *LWT-Food Science and Technology, 59*(2), 1205–1212.

Negro, V., Mancini, G., Ruggeri, B., & Fino, D., (2016). Bioresource Technology citrus waste as feedstock for bio-based products recovery : Review on limonene case study and energy valorization. *Bioresource Technology*, *214*, 806–815.

Olaya, J. M., & Méndez, J., (2003). Guía de Plantas y Productos Medicinales. (ed) Bogotá.

Otang, W. M., & Afolayan, A. J., (2016). South African journal of botany antimicrobial and antioxidant efficacy of *Citrus limon* L. peel extracts used for skin diseases by Xhosa tribe of Amathole District, Eastern Cape, South Africa. *South African Journal of Botany*, *102*, 46–49.

Pasandide, B., Khodaiyan, F., Mousavi, Z. E., & Hosseini, S. S., (2017). Optimization of aqueous pectin extraction from *Citrus medica* peel. *Carbohydrate Polymers, 178*, 27–33.

Pateiro, M., Barba, F. J., Domínguez, R., Sant'Ana, A. S., Mousavi, K. A., Gavahian, M., & Lorenzo, J. M., (2018). Essential oils as natural additives to prevent oxidation reactions in meat and meat products: A review. *Food Research International*, *113*, 156–166.

Patil, D. M., & Akamanchi, K. G., (2017). Ultrasound-assisted rapid extraction and kinetic modeling of influential factors: Extraction of camptothecin from *Nothapodytes nimmoniana* plant. *Ultrason. Sonochem., 37*, 582–591.

Périno, S., Pierson, J. T., Ruiz, K., Cravotto, G., & Chemat, F., (2016). Laboratory to pilot scale: Microwave extraction for polyphenols lettuce. *Food Chem., 204*, 108–114.

Pourbafrani, M., Forgács, G., Sárvári, I., & Niklasson, C., (2010). Bioresource Technology Production of biofuels, limonene, and pectin from citrus wastes. *Bioresource Technology, 101*(11), 4246–4250.

Quiñones, M., & Aleixandre, M. M. A., (2012). *Los Polifenoles, Compuestos De Origen Natural Con Efectos Saludables Sobre El Sistema Cardiovascular Compounds with Beneficial Effects, 27*(1), 76–89.

Rafiq, S., Kaul, R., Sofi, S. A., Bashir, N., Nazir, F., & Ahmad, G., (2016). Citrus peel as a source of functional ingredient : A review. *Journal of the Saudi Society of Agricultural Sciences*.

Redcorn, R., Fatemi, S., & Engelberth, A. S., (2018). Comparing end-use potential for industrial food-waste sources. *Engineering, 4*(3), 371–380.

Ribeiro, E. B., Reis, R., Alfonso, S., & Scarminio, I. S., (2009). Enhanced extraction yields and mobile phase separations by solvent mixtures for the analysis of metabolites in *Annona muricata* L. leaves. *J. Sep. Sci., 32*, 4176–4185.

Rivas, S. S., (2014). *Valorización De Hemicelulosas De Biomasa Vegetal*. Universidad de Vigo.

Rojas, P., Martínez, J., & Stashenko, E., (2014). Contenido de compuestos fenólicos y capacidad antioxidante de extractos de mora (RubusglaucusBenth) obtenido bajo diferentes condiciones. *Vitae., 21*(3), 218–227.

Ruiz, B. J. D. J., (2016). *Retos y Oportunidades en el Sector Citrícola* (I) (p. 1). El Economista.

SAGARPA, (2016). Agrícola nacional cítricos y toronja Mexicanos.

Saini, R. K., Shetty, N. P., & Giridhar, P., (2014). GC-FID/MS analysis of fatty acids in Indian cultivars of *Moringa oleifera*: Potential sources of PUFA. *J. Am. Oil Chem. Soc., 91*, 1029–1034.

Sánchez, R. A. M., Gutiérrez, M. A. I., Muñoz, H. J. A., & Rivera, B. C. A., (2010). Producción de bio-etanol a partir de sub productos agro-industriales lignocelulósicos bio-ethanol production from agro-industrial lignocellulosic by products. *Tumbaga*, *5*, 61–91.

Satari, B., & Karimi, K., (2018). Citrus processing wastes: Environmental impacts, recent advances, and future perspectives in total valorization. *Resources, Conservation and Recycling, 129*, 153–167.

Scalbert, A., & Williamson, G., (2000). Dietary intake and bioavailability of polyphenols. *The Journal of Nutrition, 130*(8), 2073S–2085S.

Shalaby, A., & Shanab, S. M. M., (2013). Antioxidant compounds, assays of determination and mode of action. *African Journal of Pharmacy and Pharmacology, 7*(10), 528–539.

Sharma, K., Mahato, N., Cho, M. H., & Lee, Y. R., (2017a). Converting citrus wastes into value-added products: Economic and environmentally friendly approaches. *Nutrition, 34*, 29–46.

Sharma, K., Mahato, N., Cho, M. H., & Lee, Y. R., (2017b). Converting citrus wastes into value-added products: Economic and environmentally friendly approaches. *Nutrition, 34*, 29–46.

Tchabo, W., Ma, Y., Engmann, F. N., & Zhang, H., (2015). Ultrasound-assisted enzymatic extraction (UAEE) of phytochemical compounds from mulberry (Morus nigra) must and optimization study using response surface methodology. *Ind. Crops Prod., 63*, 214–225.

Tiwari, B. K., (2015). Ultrasound: A clean, green extraction technology. *TrAC-Trends Anal. Chem., 71*, 100–109.

Toma, M., Vinatoru, M., Paniwnyk, L., & Mason, T., (2001). Investigation of the effects of ultrasound on vegetals tissues during solvent extraction. *Ultrason. Sonochem., 8*, 137–142.

Trejo, M. A., Vargas, M. G., Sánchez, A., Lira, A., Pascual, S., Granados, G., & Villavicencio, A., (2015). Extracción de compuestos bioactivos de plantas del desierto mexicano para su aplicación en envases activos para zarzamora. *Revista Iberoamericana de Tecnología Postcosecha, 16*(1), 101–107.

Valadez, A., López, E., García, R., & Ruiz, F. L., (2017). Comparación de dos técnicas de extracción de fenólicos totales y capacidad antioxidante a partir de chipilín (Crotalariamaypurensis H.B.K.). *Investigación y Desarrollo en Ciencia y Tecnología de Alimentos, 2*, 481–487.

Vieira, A. J., Beserra, F. P., Souza, M. C., Totti, B. M., & Rozza, A. L., (2018). Limonene: Aroma of innovation in health and disease. *Chemico-Biological Interactions, 283*, 97–106.

Wang, L., Xu, H., Yuan, F., Pan, Q., Fan, R., & Gao, Y., (2015). Physicochemical characterization of five types of citrus dietary fibers. *Biocatalysis and Agricultural Biotechnology, 4*(2), 250–258.

Wang, Y., Zeng, W., Xu, P., Lan, Y., Zhu, R., & Zhong, K., (2012). *Chemical Composition and Antimicrobial Activity of the Essential Oil of Kumquat (Fortunella crassifolia Swingle) Peel*, 3382–3393.

Wong, J. E., Muñiz, D. B., Martínez, G., Belmares, R. E., & Aguilar, C. N., (2015). Ultrasound-assisted extraction of polyphenols from native plants in the Mexican desert. *Ultrason. Sonochem., 22*, 474–481.

Wu, H., Zhu, J., Diao, W., & Wang, C., (2014). Ultrasound-assisted enzymatic extraction and antioxidant activity of polysaccharides from pumpkin (Cucurbita Moschata). *Carbohydr. Polym., 113*, 314–324.

Zhang, H., Tang, B., & Row, K. H., (2014). A green deep eutectic solvent-based ultrasound-assisted method to extract astaxanthin from shrimp by-products. *Anal. Lett., 47*(5), 742–749.

Zhao, Z., He, S., Hu, Y., Yang, Y., Jiao, B., Fang, Q., & Zhou, Z., (2017). Fruit flavonoid variation between and within four cultivated Citrus species evaluated by UPLC-PDA system. *Scientia Horticulturae, 224*, 93–101.

Zheng, H., Zhang, Q., Quan, J., Zheng, Q., & Xi, W., (2016). Determination of sugars, organic acids, aroma components, and carotenoids in grapefruit pulps. *Food Chemistry, 205*, 112–121.

CHAPTER 4

Use of Agro-Industrial Residues to Obtain Polyphenols with Prebiotic Effect

ANA YOSELYN CASTRO-TORRES, RAÚL RODRÍGUEZ-HERRERA, AIDÉ SÁENZ-GALINDO, JUAN ALBERTO ASCACIO-VALDÉS, JESÚS ANTONIO MORLETT-CHÁVEZ, and ADRIANA CAROLINA FLORES-GALLEGOS

Food Research Department, School of Chemistry, Universidad Autónoma de Coahuila, Blvd. Venustiano Carranza S/N, Colonia República, 25280, Saltillo, Coahuila, México, E-mail: carolinaflores@uadec.edu.mx

ABSTRACT

The exploitation of low-valued waste from natural sources to obtain biologically active compounds especially for obtaining polyphenolic compounds has had a great impact because these compounds offer a wide variety of health benefits, the use of technologies for the extraction of phenolic compounds that are friendly to the environment, apart from high yields and short extraction times compared to other traditional techniques. This chapter proposes the use of various plant materials such as pomegranate, nut and melon, soursop leaves, chili, moringa, tarbush, and creosote bush, and grains such as sorghum, among others, to obtain these compounds of interest and their potential use as a prebiotic.

4.1 INTRODUCTION

The use of natural resources referred to as waste that possesses bioactive compounds of importance for different areas such as the food industry, the pharmaceutical industry and especially in the area of health such as polyphenolic compounds, fatty acids, proteins, probiotics and prebiotics among others it has been very little studied and therefore they have not been exploited properly.

These phenolic compounds, molecules with aromatic rings, and with the presence of hydroxyl groups are the most studied phytochemical compounds. In addition to providing benefits to the plant as protection against ultraviolet rays, they improve its organoleptic properties, they are also beneficial for humans by acting as antioxidants, antitumor agents and preventing an endless number of diseases, therefore the phenolic compounds of Different plant sources, especially plants or by-products with little relevance (Ignat, Volf, and Popa, 2011) have reported that some phenolic compounds can act as prebiotics and benefit the intestinal microbiota.

The first step to obtain these bioactive compounds from plant materials and agro-industrial waste is to obtain the raw material, this has been done using traditional treatments such as maceration, percolation, soxhlet, to name a few, however, these techniques are widely used, do not achieve a satisfactory yield of bioactive compounds and affect the environment, in addition to long extraction times. This entails exposing the bioactive compounds to high temperatures; therefore, the quality of these compounds is affected (Chan et al., 2011). Due to all the aforementioned, in addition to high energy costs and the need to reduce the CO_2 emissions generated by industry action, we have chosen to look for technologies that are more friendly with the environment and that offer a sustainable growth of the chemical, pharmaceutical and food industries.

Currently, green techniques have been implemented to obtain plant extracts that offer better benefits compared to the traditional techniques used, since they provide short extraction times, better yields, and maintain the quality of the bioactive compounds. Some of these techniques are ultrasound and/or microwave-assisted extraction, high pressure, ohmic heating (Chan et al., 2011; Wang et al., 2010). These techniques are more efficient because they use fewer solvents for the extraction, in addition, the extraction times are of seconds or minutes and avoid the degradation of compounds of interest. It has been reported that these compounds have among their biological effects, the prebiotic effect, so in this chapter, we will discuss the stimulating power of polyphenolic powders extracted from various sources on probiotic bacteria.

4.2 VEGETABLE SOURCES

4.2.1 SHELLS

4.2.1.1 POMEGRANATE

Pomegranate (*Punica granatum*) is one of the most consumed and cultivated fruits worldwide. Iran occupying an important place in the production of such fruit having values of approximately 665,000 tons in 2003. Both the juice and the pomegranate peel contain a range of polyphenolic compounds such as flavonoids (anthocyanins), condensed tannins (proanthocyanidins) and hydrolyzable tannins (ellagitannins and galotannins) that can be used in different areas and that provide benefits (Loren et al., 2005). The main constituents of pomegranate are phenolic compounds including tannins and flavonoids (Al-Rawahi et al., 2014).

Punica granatum has antimicrobial, anti-inflammatory, antioxidant, and anti-diabetic properties, these properties could be related to the presence of phenolic compounds in this plant (Loizzo et al., 2019). Studies revealed that the pomegranate peel has antioxidant, antimicrobial, anticancer abilities among others (Panichayupakaranant et al., 2010). Pomegranate has a range of phenolic compounds among which tannins, anthocyanins, flavonoids, among others (Naveena, Sen, Vaithiyanathan, 2008) stand out and different parts of the plant have been used for medicinal purposes as a treatment against fever, bronchitis, urinary tract infections (Al-Zoreky, 2009).

4.2.1.2 NUT

Nutshell residues are distributed throughout the world and are rich in phenolic and tannin compounds (Ozcariz-Fermoselle et al., 2018) the nut is considered a natural source of nutrients and bioactive compounds (Pollegioni et al., 2012; Forino et al., 2016), is widely produced for its monetary value (Doğan et al., 2014).

The nut extract has reported effects on health, by curbing diabetes complications, delaying aging, and inhibiting chemical reactions among carbonyls mainly carbohydrates and thus acts as an antioxidant (Ahmad, Khan, and Wahid, 2012). Studies revealed that the phenolic compound quercetin present in this extract works as a treatment against erythema. Parts of the plant such as leaves, fruits, husks, barks, and seeds have been used by different pharmaceutical and cosmetic areas to name a few (Balasundram et al., 2006). The fruit is rich in phenolic compounds which contributes to its beneficial properties (Pérez-Gregorio et al., 2014), the leaves of this plant have been used as a traditional medicine in the treatment of inflammation among other conditions, due to its extensive range of properties that include antidiarrheal, antiseptic, antiparasitic properties, it has also been reported that the shell has natural compounds of interest because they can function as antioxidants and antimicrobials (Fernández-Agulló et al., 2013).

The seeds of this plant are a natural source of bioactive compounds among which mineral vitamins and a wide variety of phenolic compounds (flavonoids and phenolic acids) stand out, but in general, the most abundant compound in this plant is quercetin (HUO et al., 2018).

4.2.1.3 CANTALOUPE

Melon (*Cucumis melo*) is an annual life plant, it is widely consumed for its delicious flavor, being one of the most popular fruits, occupying India the 5th place in production (Revanasidda and Belavadi, 2019), it is cultivated throughout the world in tropical regions (Ibrahim, 2010). In traditional medicine, melon seeds are used as antiparasitic, antipyretic, digestive among others (De Marino et al., 2009).

According to studies reported melon seeds have a large amount of gallic acid in a concentration of (99.69 ± 23.06 mg/100 g dry weight) compared to other seeds studied in this trial (Morais et al., 2015), likewise, other phenolic compounds present in melon seed oils such as flavonoids and phenolic acids were also identified, such as the case of amentoflavone found in greater proportion (32.80 ± 0.21 μg/g dry weight) and phenolic acids such as gallic, protocathetic, caffeic, and rosmarinic acids (Mallek-Ayadi, Bahloul, and Kechaou, 2018).

4.2.2 GRAINS

Within the grains, sorghum is a crop that has a wide range of bioactive compounds with beneficial potentials, compared to other grains.

4.2.2.1 SORGHUM

Sorghum is a variable crop that supports environmental changes, tolerating high temperatures. Therefore it is a typical C4 plant from arid and hot regions of Asia, Africa, North and South America. It is positioned among the first five most produced crops worldwide in addition to having properties such as resistance to stress conditions, provides greater energy, and is rich in carbohydrates, proteins, and fats so they do better when compared to other crops such as wheat, barley, rice among others.

Sorghum, in addition to being used as a supplement in the form of grain for animal feed in different parts of the world, is also used in the food industry for its capabilities in relation to disease prevention, thanks to the bioactive compounds that it has in particular polyphenols of the genus flavonoids that contribute to improvements in glucose metabolism, lipids, fat reduction and oxidative stress proven in recent human research (Althwab and Schlegel, 2015; de Morais Cardoso et al., 2017; Awika, Ojwang, and Girard, 2018).

Despite the wide variety of bioactive compounds derived from sorghum with therapeutic activity such as anti-inflammatory, anti-cancer properties among others, this crop has been poorly valued due to its potential. However, due to their health benefits, sorghum polyphenols have been widely studied within the most relevant phenolic compounds in sorghum, phenolic acids and flavonoid derivatives are found, the presence of phenolic compounds provides protection and defense to the crop against pathogens or pest attacks, in addition to sorghum is one of the crops with the highest presence of polyphenols compared to other grains (Awika et al., 2018).

Phenolic acids especially those derived from ferulic acid are the most abundant polyphenols in sorghum, present as an essential part of the cell wall of the grains, in sorghum most of the phenolic acids are joined by covalent bonds to the polysaccharides of the cell wall (Awika et al., 2018; Chiremba et al., 2012), unbound phenolic acids are present as carbohydrate monomers (i.e., glycerol) (Svensson et al., 2010; Yang et al., 2012) and sometimes as unconjugated acids, thanks to these covalent bonds that phenolic acids have with cell wall polysaccharides, they offer a wide variety of biological benefits, since unbound forms can be better absorbed in the gastrointestinal tract, and the forms bound with the polysaccharides of the cell wall are released by the normal metabolism of the intestine and thus provide a slower systemic circulation of these compounds (Neacs et al., 2017; Vitaglione et al., 2015), however, the way in which the phenolic acids bound to the cell wall polysaccharides contribute to the improvement of the intestinal microbiota is still unknown.

Flavonoids are the most abundant polyphenols in nature, however in cereals they occupy a relatively small amount (Zamora-Ros et al., 2016), so the beneficial effects of sorghum on these compounds are not attributed, even so, there are derivatives of flavonoids such as, for example, flavones that have important biological properties including anti-inflammatory and chemoprotective functions; In addition to that flavonoids are also involved in the organoleptic properties of foods such as taste, color, texture to name a few. On the other hand, condensed tannins may be involved in the digestion and bioavailability of micronutrients in the body (Amoako and Awika, 2016; Mitaru et al., 1984; Pan et al., 2018).

Sorghum intake is related to the prevention of different diseases since this crop has bioactive compounds such as polyphenols, which are closely involved in important health benefits, thus reducing oxidative stress mechanisms and thereby decreasing aging, in addition To prevent chronic diseases and some types of cancer, they are also related to the control of glucose in the body and the balance of the intestinal microbiota. (Yang et al., 2015; Ritchie et al., 2015).

4.2.3 LEAVES

4.2.3.1 SOURSOP

Soursop (*Annona muricata*) is a popular fruit from Mexico, South America, and North America, although it is grown in different parts of the world, Mexico and Brazil are the main producers of this fruit, soursop is a continuous leaf tree and long-lasting that produces large fruits with prickly peel and a considerable amount of seeds, with a rare flavor and unusual aroma.

Studies have been conducted on the functional and medicinal properties of this fruit, although Pinto et al. (2005) mentions that the pulp of this fruit has very few nutritional benefits; however, it has considerable amounts of vitamins and minerals (Gyamfi et al., 2011), in addition to being considered a source of dietary fiber approximately 50% by dry weight (Ramírez et al., 2009), in another study, they have evaluated the other parts of the fruit such as the leaves and the peel that have presented anti-tumor properties (Gomes de Melo et al., 2010; Hamizah et al., 2012; Torres et al., 2012).

Different types of soursop *(Annona muricata, Annona squamosa,* and *Annona reticulateson*) have shown antioxidant properties, in addition to being reported to contain a large amount of dietary fiber that can function as prebiotics. Around 200 bioactive compounds have been identified in the soursop among which acetogenins, alkaloids, and phenolic compounds stand out, so the parts that have the greatest range of these phenolic compounds are the leaves and seeds so much that they are the most studied and used. Most of these compounds have been identified from organic extracts, however, aqueous extracts are currently being used.

Thirty-seven phenolic compounds have been reported in soursop, with the leaves and seeds being the one with the highest amount of these compounds, including quercetin and gallic acid, and in the pulp, there are flavonoids and antioxidant compounds such as tocotrienols and tocopherols (Correa et al., 2012). The amount of phenolic compounds obtained from aqueous extracts has been reported to have higher yields since the polyphenols are soluble in water, so the aqueous infusion is used for medicinal purposes (Figure 4.1).

4.2.3.2 MORINGA

The *Moringa oleifera* is the most used species for its relevant phytochemical and pharmacological properties (Kou et al., 2018). It is a plant native to

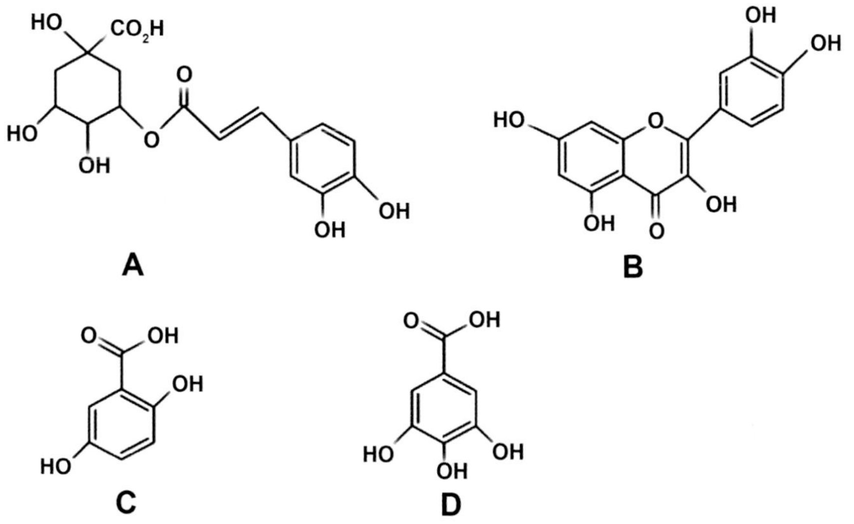

FIGURE 4.1 Chemical structures of the types of polyphenols present in the soursop: (A) chlorogenic acid type, (B) flavonoid type, (C) hydroquinone type, and (D) tannin type.

India although it has been cultivated in different countries, with wide and rapid growth, it is considered as a medicinal plant since it is a natural source of bioactive compounds with beneficial properties for health (Mahmood et al., 2010). The study of the properties of moringa has been booming, with interest in all the benefits that this plant is able to provide since all parts of the plant can be used for different purposes (Kou et al., 2018), the Moringa has been used and valued in developed countries for its nutritional, medicinal and pharmacological properties, although it can be used in underdeveloped countries because of the lack of food that is lived in these countries, the use of all parts of the moringa (leaves, flowers, pods, and seeds) has an added value, in addition to the fact that this plant can be consumed cooked or raw, providing the body with nutrients such as proteins and vitamins, thus preventing malnutrition in populations in poverty (Leone et al., 2015a).

Moringa is a plant rich in phytochemicals of the group of polyphenols (tannins, flavonoids, and phenolic acids), which given its rich source of bioactive compounds has been exploited. Thus, applying the traditional method of Folin-Ciocaleteu, it has been found that moringa leaves have the greatest amount of these phenolic compounds in a concentration around 2000 to 12,200 mg GAE/100g (Leone et al., 2015a), in comparison with flowers and seeds (Alhakmani et al., 2013); It also implies paying attention to the geographical location of the plant since it also influences the concentration of phenolic compounds that it could have, however, the Folin-Ciocaleteu method is not specific so it can react with all the phenolic compounds present in the treated sample Therefore, there are techniques that identify individual phenolic compounds (flavonoids, tannins, phenolic acids, etc.), among which HPLC, gas and/or liquid chromatography incorporated into mass spectrophotometers can be mentioned with these advanced techniques. Identify phenolic compounds (flavonoids and phenolic acids) in the different parts of the moringa tree and it has been found that flavonoids such as quercetin and glycosides (rutinosides and malonyl glycosides) are the most abundant in this plant with the exception of seeds and roots (Saini, Sivanesan, and Keum, 2016) meeting in the quercetin leaves in concentration of 0.46–16.64 mg/g dry weight and glycosides (kaempferol) in concentrations of 0.16–3.92 mg/g (Amaglo et al., 2010) (Richard N. Bennett et al., 2003) (Siddhuraju et al., 2003); significant amounts compared to other phenolic compounds found in smaller quantities such as epicatechin (Min Zhang et al., 2011) and rutin (Bajpai et al., 2005). Although in general, the phenolic compounds that make up the different parts of moringa are gallic acid (Oboh et al., 2015), cumaric acid and ellagic acid to name a few (Leone et al., 2015b). Moringa leaves also contain a considerable amount of tannins being in a concentration of 20.7 mg/g in dry weight, these phenolic compounds can bind to several molecules (protein, amino acids, etc.) and precipitate them (Teixeira et al., 2014).

4.2.3.3 CHILI

Chili (*Capsicum annuum*) is commonly used as food as it is a dietary ingredient that in addition to offering organoleptic characteristics (taste, color), has stood out because a compound with beneficial properties is a major source of capsaicin (Srinivasan, 2016), in addition to the presence of vitamins C and E, provitamin A, carotenoids and polyphenols (Marín et al., 2008), which also offer important properties, among which analgesia effects are mentioned, Anti-inflammatory, anti-litogenetic, and effects on the gastrointestinal system, the secondary metabolites of this plant especially polyphenols have shown potential beneficial effects that include antioxidant, anti-inflammatory, anti-cancer, antidiabetic, antimicrobial properties, etc. (Chen and Kang, 2013; Silva et al., 2013; Tundis et al., 2013), in addition to inhibiting metabolic enzymes such as α-amylase. A-glucosidase and lipase, involved in the hydrolysis of carbohydrates and responsible for the increase in blood glucose so they offer a regulatory effect of hyperglycemia and therefore of disorders associated with diabetes (Krentz and Bailey, 2005). This pathology causes free radicals to be generated and therefore oxidative stress is triggered (Giacco and Brownlee,

2010). Thus, the implementation of antioxidant compounds is important to counteract these oxidizing effects and therefore reduce complications from hyperglycemia (Rahimi et al., 2005), as well as prevent mutations and DNA damage due to oxidative stress (Kaur et al., 2014).

4.2.3.4 CREOSOTE BUSH

The creosote bush (*Larreta tridentata*) is a plant in the form of a shrub and an enduring leaf that is inhabiting desert areas. It is abundant in the southwestern United States and northern Mexico, it is poorly valued and there are few experimental studies on this plant (Trevino-Cueto et al., 2007). The plant as such has a spicy bitter taste, is used to prepare tea for the relief of different intestinal discomforts, this plant has a very strong and penetrating smell unpleasant for most people (Martinez-Vilalta and Pockman, 2002) including animals since it is generally toxic and can cause death (Gay and Dwyer, 1980).

The creosote bush (is a vegetable source rich in extractable compounds approximately 50% of its dry weight, about 19 phenolic compounds were found in the varnish of the creosote bush leaves among which flavonoid aglycones, lignans such as NDGA (nordihydroguayretic acid) (Konno et al., 1990) some glycosylated flavonoids, essential oils, and alkaloids.

Of the 80% of the phenolic compounds found in the creosote bush, about 5 to 10% of the dry weight of the leaves is constituted by the phenolic compound NDGA belonging to the tannin group, a study on the bioavailability of polyphenols revealed that the NDGA It is present in both leaves and flowers and stems, showing a higher concentration in leaves (38.3 mg/g) followed by stems with a concentration of (32.5 mg/g) (Hyder et al., 2002). In recent years, new phenolic compounds were isolated from the plant, including six flavonoids, three tannins and four lignans referred to as creosote compounds (Abou-Gazar et al., 2004).

NDGA is a lignan that has reported beneficial properties, has been used as a treatment against neurocutaneous syndrome due to the action of aldehyde dehydrogenase and that involves the degradation of leukotriene B4 (Willemsen et al., 2000), in addition to this compound being converted by intestinal microflora into estrogenic compounds that show effects both in vivo and in vitro (Fujimoto et al., 2004).

4.2.3.5 TARBUSH

Tarbush (*Cassia fistula*) is recommended as a medicinal plant since it has healing properties, antidiarrheal, antipyretic, antidiabetic, antimicrobial effects among others (Duraipandiyan and Ignacimuthu, 2007; Chichioco-Hernandez and Leonido, 2011).

At present, the tarbush has been studied as a potential source of novel uses, since it has compounds of interest among which phenolic compounds such as flavonoids stand out (Dillon et al., 1976; Rao, Kingston, and Spittler, 1970; Wollenweber and Dietz, 1981). Other compounds include long-chain hydrocarbons from Tetracosane 4-olide to triacotano-4-olide and lactones. *Flourensia rethinophylla* has flavonoid aglycone; 5,7,3-trihydroxy-3-iso-butyroylflavanonol; 5,7,3'-trihydroxyflvanone; 5,7-dihydroxy-3'-methoxy-flavanone; 5,7-dihydroxyphlavanone (pinocepharin) Kaempferol 3,7-dimethyl ether (Kumatakenin) (Dillon et al., 1976; Stuppner and Müller, 1994).

4.2.4 OTHERS

4.2.4.1 GRAPE POMACE AND COFFEE

Coffee is the second most commercialized product in the country followed by oil, which has gained popularity in the world since it is consumed by most people, coffee pulp is one of the main byproducts of the coffee agro-industry and it is worth mentioning that it can be a valuable key for different purposes including the extraction of polyphenols (Esquivel and Jiménez, 2012).

Coffee beans and grapes are rich in bioactive compounds generally polyphenols that have beneficial effects on the host, especially antioxidant properties (Henry-Vitrac et al., 2010; Goutzourelas et al., 2015).

Studies revealed that the grape is composed of an endless number of phenolic compounds among which catechins, epicatechin, cyanidin, mirtiline, quercetin, gallic acid can be mentioned, to name a few, likewise the study showed a total content of polyphenols in the grape of 648 mg/g of dry matter, on the other hand it was shown that coffee has a total content of phenolic compounds of 42.61 mg/GAE where the most relevant compounds found were 3-chlorogenic acid in a proportion of (16.61 mg/g) and acids 4 and 5 chlorogens at 13.62 mg/g, thanks to technologies such as HPLC and liquid chromatography incorporated into mass spectrophotometer, it was possible to detect these compounds in these plant sources (Priftis et al., 2018).

Coffee is mostly composed of phenolic acids (chlorogenic acids) (Henry-Vitrac et al., 2010; Murthy and Naidu, 2012). Chlorogenic acid has a number of benefits including anti-cancer, anti-inflammatory, antioxidant properties (Higdon and Frei, 2006; Upadhyay and Mohan Rao, 2013), it has also been reported that this compound has a neuroprotective effect counting PC12 cell lines (Cho et al., 2009; Li et al., 2008), in another study, it was reported that coffee can modify the intestinal microbiota since it has a high content of melanoidins that act as dietary fibers in the intestine (Jiménez-Zamora et al., 2015). On the other hand, the grape and its wine derivative have trans-resveratrol polyphenol belonging to the group of stilbenes (Goutzourelas et al., 2015) without neglecting that both the seed and the grape pomace also have significant amounts of this phenolic compound. that give the grape and its different parts (seeds, pomace) the antioxidant potential that it presents (Goutzourelas et al., 2014; Goutzourelas et al., 2015).

4.3 POLYPHENOLS

Polyphenols are compounds found in nature both in plants and in a wide variety of foods (Puupponen-Pimiä et al., 2002) are secondary metabolites of plants and are part of about 10,000 unique compounds that have one or more rings phenolics and substitutable groups (Pandey and Rizvi, 2009). Thus fruits, whole grains, vegetables, coffee, chocolate are rich in polyphenols, these compounds are classified according to their chemical structure, function and biological properties. Chemically they are classified into flavonoids and these, in turn, are classified into flavonols, flavones, flavonones, anthocyanidins, isoflavones, phenolic acids such as hydroxybenzoic acid and hydroxycinnamic acids, stilbenes, lignans, curcuminoids and tannins (Effects, 2000).

4.3.1 HUMAN DISEASES AND POLYPHENOLS

Epidemiological studies have repeatedly demonstrated the inverse relationship between chronic disease and the consumption of a diet high in polyphenols (Arts and Hollman, 2005). Polyphenols have phenolic groups which can couple an electron to form highly stable phenoxy radicals, thus preventing oxidation of cellular components. The antioxidant action of plasma has been established that has to do with the consumption of foods and beverages rich in polyphenols; This antioxidant action after food consumption can be said to be due to the presence of reducing polyphenols and their metabolites in the plasma, effects of polyphenols on the concentrations of other reducing agents, or for their effect on the absorption of pro components Oxidants such as iron (Scalbert et al., 2005). Low levels of oxidative damage in lymphocyte DNA have been associated thanks to the consumption of antioxidants. Likewise, the same antioxidant effects have been observed when consuming foods rich in polyphenols that indicate the protective action that these compounds have, there is certainty that polyphenols can act as antioxidants since they protect the cell from oxidative damage and therefore decrease the risk or probability of some degenerative chronic disease associated with oxidative stress (Luqman and Rizvi, 2006; Pandey and Rizvi, 2010; Pandey et al., 2009).

4.4 POLYPHENOLS AS PREBIOTICS

Studies on interactions between polyphenols and intestinal microbiota, reveal that these compounds can act as prebiotics, benefiting intestinal bacteria such as *Lactobacillus* and *Bifidobacterium*, acting as probiotics these microorganisms are able to take advantage of polyphenols and produce metabolites of interest (Bhat and Kapila, 2017; Mueller et al., 2017). However, the fermentation of prebiotics carried out by probiotic bacteria is specific and only the microorganisms that adapt to competitive conditions and produce useful metabolites are suitable for administration as probiotics (Krumbeck et al., 2016). Thus the composition of the microbiota is different in each of the organisms, according to the lifestyle (Teixeira et al., 2017).

Probiotic bacteria such as *Lactobacillus* and *Bifidobacterium* that are normally found in the intestine and take advantage of the slow absorption that certain polyphenols have, such as the case of ellagitannins and proanthocyanidins to use them in their metabolism and thus obtain nutrients (Lamuel-Raventos and Onge, 2017). The phenolic compounds around 90% that are ingested in the diet remain intact since the body does not produce enzymes that can degrade these compounds (MN Clifford, 2004), and most are stored in the intestine so they are used by microorganisms. However, the fermentation of these compounds is complex, in a study where rats were fed with grape seed extract (25% w/w) which is rich in proanthocyanidins, showed an 11% recovery of this compound in the feces.

In vitro and animal studies, the potential of polyphenols was demonstrated by stimulating the growth of intestinal microbes such as red wine extract, improved the growth of *Klebsiella*, *Alistipes*, *Cloacibacillus*, *Victivallis,* and *Akkermansia* at the same time inhibited the growth of *Bifidobacteria*, *B. coccoides* and bacteroides in a simulated environment (Kemperman et al., 2013). Similarly, the administration of black tea extract promoted and inhibited certain types of microorganisms, however, these actions to promote or inhibit were not the same, which shows that each phenolic compound has different capacities. Modifications in the intestinal microbiota due to polyphenols reduced the ratio of firmicutes: bacteroidetes (Kemperman et al., 2013).

Research has revealed the stimulating effects towards *Lactobacillus* and *Bifidobacterium* spp., Which have certain compounds present in blueberries, pomegranate, grains, skins of different vegetables, grapes, which clearly shows the prebiotic potential of these fruits (Bialonska et al., 2010; Giuseppina Mandalari et al., 2010; Molan et al., 2009; Pozuelo et al., 2012). Microbial metabolites can influence certain mechanisms that help maintain intestinal homeostasis.

The polyphenols that are ingested in the diet have to travel to certain parts of the colon to be able to stimulate or inhibit bacteria, the key part of the intestine is the lower zone since this is where the phenolic compounds exert on beneficial bacteria stimulation and pathogenic bacteria inhibition (Cardona et al., 2013; Tuohy et al., 2012; Williamson and Clifford, 2017), as well as certain phenolic compounds such as chlorogenic acids, quercetin, caffeic acid were able to stimulate beneficial bacteria such as Lactobacillus and Bifidobacterium, and decrease the amount of pathogenic bacteria (Parkar et al., 2013).

4.4.1 INTESTINAL MICROBIOTA

The intestinal microbiota is a vitally essential organ to maintain adequate health, under normal conditions beneficial microorganisms participate in the normal catabolism of nutrients and the regulation of the host's immune system (Ahern et al., 2014). These microorganisms that host the host, among which are microorganisms (commensals, pathogens and beneficial) vary throughout the gastrointestinal tract (Del Campo-Moreno et al., 2018), but bacteria generally are harmless or beneficial to humans, thanks to the fact that there is a balance in the intestinal microbiota due to the competition of complex beneficial microorganisms that eliminate pathogenic microorganisms, thus achieving homeostasis in the intestinal lumen (Buchon et al., 2013).

The development and conformation of the intestinal microbiota depends on the diet as well as on exposure to the environment, and it is different in any organism since over time it diversifies and becomes stable to provide better protection for the host (Benno and Mitsuoka, 1986; Grönlund et al., 2000; Guarner and Malagelada, 2003; Kirjavainen et al., 2001). For example, in newborns the diversity of microorganisms is deficient compared to an adult who has a wider diversity of microorganisms, the intestinal microbiota is of vital importance since in addition to protecting the host from certain pathogens by providing a barrier that prevents these microorganisms from penetrating and can penetrate inward and cause damage. It also prevents the lodging of pathogens and therefore reduces the competition of nutrients between beneficial microorganisms and pathogenic microorganisms. Therefore, benefits are provided to both the host and the microbiota creating a desirable equilibrium environment for both parties (Consortium, 2001).

Notwithstanding the differences in the intake of polyphenols by each individual and the variations in the composition of the intestinal microbiota, it can affect the availability and efficacy of polyphenols and their metabolites

(Begoña et al., 2004; Gross et al., 2010). Thus, the polyphenols-microbiota relationship seems to be complex, since studies have revealed that both the supply of phenolic compounds to probiotic bacteria and the metabolites produced can modulate, inhibit or stimulate both pathogenic and commensal bacteria, benefiting or damaging health (Tzounis et al., 2011; Hervert-Hernández et al., 2011; Oliveira et al., 2008; Queipo -Ortuño et al., 2012; Rastmanesh, 2011).

The polyphenols that are consumed in the diet are used as nutrients by the intestinal microbiota, since they are not absorbed in the intestine, the microorganisms take advantage and ferment these compounds to produce molecules of interest (Selma et al., 2009) that subsequently travel through the bloodstream (Russell and Duthie, 2011).

The microbiota, especially the one that lives in the colon, forms a fermentation environment, in which the compounds that escape or inhibit digestion such as proteins, carbohydrates and dietary fibers are metabolized, which are converted into simple metabolites such as SCFA (butyrate and propionate) and secondary bile acids, some other compounds are also fermented by the microbiota such is the case of phenolic compounds, which are converted into compounds with a more relevant potential (Levy et al., 2017).

Millions of microorganisms abound in the gastrointestinal system, which over the course of evolution have adapted to the changes; advanced species such as animals evolved in tandem with microorganisms to create a symbiosis that favors not only the intestinal flora but also goes beyond encompassing the physiology of the host, since human beings have changed throughout evolution and have adapted unfavorable lifestyles such as poor diet, industrialization, sedentary lifestyle, environmental aspects within which ultraviolet radiation pollution falls, diseases among others that damage the intestinal microbiota, leaving the host at risk, prone to be attacked by some disease (Sonnenburg and Sonnenburg, 2014; Cotillard et al., 2013).

The imbalance of the intestinal microbiota results in colonization by pathogenic microorganisms, which is why the organism is affected by suffering different conditions. At present, studies focused on the sequencing of extensions generally the 16S marker gene, metagenome sequencing (WMS) have been used to identify the intestinal microbiota (Langille et al., 2013).

Metagenome sequencing (WMS) provides valuable information as it includes prokaryotic species such as eukaryotes (bacteria, archaea, viruses), as well as allowing you to delve deeper into the functions and metabolic pathways of these species. Current information suggests working with intestinal bacteria profiles to predict a possible disease and the management of intestinal flora as an approach to disease treatment (Turnbaugh et al., 2006).

ACKNOWLEDGMENTS

Castro-Torres thanks SAGARPA-CONACYT for the financial support given to her studies at DIA-FCQ (UAdeC) within the project 2015–4-266936 "*Obtención, purificación y escalado de compuestos de extractos bioactivos con valor industrial, obtenidos usando tecnología avanzadas de extracción y a partir de cultivos, subproductos y recursos naturales poco valorados.*"

KEYWORDS

- **grains**
- **husks**
- **leaves**
- **polyphenols**
- **prebiotics**
- **residues**

REFERENCES

Abou-Gazar, H., Bedir, E., Takamatsu, S., Ferreira, D., & Khan, I. A., (2004). Antioxidant lignans from *Larrea tridentata*. *Phytochemistry, 65*(17), 2499–2505. https://doi.org/10.1016/J.PHYTOCHEM.2004.07.009 (accessed on 24 July 2020).

Ahern, P. P., Faith, J. J., & Gordon, J. I., (2014). Mining the human gut micro biota for effector strains that shape the immune system. *Immunity, 40*(6), 815–823. https://doi.org/10.1016/J.IMMUNI.2014.05.012 (accessed on 24 July 2020).

Ahmad, H., Khan, I., & Wahid, A., (2012). Antiglycation and antioxidation properties of *Juglans regia* and *Calendula officinalis*: Possible role in reducing diabetic complications and slowing down aging. *Journal of Traditional Chinese Medicine, 32*(3), 411–414. https://doi.org/10.1016/S0254–6272(13)60047–3 (accessed on 24 July 2020).

Alhakmani, F., Kumar, S., & Khan, S. A., (2013). Estimation of total phenolic content, in–vitro antioxidant and anti–inflammatory activity of flowers of *Moringa oleifera*. *Asian Pacific Journal of Tropical Biomedicine, 3*(8), 623–627. https://doi.org/10.1016/S2221–1691(13)60126–4 (accessed on 24 July 2020).

Al-Rawahi, A., Edwards, G., & Al-Sibani, M. (2014). Phenolic Constituents of Pomegranate Peels (*Punica granatum* L) Cultivated in Oman. *European Journal of Medicinal Plants 4*(3), 315–331. https://doi.org/10.9734/EJMP/2014/6417

Althwab, S., Carr, T. P., Weller, C. L., Dweikat, I. M., & Schlegel, V., (2015). Advances in grain sorghum and its co-products as a human health promoting dietary system. *FRIN, 77*, 349–359. https://doi.org/10.1016/j.foodres.2015.08.011 (accessed on 24 July 2020).

Al-Zoreky, N. S., (2009). Antimicrobial activity of pomegranate (Punica granatum L.) fruit peels. *International Journal of Food Microbiology, 134*(3), 244–248. https://doi.org/10.1016/J.IJFOODMICRO.2009.07.002 (accessed on 24 July 2020).

Amaglo, N. K., Bennett, R. N., Lo Curto, R. B., Rosa, E. A. S., Lo Turco, V., Giuffrida, A., & Timpo, G. M., (2010). Profiling selected phytochemicals and nutrients in different tissues of the multipurpose tree *Moringa oleifera* L, grown in Ghana. *Food Chemistry, 122*(4), 1047–1054. https://doi.org/10.1016/J.FOODCHEM.2010.03.073 (accessed on 24 July 2020).

Amoako, D. B., & Awika, J. M., (2016). Polymeric tannins significantly alter properties and in vitro digestibility of partially gelatinized intact starch granule. *Food Chemistry*, *208*, 10–17. https://doi.org/10.1016/J.FOODCHEM.2016.03.096 (accessed on 24 July 2020).

Arts, I. C. W., & Hollman, P. C. H., (2005). Polyphenols and disease risk in epidemiologic studies 1–4, *81*.

Awika, J. M., Rose, D. J., & Simsek, S., (2018). Complementary effects of cereal and pulse polyphenols and dietary fiber on chronic inflammation and gut health. *Food and Function, 9*(3), 1389–1409. https://doi.org/10.1039/C7FO02011B (accessed on 24 July 2020).

Awika, J., Ojwang, L., & Girard, A., (2018). Chapter 14. *Anthocyanins, Deoxyanthocyanins and Proanthocyanidins as Dietary Constituents in Grain Products* (pp. 305–331). https://doi.org/10.1039/9781788012799–00305 (accessed on 24 July 2020).

Bajpai, M., Pande, A., Tewari, S. K., & Prakash, D., (2005). Phenolic contents and antioxidant activity of some food and medicinal plants. *International Journal of Food Sciences and Nutrition, 56*(4), 287–291. https://doi.org/10.1080/09637480500146606 (accessed on 24 July 2020).

Balasundram, N., Sundram, K., & Samman, S. (2006). Phenolic compounds in plants and agri-industrial by-products: Antioxidant activity, occurrence, and potential uses. *Food Chemistry 99*(1), 191–203. https://doi.org/10.1016/j.foodchem.2005.07.042.

Begoña, C., Tomás-Barberán, F. A., & Espín, J. C., (2004). *Metabolism of Antioxidant and Chemo Preventive Ellagitannins from Strawberries, Raspberries, Walnuts, and Oak-Aged Wine in Humans: Identification of Biomarkers and Individual Variability*. https://doi.org/10.1021/JF049144D (accessed on 24 July 2020).

Benno, Y., & Mitsuoka, T., (1986). Development of intestinal microflora in humans and animal. *Bifidobacteria and Microflora 5*(1), 13–25. https://doi.org/10.12938/bifidus1982.5.1_13.

Bhat, M. I., & Kapila, R., (2017). Dietary metabolites derived from gut microbiota: Critical modulators of epigenetic changes in mammals. *Nutrition Reviews, 75*(5), 374–389. https://doi.org/10.1093/nutrit/nux001 (accessed on 24 July 2020).

Bialonska, D., Ramnani, P., Kasimsetty, S. G., Muntha, K. R., Gibson, G. R., & Ferreira, D., (2010). The influence of pomegranate by-product and punicalagins on selected groups of human intestinal microbiota. *International Journal of Food Microbiology, 140*(2, 3), 175–182. https://doi.org/10.1016/J.IJFOODMICRO.2010.03.038 (accessed on 24 July 2020).

Buchon, N., Broderick, N. A. & Lemaitre, B., (2013). Gut homeostasis in a microbial world: insights from *Drosophila melanogaster*. *Nature Reviews Microbiology 11*, 615–626. https://doi.org/10.1038/nrmicro3074.

Cardona, F., Andrés-Lacueva, C., Tulipani, S., Tinahones, F. J., & Queipo-Ortuño, M. I., (2013). Benefits of polyphenols on gut micro biota and implications in human health. *The Journal of Nutritional Biochemistry, 24*(8), 1415–1422. https://doi.org/10.1016/J.JNUTBIO.2013.05.001 (accessed on 24 July 2020).

Chen, L., & Kang, Y. H., (2013). Anti-inflammatory and antioxidant activities of red pepper (Capsicum annuum L.) stalk extracts: Comparison of pericarp and placenta extracts. *Journal of Functional Foods, 5*(4), 1724–1731. https://doi.org/10.1016/J.JFF.2013.07.018 (accessed on 24 July 2020).

Chichioco-Hern, C. L., & Leonido, F. M. G., (2011). Journal of medicinal plant research. *Journal of Medicinal Plants Research* (Vol. 5). Academic Journals. Retrieved from https://academicjournals.org/journal/JMPR/article-abstract/911EEE525331 (accessed on 24 July 2020).

Chiremba, C., Taylor, J. R. N., & Rooney, L. W., (2012). Phenolic acid content of sorghum and maize cultivars varying in hardness. *Food Chemistry*, *134*, 81–88. https://doi.org/10.1016/j.foodchem.2012.02.067 (accessed on 24 July 2020).

Cho, E. S., Jang, Y. J., Hwang, M. K., Kang, N. J., Lee, K. W., & Lee, H. J., (2009). Attenuation of oxidative neuronal cell death by coffee phenolic phytochemicals. *Mutation Research/Fundamental and Molecular Mechanisms of Mutagenesis, 661*(1, 2), 18–24. https://doi.org/10.1016/J.MRFMMM.2008.10.021 (accessed on 24 July 2020).

Consortium, I. H. G. S., (2001). Initial sequencing and analysis of the human genome. *Nature, 409*(6822), 860–921. https://doi.org/10.1038/35057062 (accessed on 24 July 2020).

Correa, J., Ortiz, D., Larrahondo, J., Sanchez, M., & Pachon, H., (2012) Actividad antioxidante en guanábana (*Annona muricata* L.): una revisión bibliográfica. *BLACPMA 1*(2). Retrieved from http://www.revistas.usach.cl/ojs/index.php/blacpma/article/view/553 (August 12th, 2020).

Cotillard, A., Kennedy, S. P., Kong, L. C., Prifti, E., Pons, N., Le Chatelier, E., & Ehrlich, S. D., (2013). Dietary intervention impact on gut microbial gene richness. *Nature, 500*(7464), 585–588. https://doi.org/10.1038/nature12480 (accessed on 24 July 2020).

De Marino, S., Festa, C., Zollo, F., & Iorizzi, M., (2009). Phenolic glycosides from *Cucumis* melo var. inodorus seeds. *Phytochemistry Letters, 2*(3), 130–133. https://doi.org/10.1016/J.PHYTOL.2009.04.001 (accessed on 24 July 2020).

De Morais, C. L., Pinheiro, S. S., Martino, H. S. D., & Pinheiro-Sant'Ana, H. M., (2017). Sorghum (*Sorghum bicolor* L.): Nutrients, bioactive compounds, and potential impact on human health. *Critical Reviews in Food Science and Nutrition, 57*(2), 372–390. https://doi.org/10.1080/10408398.2014.887057 (accessed on 24 July 2020).

Del Campo-Moreno, R., Alarcón-Cavero, T., D'Auria, G., Delgado-Palacio, S., & Ferrer-Martínez, M. (2018). Microbiota and Human Health: characterization techniques and transference. *Enfermedades Infecciosas y Microbiología Clínica*, *36*(4), 241–245. https://doi.org/10.1016/J.EIMC.2017.02.007.

Dillon, M. O., Mabry, T. J., Besson, E., Bouillant, M. L., & Chopin, J., (1976). New flavonoids from *Flourensia cernua*. *Phytochemistry*. Retrieved from http://agris.fao.org/agris-search/search.do?recordID=US201303043102 (accessed on 24 July 2020).

Doğan, Y., Kafkas, S., Sütyemez, M., Akça, Y., & Türemiş, N., (2014). Assessment and characterization of genetic relationships of walnut (Juglans regia L.) genotypes by three types of molecular markers. *Scientia Horticulturae*, *168*, 81–87. https://doi.org/10.1016/J.SCIENTA.2014.01.024 (accessed on 24 July 2020).

Duraipandiyan, V., & Ignacimuthu, S., (2007). Antibacterial and antifungal activity of *Cassia fistula* L.: An ethno medicinal plant. *Journal of Ethno Pharmacology, 112*(3), 590–594. https://doi.org/10.1016/J.JEP.2007.04.008 (accessed on 24 July 2020).

Effects, A., (2000). *Polyphenols: Extraction Methods, Anti-oxidative Action, Bioavailability and Anticarcinogenic Effects*. https://doi.org/10.3390/molecules21070901 (accessed on 24 July 2020).

Esquivel, P., & Jiménez, V. M., (2012). Functional properties of coffee and coffee by-products. *Food Research International, 46*(2), 488–495. https://doi.org/10.1016/j.foodres.2011.05.028 (accessed on 24 July 2020).

Fernández-Agulló, A., Pereira, E., Freire, M. S., Valentão, P., Andrade, P. B., González-Álvarez, J., & Pereira, J. A., (2013). Influence of solvent on the antioxidant and antimicrobial properties of walnut (Juglans regia L.) green husk extracts. *Industrial Crops and Products*, *42*, 126–132. https://doi.org/10.1016/J.INDCROP.2012.05.021 (accessed on 24 July 2020).

Forino, M., Stiuso, P., Lama, S., Ciminiello, P., Tenore, G. C., Novellino, E., & Taglialatela-Scafati, O., (2016). Bioassay-guided identification of the antihyperglycaemic constituents of walnut (Juglans regia) leaves. *Journal of Functional Foods*, *26*, 731–738. https://doi.org/10.1016/J.JFF.2016.08.053 (accessed on 24 July 2020).

Fujimoto, N., Kohta, R., Kitamura, S., & Honda, H., (2004). Estrogenic activity of an antioxidant, nordihydroguaiaretic acid (NDGA). *Life Sciences, 74*(11), 1417–1425. https://doi.org/10.1016/J.LFS.2003.08.012 (accessed on 24 July 2020).

Gay, C., & Dwyer, D., (1980). *New Mexico Range Plants*. Retrieved from http://agris.fao.org/agris-search/search.do?recordID=US201300395306 (accessed on 24 July 2020).

Giacco, F., & Brownlee, M., (2010). Oxidative stress and diabetic complications. *Circulation Research, 107*(9), 1058–1070. https://doi.org/10.1161/CIRCRESAHA.110.223545 (accessed on 24 July 2020).

Gomes, D. M. J., De Sousa, A. T. A., Thijan, N. D. A. E. C. V., Lyra, D. V. C. D., Do Desterro, R. M., Carneiro, D. N. S., & De Albuquerque, U. P., (2010). Anti-proliferative activity, antioxidant capacity and tannin content in plants of semi-arid Northeastern Brazil. *Molecules, 15*(12), 8534–8542. https://doi.org/10.3390/molecules15128534 (accessed on 24 July 2020).

Goutzourelas, N., Stagos, D., Demertzis, N., Mavridou, P., Karterolioti, H., Georgadakis, S., & Kouretas, D., (2014). Effects of polyphenolic grape extract on the oxidative status of muscle and endothelial cells. *Human and Experimental Toxicology, 33*(11), 1099–1112. https://doi.org/10.1177/0960327114533575 (accessed on 24 July 2020).

Goutzourelas, N., Stagos, D., Spanidis, Y., Liosi, M., Apostolou, A., Priftis, A., & Kouretas, D., (2015). Polyphenolic composition of grape stem extracts affects antioxidant activity in endothelial and muscle cells. *Molecular Medicine Reports, 12*(4), 5846–5856. https://doi.org/10.3892/mmr.2015.4216 (accessed on 24 July 2020).

Grönlund, M., Arvilommi, H., Kero, P., Lehtonen, O-P., & Isolauri, E., (2000). Importance of intestinl colonisation in the maturation of humoral immunity in early infancy: a prospective follow up study of healthy infants aged 0-6 monts. *Archives of Disease in Childhood-Fetal and Neonatal Edition 84*(2), 92–95. http://dx.doi.org/10.1136/fn.83.3.F186.

Gross, G., Jacobs, D. M., Peters, S., Possemiers, S., Van, D. J., Vaughan, E. E., & Van, D. W. T., (2010). *In Vitro* bioconversion of polyphenols from black tea and red wine/grape Juice by human intestinal microbiota displays strong interindividual variability. *Journal of Agricultural and Food Chemistry, 58*(18), 10236–10246. https://doi.org/10.1021/jf101475m (accessed on 24 July 2020).

Guarner, F., & Malagelada, J. (2003). Gut flora in health and disease. *Lancet 8*(361), 512–519. https://doi.org/10.1016/S0140-6736(03)12489-0.

Gyamfi, K., Sarfo, D., Nyarko, B., Akaho, E., Serfor-Armah, Y., & Ampomah-Amoako, E. (2011). Assessment of elemental content in the fruit of graviola plant, *Annona muricata*, from some selected communities in Ghana by instrumental neutron activation analysis. *Elixir Food Science 41*, 5671–5675.

Hamizah, S., Roslida, A. H., Fezah, O., Tan, K. L., Tor, Y. S., & Tan, C. I., (2012). Chemo preventive potential of *Annona muricata* L., leaves on chemically-induced skin papilloma genesis in mice. *Asian Pacific Journal of Cancer Prevention, 13*(6), 2533–2539. https://doi.org/10.7314/APJCP.2012.13.6.2533 (accessed on 24 July 2020).

Henry-Vitrac, C., Ibarra, A., Roller, M., Mérillon, J. M., & Vitrac, X., (2010). Contribution of chlorogenic acids to the inhibition of human hepatic glucose-6-phosphatase activity in vitro by svetol, a standardized decaffeinated green coffee extract. *Journal of Agricultural and Food Chemistry, 58*(7), 4141–4144. https://doi.org/10.1021/jf9044827 (accessed on 24 July 2020).

Hervert-Hernández, D., & Goñi, I., (2011). Dietary polyphenols and human gut micro biota: A review. *Food Reviews International, 27*(2), 154–169. https://doi.org/10.1080/87559129.2010.535233 (accessed on 24 July 2020).

Higdon, J. V., & Frei, B., (2006). Coffee and health: A review of recent human research. *Critical Reviews in Food Science and Nutrition, 46*(2), 101–123. https://doi.org/10.1080/10408390500400009 (accessed on 24 July 2020).

Huo, J. H., Du, X. W., Sun, G. D., Dong, W. T., & Wang, W. M., (2018). Identification and characterization of major constituents in *Juglans mandshurica* using ultra performance liquid chromatography coupled with time-of-flight mass spectrometry (UPLC-ESI-Q-TOF/MS). *Chinese Journal of Natural Medicines, 16*(7), 525–545. https://doi.org/10.1016/S1875–5364(18)30089-X (accessed on 24 July 2020).

Hyder, P. W., Fredrickson, E., Estell, R. E., Tellez, M., & Gibbens, R. P., (2002). Distribution and concentration of total phenolics, condensed tannins, and nordihydroguaiaretic acid (NDGA) in creosote bush (*Larrea tridentata*). *Biochemical Systematics and Ecology, 30*(10), 905–912. https://doi.org/10.1016/S0305–1978(02)00050–9 (accessed on 24 July 2020).

Ibrahim, S. R. M., (2010). New 2-(2-phenylethyl)chromone derivatives from the seeds of *Cucumis melo* L var. reticulatus. *Natural Product Communications, 5*(3), 403–406. Retrieved from http://www.ncbi.nlm.nih.gov/pubmed/20420317 (accessed on 24 July 2020).

Ignat, I., Volf, I., & Popa, V. I., (2011). A critical review of methods for characterization of polyphenolic compounds in fruits and vegetables. *Food Chemistry, 126*(4), 1821–1835. https://doi.org/10.1016/J.FOODCHEM.2010.12.026 (accessed on 24 July 2020).

Jiménez-Zamora, A., Pastoriza, S., & Rufián-Henares, J. A., (2015). Revalorization of coffee by-products. Prebiotic, antimicrobial and antioxidant properties. *LWT-Food Science and Technology, 61*(1), 12–18. https://doi.org/10.1016/J.LWT.2014.11.031 (accessed on 24 July 2020).

Kaur, R., Kaur, J., Mahajan, J., Kumar, R., & Arora, S., (2014). Oxidative stress—implications, source and its prevention. *Environmental Science and Pollution Research, 21*(3), 1599–1613. https://doi.org/10.1007/s11356–013–2251–3 (accessed on 24 July 2020).

Kemperman, R. A., Gross, G., Mondot, S., Possemiers, S., Marzorati, M., Van, D. W. T., & Vaughan, E. E., (2013). Impact of polyphenols from black tea and red wine/grape juice on a gut model microbiome. *Food Research International, 53*(2), 659–669. https://doi.org/10.1016/J.FOODRES.2013.01.034.

Kirjavainen, P., Apostolou, E., & Arvola, T. (2001). Characterizing the composition of intestinal microflora as a prospective treatment target in infant allergic disease. *FEMS Immunology & Medical Microbiology 32*(1), 1–7. https://doi.org/10.1111/j.1574-695X.2001.tb00526.x.

Konno, C., Lu, Z. Z., Xue, H. Z., Erdelmeier, C. A. J., Meksuriyen, D., Che, C. T., & Fong, H. H. S., (1990). Furanoid lignans from *Larrea tridentata*. *Journal of Natural Products, 53*(2), 396–406. https://doi.org/10.1021/np50068a019 (accessed on 24 July 2020).

Kou, X., Li, B., Olayanju, J., Drake, J., Chen, N., Kou, X., & Chen, N., (2018). Nutraceutical or pharmacological potential of *Moringa oleifera* Lam. *Nutrients, 10*(3), 343. https://doi.org/10.3390/nu10030343 (accessed on 24 July 2020).

Krentz, A. J., & Bailey, C. J., (2005). Oral antidiabetic agents. *Drugs, 65*(3), 385–411. https://doi.org/10.2165/00003495–200565030–00005 (accessed on 24 July 2020).

Krumbeck, J. A., Maldonado-Gomez, M. X., Ramer-Tait, A. E., & Hutkins, R. W., (2016). Prebiotics and synbiotics. *Current Opinion in Gastroenterology, 32*(2), 110–119. https://doi.org/10.1097/MOG.0000000000000249 (accessed on 24 July 2020).

Lamuel-Raventos, R. M., & Onge, M. P. S., (2017). Prebiotic nut compounds and human micro biota. *Critical Reviews in Food Science and Nutrition, 57*(14), 3154–3163. https://doi.org/10.1080/10408398.2015.1096763 (accessed on 24 July 2020).

Langille, M. G. I., Zaneveld, J., Caporaso, J. G., McDonald, D., Knights, D., Reyes, J. A., & Huttenhower, C., (2013). Predictive functional profiling of microbial communities using 16S rRNA marker gene sequences. *Nature Biotechnology, 31*(9), 814–821. https://doi.org/10.1038/nbt.2676 (accessed on 24 July 2020).

Leone, A., Spada, A., Battezzati, A., Schiraldi, A., Aristil, J., Bertoli, S., & Bertoli, S., (2015a). Cultivation, genetic, ethnopharmacology, phytochemistry and pharmacology of *Moringa oleifera* leaves: An overview. *International Journal of Molecular Sciences, 16*(12), 12791–12835. https://doi.org/10.3390/ijms160612791 (accessed on 24 July 2020).

Leone, A., Spada, A., Battezzati, A., Schiraldi, A., Aristil, J., Bertoli, S., & Bertoli, S., (2015b). Cultivation, genetic, ethnopharmacology, phytochemistry and pharmacology of *Moringa oleifera* leaves: An overview. *International Journal of Molecular Sciences, 16*(12), 12791–12835. https://doi.org/10.3390/ijms160612791 (accessed on 24 July 2020).

Levy, M., Kolodziejczyk, A. A., Thaiss, C. A., & Elinav, E., (2017). Dysbiosis and the immune system. *Nature Reviews Immunology, 17*(4), 219–232. https://doi.org/10.1038/nri.2017.7 (accessed on 24 July 2020).

Li, Y., Shi, W., Li, Y., Zhou, Y., Hu, X., Song, C., & Li, Y., (2008). Neuro protective effects of chlorogenic acid against apoptosis of PC12 cells induced by methylmercury. *Environmental Toxicology and Pharmacology, 26*(1), 13–21. https://doi.org/10.1016/J.ETAP.2007.12.008 (accessed on 24 July 2020).

Loizzo, M. R., Aiello, F., Tenuta, M. C., Leporini, M., Falco, T., & Tundis, R., (2019). Pomegranate (*Punica granatum* L.). *Nonvitamin and Nonmineral Nutritional Supplements*, 467–472. https://doi.org/10.1016/B978–0-12–812491–8.00062-X (accessed on 24 July 2020).

Loren, D. J., Seeram, N. P., Schulman, R. N., & Holtzman, D. M., (2005). *Maternal Dietary Supplementation with Pomegranate Juice is Neuro protective in an Animal Model of Neonatal Hypoxic-Ischemic Brain Injury*. https://doi.org/10.1203/01.PDR.0000157722.07810.15 (accessed on 24 July 2020).

Luqman, S., & Rizvi, S. I., (2006). Protection of lipid peroxidation and carbonyl formation in proteins by capsaicin in human erythrocytes subjected to oxidative stress. *Phytotherapy Research, 20*(4), 303–306. https://doi.org/10.1002/ptr.1861 (accessed on 24 July 2020).

Mahmood, K., Mugal, T., & Haq, I., (2010). *Moringa oleifera*: Una revisión natural del regalo-A. *Revista De Ciencias Farmacéuticas Y*. Retrieved from http://citeseerx.ist.psu.edu/viewdoc/download?doi=10.1.1.193.7882&rep=rep1&type=pdf (accessed on 24 July 2020).

Mallek-Ayadi, S., Bahloul, N., & Kechaou, N., (2018). Chemical composition and bioactive compounds of *Cucumis melo* L. seeds: Potential source for new trends of plant oils. *Process Safety and Environmental Protection, 113*, 68–77. https://doi.org/10.1016/J.PSEP.2017.09.016 (accessed on 24 July 2020).

Marín, A., Gil, M. I., Flores, P., Hellín, P., & Selma, M. V., (2008). Microbial quality and bioactive constituents of sweet peppers from sustainable production systems. *Journal of Agricultural and Food Chemistry, 56*(23), 11334–11341. https://doi.org/10.1021/jf8025106 (accessed on 24 July 2020).

Martinez-Vilalta, J., & Pockman, W. T., (2002). The vulnerability to freezing-induced xylem cavitation of *Larrea tridentata* (Zygophyllaceae) in the Chihuahuan desert. *American*

Journal of Botany, 89(12), 1916–1924. https://doi.org/10.3732/ajb.89.12.1916 (accessed on 24 July 2020).

Min Zhang, Hettiarachchy, N. S., Horax, R., Kannan, A., Apputhury, P. M. D., Muhundan, A., & Mallangi, R. C. (2011). Phytochemicals, antioxidant and antimicrobial activity of *Hibiscus sabdariffa, Centella asiatica, Moringa oleifera* and *Murraya koenigi*i leaves. *Journal of Medicinal Plants Research*, *5*(30), 6672–6680. https://doi.org/10.5897/JMPR11.621

Mitaru, B. N., Reichert, R. D., & Blair, R., (1984). The binding of dietary protein by sorghum tannins in the digestive tract of pigs. *The Journal of Nutrition, 114*(10), 1787–1796. https://doi.org/10.1093/jn/114.10.1787 (accessed on 24 July 2020).

Molan, A. L., Lila, M. A., Mawson, J., & De, S., (2009). *In vitro* and *in vivo* evaluation of the prebiotic activity of water-soluble blueberry extracts. *World Journal of Microbiology and Biotechnology, 25*(7), 1243–1249. https://doi.org/10.1007/s11274–009–0011–9 (accessed on 24 July 2020).

Morais, D. R., Rotta, E. M., Sargi, S. C., Schmidt, E. M., Bonafe, E. G., Eberlin, M. N., & Visentainer, J. V., (2015). Antioxidant activity, phenolics and UPLC–ESI(–)–MS of extracts from different tropical fruits parts and processed peels. *Food Research International*, *77*, 392–399. https://doi.org/10.1016/J.FOODRES.2015.08.036 (accessed on 24 July 2020).

Mueller, D., Jung, K., Winter, M., Rogoll, D., Melcher, R., & Richling, E., (2017). Human intervention study to investigate the intestinal accessibility and bioavailability of anthocyanins from bilberries. *Food Chemistry*, *231*, 275–286. https://doi.org/10.1016/J.FOODCHEM.2017.03.130 (accessed on 24 July 2020).

Murthy, P. S., & Naidu, M. M., (2012). Recovery of phenolic antioxidants and functional compounds from coffee industry by-products. *Food and Bioprocess Technology, 5*(3), 897–903. https://doi.org/10.1007/s11947–010–0363-z (accessed on 24 July 2020).

Naveena, B. M., Sen, A. R., Vaithiyanathan, S., Babji, Y., & Kondaiah, N., (2008). Comparative efficacy of pomegranate juice, pomegranate rind powder extract and BHT as antioxidants in cooked chicken patties. *Meat Science, 80*(4), 1304–1308. https://doi.org/10.1016/J.MEATSCI.2008.06.005 (accessed on 24 July 2020).

Neacsu, M., McMonagle, J., Fletcher, R. J., Hulshof, T., Duncan, S. H., Scobbie, L., & Russell, W. R., (2017). Availability and dose response of phytophenols from a wheat bran rich cereal product in healthy human volunteers. *Molecular Nutrition and Food Research, 61*(3), 1600202. https://doi.org/10.1002/mnfr.201600202 (accessed on 24 July 2020).

Oboh, G., Ademiluyi, A. O., Ademosun, A. O., Olasehinde, T. A., Oyeleye, S. I., Boligon, A. A., & Athayde, M. L., (2015). Phenolic extract from *Moringa oleifera* Leaves inhibits key enzymes linked to erectile dysfunction and oxidative stress in rats' penile tissues. *Biochemistry Research International*, *2015*, 1–8. https://doi.org/10.1155/2015/175950 (accessed on 24 July 2020).

Oliveira, I., Sousa, A., Morais, J., & Ferreira, I. (2008). Chemical composition, and antioxidant and antimicrobial activities of three hazelnut (*Corylus avellana* L.) cultivars. *Food and Chemical Toxicology 46*(5), 1801–1807. https://doi.org/10.1016/j.fct.2008.01.026.

Ozcariz-Fermoselle, M. V., Fraile-Fabero, R., Girbés-Juan, T., Arce-Cervantes, O., Rueda-Salgueiro, J. A. O. D., & Azul, A. M., (2018). Use of lignocellulosic wastes of pecan (*Carya illinoinensis*) in the cultivation of *Ganoderma* lucidum. *Revista Iberoamericana de Micología, 35*(2), 103–109. https://doi.org/10.1016/J.RIAM.2017.09.005 (accessed on 24 July 2020).

Pan, L., Ma, X., Hu, J., Liu, L., Yuan, M., Liu, L., & Piao, X., (2018). Low-tannin white sorghum contains more digestible and metabolizable energy than high-tannin red sorghum if fed to growing pigs. *Animal Production Science*. https://doi.org/10.1071/AN17245 (accessed on 24 July 2020).

Pandey, K. B., & Rizvi, S. I., (2009). Plant polyphenols as dietary antioxidants in human health and disease. *Oxidative Medicine and Cellular Longevity, 2*(5), 270–278. https://doi.org/10.4161/oxim.2.5.9498 (accessed on 24 July 2020).

Pandey, K. B., & Rizvi, S. I., (2010). Protective effect of resveratrol on markers of oxidative stress in human erythrocytes subjected to in vitro oxidative insult. *Phytotherapy Research, 24*(S1), S11–S14. https://doi.org/10.1002/ptr.2853 (accessed on 24 July 2020).

Pandey, Mishra, N., & Rizvi, S. I., (2009). Protective role of myricetin on markers of oxidative stress in human erythrocytes subjected to oxidative stress. *Nat. Prod. Commun., 4*, 221–226.

Panichayupakaranant, P., Tewtrakul, S., & Yuenyongsawad, S., (2010). Antibacterial, anti-inflammatory and anti-allergic activities of standardized pomegranate rind extract. *Food Chemistry, 123*(2), 400–403. https://doi.org/10.1016/J.FOODCHEM.2010.04.054 (accessed on 24 July 2020).

Parkar, S. G., Trower, T. M., & Stevenson, D. E., (2013). Fecal microbial metabolism of polyphenols and its effects on human gut micro biota. *Anaerobe, 23*, 12–19. https://doi.org/10.1016/J.ANAEROBE.2013.07.009 (accessed on 24 July 2020).

Pérez-Gregorio, M. R., Regueiro, J., Simal-Gándara, J., Rodrigues, A. S., & Almeida, D. P. F., (2014). Increasing the added-value of onions as a source of antioxidant flavonoids: A critical review. *Critical Reviews in Food Science and Nutrition, 54*(8), 1050–1062. https://doi.org/10.1080/10408398.2011.624283 (accessed on 24 July 2020).

Pollegioni, P., Van, D. L. G., Belisario, A., Gras, M., Anselmi, N., Olimpieri, I., & Malvolti, M. E., (2012). Mechanisms governing the responses to anthracnose pathogen in *Juglans* spp. *Journal of Biotechnology, 159*(4), 251–264. https://doi.org/10.1016/J.JBIOTEC.2011.08.020 (accessed on 24 July 2020).

Pozuelo, M. J., Agis-Torres, A., Hervert-Hernández, D., López-Oliva, M. E., Muñoz-Martínez, E., Rotger, R., & Goñi, I., (2012). Grape antioxidant dietary fiber stimulates lactobacillus growth in rat cecum. *Journal of Food Science, 77*(2), H59–H62. https://doi.org/10.1111/j.1750–3841.2011.02520.x (accessed on 24 July 2020).

Priftis, A., Goutzourelas, N., Halabalaki, M., Ntasi, G., Stagos, D., Amoutzias, G. D., & Kouretas, D., (2018). *Effect of Polyphenols From Coffee and Grape on Gene Expression in Myoblasts.* https://doi.org/10.1016/j.mad.2017.11.015 (accessed on 24 July 2020).

Puupponen-Pimiä, R., Aura, A. M., Oksman-Caldentey, K. M., Myllärinen, P., Saarela, M., Mattila-Sandholm, T., & Poutanen, K., (2002). Development of functional ingredients for gut health. *Trends in Food Science and Technology*. Elsevier. https://doi.org/10.1016/S0924–2244(02)00020–1 (accessed on 24 July 2020).

Queipo-Ortuño, M. I., Boto-Ordóñez, M., Murri, M., Gomez-Zumaquero, J. M., Clemente-Postigo, M., Estruch, R., & Tinahones, F. J., (2012). Influence of red wine polyphenols and ethanol on the gut micro biota ecology and biochemical biomarkers. *The American Journal of Clinical Nutrition, 95*(6), 1323–1334. https://doi.org/10.3945/ajcn.111.027847 (accessed on 24 July 2020).

Rahimi, R., Nikfar, S., Larijani, B., & Abdollahi, M. (2005). A review on the role of antioxidants in the management of diabetes and its complications. *Biomedicine & Pharmacotherapy 59*(7), 365-373. https://doi.org/10.1016/j.biopha.2005.07.002.

Ramírez, A., & Pacheco de Delahaye, E. (2009). Functional properties of starches with high dietetic fiber content obtained from pineapple, guava and soursop. *Interciencia, 34*(4), 293–298. Retrieved from http://ve.scielo.org/scielo.php?script=sci_arttext&pid=S0378-18442009000400014&lng=es&tlng=en (accessed on August 11th, 2020).

Rao, M. M., Kingston, D. G. I., & Spittler, T. D., (1970). Flavonoids from *Flourensia cernua. Phytochemistry*. Retrieved from http://agris.fao.org/agris-search/search.do?recordID=US201301161778 (accessed on 24 July 2020).

Rastmanesh, R., (2011). High polyphenol, low probiotic diet for weight loss because of intestinal micro biota interaction. *Chemico-Biological Interactions, 189*(1, 2), 1–8. https://doi.org/10.1016/J.CBI.2010.10.002 (accessed on 24 July 2020).

Revanasidda, & Belavadi, V. V., (2019). Floral biology and pollination in *Cucumis melo* L, a tropical andromonoecious cucurbit. *Journal of Asia-Pacific Entomology, 22*(1), 215–225. https://doi.org/10.1016/J.ASPEN.2019.01.001 (accessed on 24 July 2020).

Richard, N. B., Fred, A. M., Nikolaus, F., John, H. P., Susan, D. M., Lionel, P., & Kroon, P. A., (2003). *Profiling Glucosinolates and Phenolics in Vegetative and Reproductive Tissues of the Multi-Purpose Trees Moringa oleifera L. (Horseradish Tree) and Moringa stenopetala L.* https://doi.org/10.1021/JF0211480 (accessed on 24 July 2020).

Ritchie, L. E., Sturino, J. M., Carroll, R. J., Rooney, L. W., Azcarate-Peril, M. A., & Turner, N. D., (2015). Polyphenol-rich sorghum brans alter colon microbiota and impact species diversity and species richness after multiple bouts of dextran sodium sulfate-induced colitis. *FEMS Microbiology Ecology, 91*(3). https://doi.org/10.1093/femsec/fiv008 (accessed on 24 July 2020).

Russell, W., & Duthie, G., (2011). Plant secondary metabolites and gut health: The case for phenolic acids. *Proceedings of the Nutrition Society, 70*(3), 389–396. https://doi.org/10.1017/S0029665111000152 (accessed on 24 July 2020).

Scalbert, A., Manach, C., Morand, C., & Em, C. R., (2005). *Dietary Polyphenols and the Prevention of Diseases, 306*, 287–306. https://doi.org/10.1080/1040869059096 (accessed on 24 July 2020).

Selma, M. V., Espín, J. C., & Tomás-Barberán, F. A., (2009). Interaction between phenolics and gut micro biota: Role in human health. *Journal of Agricultural and Food Chemistry, 57*(15), 6485–6501. https://doi.org/10.1021/jf902107d (accessed on 24 July 2020).

Siddhuraju, P., & Becker, K. (2003). Antioxidant Properties of Various Solvent Extracts of Total Phenolic Constituents from Three Different Agroclimatic Origins of Drumstick Tree (*Moringa oleifera* Lam.) Leaves. *Journal of Agricultural and Food Chemistry 51*(8), 2144–2155. https://doi.org/10.1021/jf020444+.

Silva, L., Azevedo, J., Pereira, M., & Valentão, P. (2013). Chemical assessment and antioxidant capacity of pepper (*Capsicum annuum* L.) seeds. *Food and Chemical Toxicology 53*, 240–248. https://doi.org/10.1016/j.fct.2012.11.036.

Sonnenburg, E. D., & Sonnenburg, J. L., (2014). Starving our microbial self: The deleterious consequences of a diet deficient in micro biota-accessible carbohydrates. *Cell Metabolism, 20*(5), 779–786. https://doi.org/10.1016/j.cmet.2014.07.003 (accessed on 24 July 2020).

Srinivasan, K., (2016). Biological activities of red pepper (*Capsicum annuum*) and its pungent principle capsaicin: A review. *Critical Reviews in Food Science and Nutrition, 56*(9), 1488–1500. https://doi.org/10.1080/10408398.2013.772090 (accessed on 24 July 2020).

Stuppner, H., & Müller, E. P., (1994). Rare flavonoid aglycones from *Flourensia retinophylla*. *Phytochemistry, 37*(4), 1185–1187. https://doi.org/10.1016/S0031–9422(00)89554–0 (accessed on 24 July 2020).

Svensson, L., Sekwati-Monang, B., Lutz, D. L., Schieber, A., & Gänzle, M. G., (2010). Phenolic acids and flavonoids in nonfermented and fermented red sorghum (*Sorghum bicolor* (L.) Moench). *Journal of Agricultural and Food Chemistry, 58*(16), 9214–9220. https://doi.org/10.1021/jf101504v (accessed on 24 July 2020).

Teixeira, E. M. B., Carvalho, M. R. B., Neves, V. A., Silva, M. A., & Arantes-Pereira, L., (2014). Chemical characteristics and fractionation of proteins from *Moringa oleifera Lam.* leaves. *Food Chemistry*, *147*, 51–54. https://doi.org/10.1016/J.FOODCHEM.2013.09.135 (accessed on 24 July 2020).

Teixeira, L. L., Costa, G. R., Dörr, F. A., Ong, T. P., Pinto, E., Lajolo, F. M., & Hassimotto, N. M. A., (2017). Potential anti-proliferative activity of polyphenol metabolites against human breast cancer cells and their urine excretion pattern in healthy subjects following acute intake of a polyphenol-rich juice of grumixama (*Eugenia brasiliensis* Lam.). *Food and Function, 8*(6), 2266–2274. https://doi.org/10.1039/C7FO00076F (accessed on 24 July 2020).

Torres, M. P., Rachagani, S., Purohit, V., Pandey, P., Joshi, S., Moore, E. D., & Batra, S. K., (2012). Graviola: A novel promising natural-derived drug that inhibits tumorigenicity and metastasis of pancreatic cancer cells *in vitro* and *in vivo* through altering cell metabolism. *Cancer Letters, 323*(1), 29–40. https://doi.org/10.1016/J.CANLET.2012.03.031 (accessed on 24 July 2020).

Treviño-Cueto, B., Luis, M., Contreras-Esquivel, J. C., Rodríguez, R., Aguilera, A., Aguilar, C.N. (2007) Gallic acid and tannase accumulation during fungal solid state culture of a tannin-rich desert plant (*Larrea tridentata* Cov.). *Bioresource Technology 98*(3), 721–724. https://doi.org/10.1016/j.biortech.2006.02.015.

Tundis, R., Menichini, F., Bonesi, M., Conforti, F., Statti, G., Menichini, F., & Loizzo, M. R., (2013). Antioxidant and hypoglycaemic activities and their relationship to phytochemicals in *Capsicum annuum* cultivars during fruit development. *LWT-Food Science and Technology, 53*(1), 370–377. https://doi.org/10.1016/J.LWT.2013.02.013 (accessed on 24 July 2020).

Tuohy, K. M., Conterno, L., Gasperotti, M., & Viola, R., (2012). Up-regulating the human intestinal micro biome using whole plant foods, polyphenols, and/or fiber. *Journal of Agricultural and Food Chemistry, 60*(36), 8776–8782. https://doi.org/10.1021/jf2053959 (accessed on 24 July 2020).

Turnbaugh, P. J., Ley, R. E., Mahowald, M. A., Magrini, V., Mardis, E. R., & Gordon, J. I., (2006). An obesity-associated gut micro biome with increased capacity for energy harvest. *Nature, 444*(7122), 1027–1031. https://doi.org/10.1038/nature05414 (accessed on 24 July 2020).

Tzounis, X., Rodriguez-Mateos, A., Vulevic, J., Gibson, G. R., Kwik-Uribe, C., & Spencer, J. P., (2011). Prebiotic evaluation of cocoa-derived flavanols in healthy humans by using a randomized, controlled, double-blind, crossover intervention study. *The American Journal of Clinical Nutrition, 93*(1), 62–72. https://doi.org/10.3945/ajcn.110.000075 (accessed on 24 July 2020).

Upadhyay, R., & Mohan, R. L. J., (2013). An outlook on chlorogenic acids—occurrence, chemistry, technology, and biological activities. *Critical Reviews in Food Science and Nutrition, 53*(9), 968–984. https://doi.org/10.1080/10408398.2011.576319 (accessed on 24 July 2020).

Vitaglione, P., Mennella, I., Ferracane, R., Rivellese, A. A., Giacco, R., Ercolini, D., & Fogliano, V., (2015). Whole-grain wheat consumption reduces inflammation in a randomized controlled trial on overweight and obese subjects with unhealthy dietary and lifestyle behaviors: Role of polyphenols bound to cereal dietary fiber. *The American Journal of Clinical Nutrition, 101*(2), 251–261. https://doi.org/10.3945/ajcn.114.088120 (accessed on 24 July 2020).

Wang, W., He, J., Pan, D., Wu, Z., Guo, Y., Zeng, X., & Lian, L., (2018). Metabolomics analysis of *Lactobacillus plantarum* ATCC 14917 adhesion activity under initial acid and alkali stress. *Plos One, 13*(5), e0196231. https://doi.org/10.1371/journal.pone.0196231 (accessed on 24 July 2020).

Willemsen, M. A. A. P., Rotteveel, J. J., Steijlen, P. M., Heerschap, A., & Mayatepek, E., (2000). 5-lipoxygenase inhibition: A new treatment strategy for the sjögren-larsson syndrome. *Neuropediatrics, 31*(1), 1–3. https://doi.org/10.1055/s-2000–15288 (accessed on 24 July 2020).

Williamson, G., & Clifford, M. N., (2017). Role of the small intestine, colon and microbiota in determining the metabolic fate of polyphenols. *Biochemical Pharmacology*, *139*, 24–39. https://doi.org/10.1016/J.BCP.2017.03.012 (accessed on 24 July 2020).

Wollenweber, E., & Dietz, V. H., (1981). Occurrence and distribution of free flavonoid aglycones in plants. *Phytochemistry, 20*(5), 869–932. https://doi.org/10.1016/0031–9422(81)83001–4 (accessed on 24 July 2020).

Yang, L., Allred, K. F., Dykes, L., Allred, C. D., & Awika, J. M., (2015). Enhanced action of apigenin and naringenin combination on estrogen receptor activation in non-malignant colonocytes: Implications on sorghum-derived phytoestrogens. *Food and Function, 6*(3), 749–755. https://doi.org/10.1039/C4FO00300D (accessed on 24 July 2020).

Yang, L., Allred, K. F., Geera, B., Allred, C. D., & Awika, J. M., (2012). *Sorghum* phenolics demonstrate estrogenic action and induce apoptosis in nonmalignant colonocytes. *Nutrition and Cancer, 64*(3), 419–427. https://doi.org/10.1080/01635581.2012.657333 (accessed on 24 July 2020).

Zamora-Ros, R., Knaze, V., Rothwell, J. A., Hémon, B., Moskal, A., Overvad, K., & Scalbert, A. (2016). Dietary polyphenol intake in Europe: the European Prospective Investigation into Cancer and Nutrition (EPIC) study. *European Journal of Nutrition*, *55*(4), 1359–1375. https://doi.org/10.1007/s00394-015-0950-x.

CHAPTER 5

Valorization of Pomegranate Residues

PALOMA ALMANZA-TOVANCHE, RAÚL RODRÍGUEZ-HERRERA, AIDÉ SÁENZ-GALINDO, JUAN ALBERTO ASCACIO-VALDÉS, CRISTÓBAL N. AGUILAR, and ADRIANA CAROLINA FLORES-GALLEGOS

Food Research Department, School of Chemistry, Universidad Autónoma de Coahuila, Blvd. Venustiano Carranza S/N, Colonia República, 25280, Saltillo, Coahuila, México, E-mail: carolinaflores@uadec.edu.mx

ABSTRACT

Billions of tons of agro-industrial wastes are generated every year, which could be used to obtain value-added resources. In addition to that, it would reduce pollution and environmental problems. That is why the efficient use of agroindustrial waste to obtain value-added resources is an economically, environmentally, and socially acceptable alternative. The recovery of these materials implies their conversion to a wide range of products within which their antioxidant and pharmacological properties have been exploited. In recent years, numerous investigations have focused on the study of bioactive compounds of phenolic origin, such as polyphenols, flavonoids, and phenolic acids. Recent works have evidenced the presence of these compounds in plant species, such as pomegranate (*Punica granatum*), appearing in greater quantity in the husk/shell and the seeds. That is why this chapter focuses on the pomegranate husk as a source of bioactive compounds.

5.1 INTRODUCTION

Agricultural and agro-industrial wastes are potential sources for products and molecules of industrial interest; however, only a small fraction of these natural sources are used (Yemiş and Mazza, 2012). These natural sources have a large number of compounds with multiple biological properties. An example of these are polyphenols, which can be found naturally in plant tissues and are necessary for pigmentation, growth, reproduction, resistance to pathogens, and many other functions. Martínez-Ramírez et al., (2010) managed to extract different polyphenolic compounds from the pomegranate shell using microwave technology. In Mexico, in general, the pulp of this fruit is used in the preparation of typical dishes, desserts, or is freshly consumed, while the husk is not discarded, and it is used in traditional herbalism.

Pomegranate contains antioxidant substances such as phenols, polyphenols, and anthocyanins. Compounds such as gallic acid and ursolic acid are found in the husk (Huang et al., 2005). There are several studies, which report the antioxidant capacity of pomegranate compounds. Iqbal et al., (2008) obtained methanolic extracts from the shell of pomegranate and mention that these compounds are potent antioxidants, compared to BHT (butyl-hydroxy-toluene), synthetic antioxidant of conventional use in the preservation of fats and oils.

In recent years, alternative technologies have been developed to obtain plant extracts, better known as "green techniques" (Ballard et al., 2010) because they allow the recovery of a greater amount of bioactive compounds of interest, use shorter extraction periods (seconds and in some cases minutes), increase the quality of extracts with lower processing costs and improve the extraction of compounds that are difficult to extract. Some of these techniques are ultrasound -assisted extraction, ohmic heating, high pressure, supercritical fluid extraction, and microwave-assisted extraction (Xia et al., 2011). Green techniques have the advantage of being more efficient and environmentally friendly since these techniques use fewer solvents and energy, consume less extraction time, they are simpler compared to conventional techniques, there is less degradation of thermolabile compounds, better products and higher yields (Hao et al., 2002).

5.2 USE OF VEGETABLE SOURCES FOR OBTAINING BIOACTIVE COMPOUNDS

Plants have two types of metabolism, the primary one which is responsible for performing all the essential chemical processes for life and the secondary metabolism whose function is to produce and accumulate natural products, which have various uses and applications such as in medicines, insecticides, herbicides, perfumes or dyes. Within the primary metabolites, we can find sugars, amino acids, fatty acids, nucleotides, polysaccharides, proteins, lipids, RNA and DNA can be found, while as secondary metabolites we can mention pigments, natural pesticides, vitamins, and bioactive compounds to name a few (Ávalos and Pérez, 2009).

A bioactive compound is that essential or non-essential compound found within nature and that influences cellular activity, as well as different physiological mechanisms (Biesalski et al., 2009). The National Cancer Institute mentions that bioactive substances can provide and promote good health. These substances can be found in small quantities in plants and some foods such as fruits, vegetables, nuts, oils, and grains. Although plants are a source of obtaining these types of molecules, their great diversity, the little information that exists on their structure and the mode of action of some of their phytochemicals, make it complex to take advantage of their full potential, since not all bioactive compounds can be found in all plant groups. They are also synthesized in small quantities and not in a generalized way; that is, its production becomes restricted to certain areas of the plant or even to certain genera, families, or species (Ávalos and Pérez, 2009).

Bioactive compounds have a very important value in industries such as cosmetics, pharmaceuticals, and food. In ancient times a large number of these products were used for the treatment of various diseases. As examples we have the pomegranate leaves, which were used in traditional Mexican herbalism for the treatment of diarrhea or as an antimicrobial against different microorganisms (Digital Library of Traditional Mexican Medicine, 2009), meanwhile, the leaves of the serrano pepper are used for the treatment of areas of the body affected by ulcers, abscesses or skin infections.

Ávalos and Pérez (2009) classify secondary metabolites into terpenes, phenolic compounds, glycosides, and alkaloids. Terpenes, or also known as terpenoids, represent the largest group of secondary metabolites with more than 40,000 different molecules; these have the characteristic of being insoluble in water and come from the union of isoprene units (5 C atoms). Terpenes are classified based on the number of isoprene units they contain: (1) monoterpenes with 10 C containing two units of 5 C each, (2) sesquiterpenes with 15 C with three units and, followed by those (3) diterpenes, which have 20 C and four 5 C units. On the other hand, the triterpenes have 30 C, the tetraterpenes have 40 C and finally, the polyterpenes are those that have up to 8 isoprene units (Ávalos and Pérez, 2009).

In the group of terpenes, it can be found hormones, pigments, sterols, derivatives of sterols, and essential oils. Many of these compounds have a high physiological and commercial value since they are used for their ability to generate aromas and fragrances in some foods and the cosmetic industry; while others are used in medicine for its anticarcinogenic, antiulcer, antimicrobial, etc. (Ávalos and Pérez, 2009). Limonene, for example, is the second most used terpene in the industry and we can obtain it from the skin of lemon or other citrus fruits (Terpenes | CANNA Foundation: Cannabis Research and Analysis).

For its part, a phenolic compound is an aromatic ring with a hydroxyl group from the synthesis of a phenol (Ávalos and Pérez, 2009). In this group, we can find polyphenols or phenylpropanoids. Polyphenols are molecules with an antioxidant capacity which contain acidic phenols and flavonoids. A variety of acidic phenols can be found in products of plant origin, such as caffeic, ferulic, chlorogenic acids present in the fruit and seeds of coffee and soy (Milner, 2004). Flavonoids are polyphenolic compounds characterized by having a three-ring structure with two aromatic centers and an oxygenated heterocycle in the central part. Within these, flavones, flavanones, catechins and anthocyanins are included. In onion and lettuce, we can find quercetin (flavone), while in citrus we find a flavonone called fisetin (Nijveldt et al., 2001).

Taking into account its chemical structure, the group of phenolic compounds is very varied, as there are simple molecules such as phenolic acids and very complex molecules such as tannins and lignin. Tannins are polymeric phenolic compounds that bind to proteins by denaturing them. Its name derives from the practice of using plant extracts to convert the skin into leather and these are divided into condensed tannins and hydrolyzable tannins (Ávalos and Pérez 2009).

Condensed tannins are those polymers of flavonoids linked by C-C bonds, which cannot be hydrolyzed, but can be oxidized by a strong acid to produce anthocyanidins. Hydrolyzable tannins are heterogeneous polymers, contain phenolic acids and simple sugars; they are smaller compared to condensed and hydrolyze more easily (Ávalos and Pérez, 2009). They can be considered as toxins because they show a great capacity for protein binding. In plants, terpenes serve as protection against insects or herbivorous animals, or against high temperatures (Terpenes | CANNA Foundation: Cannabis Research and Analysis). In the case of protection against animals, the immature fruits contain the tannins in the skin so they block the attack by the animals since generating bitter flavors in the plant. However, not all tannins have toxic effects, for example, the tannins contained in red wine cause a beneficial effect on health by blocking the formation of endothelin-1, a signal molecule that causes vasoconstriction (Ávalos and Pérez, 2009).

Glycosides owe their name to the glycosidic bond that is generated through the condensation of a sugar molecule with a molecule that contains a hydroxyl group. Glycosides can be classified into three groups: saponins, cardiac glycosides, and cyanogenic glycosides. Although a fourth group, glucosinolates, can also be included, this is because their structure is very similar to that of glycosides (Ávalos and Pérez, 2009). Jaramillo et al. (2016) evaluated the presence of alkaloids, cyanogenic glycosides, and saponins in five of the most commonly used medicinal plants in Ecuador: *Artemisia absinthium, Cnidoscolus aconitifolius, Parthenium hysterophorus* Linn, *Piper carpunya* Ruiz & Pav, and *Taraxacum officinale*. They found that all plant leaf extracts tested positive for the qualitative test for saponins and the species with the highest content was *P. hysterophorus*, while the one with the lowest content was *P. carpunya*. On the other hand, in the quantification of cyanogenic glycosides it was discovered that the plant that showed the highest concentration was *C. aconitifolius* and the one with the lowest concentration was *A. absinthium*.

The alkaloids generate more than 15,000 secondary metabolites which share three characteristics:

1. they are water-soluble;
2. they contain at least one nitrogen atom in the molecule; and
3. exhibit biological activity.

Most of them are heterocycles, although some are non-cyclic such as mescaline or colchicine. They can be found approximately 20% of vascular plants. In humans, alkaloids can cause physiological or psychological responses since they can interact with neurotransmitters (Ávalos and Pérez, 2009). If alkaloids are used in high doses, they can cause toxicity; on the contrary, if they are used in low or moderate doses, they can generate a therapeutic value as muscle relaxants, tranquilizers or analgesics. The classification of the alkaloids is based on the rings presented by the molecule, so we have quinoline, isoquinoline, indole, tropane, quinolizidine, piperidine, purine and pyrrolizidine (Serrano, López and Espuñes, 2006).

Caffeine is an example of alkaloids, it belongs to the group of xanthines and more than approximately 120,000 tons per year is consumed; within the food industry, caffeine is used in the formulation of carbonated drinks, bakery and confectionery. Another use of caffeine is in the treatment of childhood apnea, acne and migraine. In the pharmaceutical industry, caffeine is used in products such as analgesics and diuretics, to name a few.

5.3 POMEGRANATE

5.3.1 GENERALITIES OF THE POMEGRANATE (PUNICA GRANATUM)

The *Punica granatum* species belongs to the Punicaceae family (Mohseni, 2009; Sheets et al., 2009). This plant generates the fruit we know in Mexico as pomegranate. The term "punica" is derived from the Roman name that was coined to Cartagena, where the best grenades came to Italy (Morton, 1987). Pomegranate is a fruit of Iranian origin, which arrived in America during the Spanish conquest, being able to adapt mainly in areas of warm and arid climate of Mexico and the United States (López-Mejía, López-Malo and Palou, 2010).

According to López-Mejía et al. (2010) the fruit of the *Punica granatum* is known by various names around the world. In Mexico, it is usually called by common names such as pomegranate or pomogranada, while in countries like Brazil it is known as Roman, Romaeira or Romazeira; on the other hand, in Italy it is named as melogranato, grano grain melo, granato pommel or punic pommel. The pomegranate tree can reach a height of 3 to 6 meters, has a semi-woody trunk and very branched with some thorns. Its leaves are green, bright, elongated and slightly wavy (Digital Library of Traditional Mexican Medicine, 2009). It has bright flowers whose colors range between orange and red, usually contain 5 to 8 petals and are full of hairs on the inside of the flower similar to stamens (United States Department of Agriculture, 2009). Its fruit is the pomegranate which has a globose shape with a diameter of approximately 6-12 centimeters, has a thick and strong crust that ranges from yellow-reddish to green and can reach a scarlet red color, with a chalice similar to a crown at one of its ends (Figure 5.1) (Royal Spanish Academy, 2010). Pomegranate contains many fleshy and edible grains, which have white seeds inside and are separated by an astringent membrane (López-Mejía, López-Malo and Palou, 2010).

5.3.2 WORLD PRODUCTION OF POMEGRANATE

Currently, the cultivation of pomegranate occurs in various parts of the world and has taken a lot of global relevance for its antioxidant properties which, in turn, confer pharmacological properties such as anti-cancer, anti-tumor, anti-microbial and anti-diarrheal, to mention some (López-Mejía, López-Malo and Palou, 2010). The Daily Records (2017) generated a list of the top 10 pomegranate producing countries in the world: in the first place is Iran, its origin country. Mohseni (2009) reported that 56,329 hectares were cultivated in that country and they generated around 705,166 tons of pomegranate per year. The majority of the product generated in Iran is destined for export to countries in the Persian Gulf, Europe and China (The Daily Records, 2017).

In the second place of the list we find the United States, mainly the state of California where the Spanish Ruby and Sweet Fruited varieties were grown, but then from the 20th century they were replaced by the Wonderful variety (Morton, 1987). Pomegranates produced in the United States are recognized for their large sizes and sweet taste (The Daily Records, 2017). Since 2016,

FIGURE 5.1 Branches, leaves and pomegranate fruit.

Snoei reported a decrease in the production of pomegranate from 10 to 15%. This decline continues to increase as producers have chosen to decrease the area available for said crop and rotate the crop towards more profitable products, about 40 % of the pomegranate crop is exported to countries such as South Korea, Canada, Taiwan, Brazil, Australia, Japan and Mexico (Snoei, 2016; The Daily Records, 2017).

China ranks third in pomegranate production worldwide (The Daily Records, 2017), the most popular variety is Tunisian because its flavor is very sweet. Most pomegranates are grown in provinces such as Yunnan, Henan and Sichuan (Snoei, 2016). The fourth place is occupied by India where the fruit is usually sold fresh and the most sown varieties have the characteristic of having a large amount of pulp around the seed, which is why it is called "seedless". Some of the best varieties are Bedana, which has as its characteristics a pink pulp, sweet flavor and soft seeds, and Kandhari, which has a strong red pulp with an acid flavor and hard seeds (Morton, 1987). It is estimated that the production of pomegranate in India increases every year between 20 and 25%, in addition to be the only country with the ability to produce pomegranate throughout the year due to its climate (Snoei, 2016). Despite its high production, India is not considered a pomegranate exporting country because in recent years there are almost zero export records, since India is its main consumer (The Daily Records, 2017).

The fifth and sixth place is occupied by Israel and Egypt respectively; in Israel the fruit is grown in the central and southern regions of the country (The Daily Records, 2017), where the main cultivated varieties are Wonderful and Red Loufani (Morton, 1987). Most of the Israeli product is exported to Europe where 3.5 kilograms of pomegranate have a cost of 5 euros so that farmers have high yields. In the case of Egypt, the best-known varieties are the Wonderful and the Baladi. Baladi has weaker skin than Wonderful, while Wonderful has a less sweet taste than Baladi. All pomegranates generated in Egypt are exported, approximately 700 tons per year (The Daily Records, 2017).

In seventh place is Spain, which is considered the country with the most important production in the European Union. Melgarejo (2008) and Özgüven et al. (2009) mentioned that the cultivation of pomegranates in Spain was approximately 2,325 hectares, producing around 27,389 tons of pomegranate. They also mentioned the province of Alicante as the main producer since only 2,047 hectares were cultivated there and 25,104 tons of pomegranate per year were produced. Snoei (2016) says that the most cultivated variety is Mollar, which is grown mainly in the Elche region. Although he also mentions that lately, producers have chosen to try new varieties such as Smith and Acco. Approximately 30% of the production of pomegranate in Spain is destined for export; initially Russia was the key market for Spanish pomegranates. However, currently the pomegranates are aimed at countries such as the Far and the Middle East as well as Europe, Italy and Ireland (The Daily Records, 2017).

In the eighth and ninth places we have Turkey and Afghanistan. López-Mejía et al. (2010) mentioned that in Turkey about 80,000 tons were produced per year. Turkey is currently considered as a growing producer in the pomegranate market, the fruit can be found in the months of October to March having as the most cultivated varieties of Hicaz, Wonderful and Caner. The pomegranate of the Hicaz variety is the most produced and consumed in Turkey. This and the Wonderful variety are very similar, however, the Wonderful is sweeter. On the other hand, the pomegranate of the Caner variety is the cheapest of the three (Snoei, 2016). In Afghanistan, the production of pomegranate dates back several decades, although in the 1970s it was interrupted as there was political instability which reduced exports. After the war, the cultivation of pomegranate was resumed in this country, although its rebirth can be considered as of the year 2009, the year in which thousands of pomegranate plants were cultivated in the country and a juice company was generated which opened A market for farmers. With this, not only has an increase in the local cost of the fruit been achieved but also a great boost has been generated for the fruit industry (The Daily Records, 2017).

Belgium has the last place, a country in which since 2016 there has been a growth in the production of pomegranate. The high consumption of this fruit in the country has generated that local agriculture is unable to meet the needs of the population, so Belgium has been in need of importing pomegranate from countries such as Spain, Turkey, Israel, Iran, Egypt, India, South Africa, Peru, Chile and the United States. Despite this, farmers are still looking for ways to increase crops and yields of pomegranate to feed the growing demand. The most cultivated variety in Belgium is the Wonderful since it has been in the market for more than ten years; however, other varieties such as Mollar, Baghwa, Hershkovitz, Akko, Hicaz and Emek are also found (Snoei, 2016; The Daily Records, 2017).

5.3.3 POMEGRANATE PRODUCTION IN MEXICO

The fruit of the pomegranate is not native to Mexico, it was introduced by the Spaniards and its main crop was located in Tehuacán, Puebla, this until the 80s (Morton, 1987). After this, its production suddenly declined due to urbanization. Jiménez (2007) mentioned that according to government statistics in the state of Puebla, 5 hectares were sown, and 30 tons of pomegranate were produced per year.

In 2016, the world production of pomegranate was estimated at over 2 million tons, where Mexico was ranked 11th (Fresh plaza, 2016). In Mexico, pomegranate is grown mainly in places with warm, semi-warm, semi-dry and temperate climate (de la Cruz Quiroz et al., 2011). According to data from SAGARPA (2014), the season for the harvest of pomegranate is between August and October.

Within the main pomegranate producing states in Mexico, we find Hidalgo, Guanajuato and Oaxaca as the first three producers according to SIAP (2015), followed by Chihuahua, Coahuila, Zacatecas, Aguascalientes, Durango, Oaxaca and Jalisco (SAGARPA, 2014). In states that are highly produced like Hidalgo and Guanajuato, it is mainly for commercial purposes. For example, in the municipalities of Apaseo el Alto and Apaseo el Grande in Guanajuato, pomegranate is marketed in different restaurants in the cities of Mexico, Monterrey, Guadalajara and Puebla (López-Mejía, López-Malo and Palou, 2010).

5.4 USES OF THE PLANT AND FRUIT OF THE POMEGRANATE

Although currently the main use of pomegranate is its consumption in fresh, during recent years has gained relevance in the industry, mainly for obtaining juices and extracts from its different parts (Moreno, 2010). Around the world, pomegranate juice has been used to make grenadine, which is a soda, which can be consumed alone or in combination with other drinks (Royal Spanish Academy, 2010). In India, pomegranate has been used as a toothpaste or for dyeing fabrics, to manufacture sodium citrate or citric acid, or in the production of carbonated or non-carbonated beverages (Maharashtra Pomegranate Growers Research Association, 2009).

In Turkey, it has been used mainly in the production of juices; however, this is not its only application because it has also been used for the manufacture of citric acid, in the preparation of canned food and for animal feed (Özgüven et al., 2009). In the case of Mexico, the fruit of the pomegranate is consumed fresh or used for the preparation of desserts or typical dishes such as chili in nogada, in addition the pomegranate has various applications in the traditional Mexican herbalist where it is used the bark of the tree, the leaves, the flowers and the fruit (juice, seed, peel) (López-Mejía, López-Malo, and Palou, 2010).

In Mexico, tree bark has been used to treat diarrhea or dysentery in Mexico City, Guerrero, Guanajuato, Hidalgo, Jalisco, Morelos, Oaxaca, Puebla, Quintana Roo, Sonora, Tabasco, Veracruz and Tlaxcala. It has also been used for treatment against intestinal parasites and skin problems such as measles or rubella (Digital Library of Traditional Mexican Medicine, 2009). In the case of pomegranate leaves, these are used for antidisenteric and antidiarrheal treatments, in addition to their high antimicrobial activity against different microorganisms has been demonstrated (United States Department of Agriculture, 2009, Mexican Traditional Medicines Digital Library, 2009).

Pomegranate flowers are used in the treatment of oral or stomach infections and against intestinal parasites. While from the fruit it can be obtained juice, seeds and the peel, the fresh and dried fruit is used for ornamental use (Moreno, 2010). The seeds and juice help in throat problems; meanwhile the shell has been used against bronchial conditions (Digital Library of Traditional Mexican Medicine, 2009). Pomegranate can also be used to treat indigestion, vomiting, mouth lesions such as cold sores or grains. Through the passage of human history, this fruit has been used as a nutraceutical food (Longtin, 2003), as it has the capacity to provide health benefits due to its high content of bioactive compounds (Caligiani et al., 2010).

The pomegranate tree has been used to obtain wood. From its bark, five alkaloids can be obtained, such as pelletierin, pseudopeletierine, isopeletierine, pseudopeletierine and methyl pelletierine, which can be found between 0.5 and 0.9%. These alkaloids can also be found in the trunk and branches of the pomegranate only in a smaller amount. The use of the bark is very wide since it not only has the presence of alkaloids but also has tannins which, when mixed, reduce intestinal absorption (Moreno, 2010).

In addition to the pomegranate fruit, different food, pharmaceutical or cosmetic products can be obtained, such as whole seeds for fresh consumption, juices, syrups, jams, jams, jellies, food fiber and pomegranate oil. Using bark and pomegranate seeds, products can be obtained that help prevent diseases (Moreno, 2010). Pomegranate has antidiarrheal, antimicrobial, anticancer properties and aids in cardiovascular diseases. It contains minerals such as potassium and phosphorus which are essential in the body, in addition to providing vitamins such as K. Another property, which has gained importance in recent years is that it has a high content of antioxidants such as polyphenols (SAGARPA, 2014).

5.5 COMPOSITION AND USES OF THE POMEGRANATE HUSK

The pomegranate peel is considered an agro-industrial residue; however, it is a rich source of polyphenolic and ellagitannin compounds. Ellagitannins can be degraded to obtain ellagic acid, which can provide various health benefits. In the shell of pomegranate are various compounds which can be useful in the treatment and prevention of a large number of diseases or can give different benefits in the nutrition of people. Some of these products are fatty acids, anthocyanins, tannins, punicalagin, ellagic acid, vitamins, minerals, crude fiber, sugars and organic acids (Moreno, 2010).

The pomegranate contains carpels in which the seeds are found, which are considered the edible portion of the fruit and represent between 58 and 75% of the total fruit; depending on the variety will be the percentage represented by the carpellary membranes and the shell, but approximately these represent between 25 and 42% of the weight of the fruit. The fruit of the pomegranate is composed of approximately 80% water (Moreno, 2010). As mentioned before, the pomegranate peel is rich in hydrolyzable tannins or also called ellagitannins. Aguilar et al. (2008) and Haidari et al. (2009) mentioned that some of these ellagitannins were punicalagin (15.7%) in conjunction with its isomers, punicalin which represents 2% of the husk, gallic and ellagic acids in 3% and the rest of the pomegranate peel are oligomers (77%).

Moreno (2010) mentions that by degrading the ellagitannins present in the pomegranate peel, ellagic acid is obtained, a substance that causes cell death (apoptosis) in cells such as cancer cells without damaging normal cells, since ellagic acid is a potent antioxidant and anticarcinogen, and manages to protect cells from damage that can cause free radicals. Sánchez (2009) reports other functions of ellagic acid; for example, in the hair it improves its capillary activity, as well as strengthens the capillary membranes, in the case of the skin it softens it and improves its elasticity; ellagic acid also reduces diabetic retinopathy, improves vision, reduces varicose veins and helps improve brain function and in diseases such as arthritis reduces inflammation.

Peris et al. (1996) evaluated the composition of the seeds and the pomegranate shell on a dry basis and the results are presented in Table 5.1.

Currently, products derived from the husk of the pomegranate can be found on the market, which is made from extracts with functional properties. Most of them contain punicalagin and ellagic acid, these extracts are found in concentrations of 40 to 90% in the products (Moreno, 2010). Punicalagin is one of the most important compounds in the pomegranate peel. Lu, Ding, and Yuan (2008) determined 82.4 mg of this compound in a gram of pomegranate peel. Huang et al. (2005) mention that compounds such as gallic acid and ursolic acid can be found in flowers, seeds, and pomegranate peel.

The pomegranate peel has been used in traditional herbalism, to treat diseases such as diarrhea, dysentery, cough, or against intestinal parasites such as *Ascaris lumbricoides* (Digital Library of Traditional Mexican Medicine, 2009). Iqbal et al. (2008) used methanolic extracts of pomegranate husk added to sunflower oil, and demonstrated that these extracts are potent

TABLE 5.1 Pomegranate Seeds and Husk Composition

Dry based analysis	**Quantity per 100 g of sample**	
	Seed (g)	**Husk (g)**
Dry material	74.33 ± 0.09	94.45 ± 1.25
Humidity	25.66 ± 0.09	5.5 ± 1.25
Ashes	3.62 ± 0.13	3.59 ± 0.08
Raw fat	10.33 ± 0.17	3.57 ± 0.38
Total protein	10.42 ± 2.61	1.26 ± 0.17
Raw fiber	12.12 ± 2.10	17.75 ± 1.61
Total Sugars	10.13 ± 0.33	16.08 ± 0.61
Reducing sugars	4.67 ± 0.02	4.34 ± 0.01
Total hydrolyzable phenols	0.002 ± 0.00015	6.11 ± 1.83
Ellagic acid	0.0048 ± 0.00002	0.26 ± 0.08
Water absorption index (WAI)	1.62 ± 0.020	4.84 ± 0.006
Critical Humidity Point (CHP)	64.55 ± 2.170	22.08 ± 0.630
pH	5.37 ± 0.354	5.70 ± 0.420
Water activity (a_w)	0.156 ± 0.015	0.189 ± 0.018

antioxidants of this oil. On the other hand, Voravuthikunchai et al. (2004) reported that aqueous extracts of pomegranate peel were highly effective against *Escherichia coli* serotype O157:H7, which could help in the creation of alternative medicines for the treatment of this infection.

Another potential use of the pomegranate peel is in the area of cosmetology since it has a high astringent capacity. Aslam, Lansky, and Varani (2006) used aqueous extracts of pomegranate husks to determine their functionality on the skin; and showed that these extracts promoted skin repair.

5.6 POMEGRANATE WASTE AS A SOURCE OF BIOACTIVE COMPOUNDS

Moreno (2010) cited that compounds such as anthocyanins, fatty acids, crude fiber, tannins, ellagitannins and punicalagin are present in the pomegranate and are among the most important. The anthocyanins are the compounds responsible for granting the red color to the pomegranates, the red coloration depends directly on the concentration and type of anthocyanins that the fruit contains. It has been possible to identify six types of anthocyanins responsible for the coloring of pomegranate juice, which are delphinidin 3-glucoside and 3,5-diglucoside, cyanidin 3-glucoside and 3,5-diglucoside and pelargonidin 3-glucoside and 3, 5-diglucoside (Du et al., 1975).

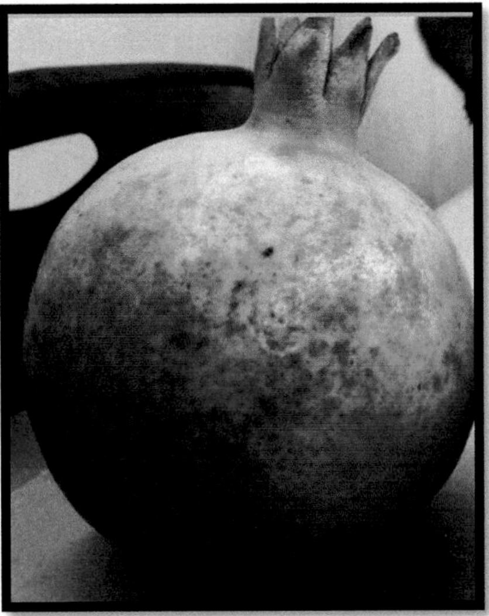

FIGURE 5.2 Pomegranate (fruit).

In addition to coloring the fruit of the pomegranate, anthocyanins are antioxidant compounds that protect the skin from the aging process because they delay the action of free radicals on cells (Moreno, 2010). The antioxidant capacity in a pomegranate juice is three times higher than the antioxidant capacity of red wine or a green tea (Gil et al., 2000). Anthocyanins have a high application in the textile and food industry as natural dyes (Moreno, 2010). In the case of fatty acids, these are important in the diet of people because they grant different benefits, for example, they help prevent coronary heart disease, reduce blood cholesterol levels, inhibit angiogenesis and possess cytotoxic activity on tumor cells. In general, essential fatty acids (n–3 and n–6) are provided by fish such as salmon, trout and mackerel, although they can also be obtained through vegetable sources (Serrano, López, and Espuñes, 2006).

Pomegranate contains its highest percentage of fatty acids in the seeds, being approximately between 37 and 143 g/kg of fruit (Moreno, 2010). It contains both saturated (or essential) and polyunsaturated fatty acids; among the essential fatty acids are linoleic, linolenic and arachidonic. The consumption of this type of acids is important to prevent cardiovascular diseases and other problems in the heart, as it has been shown that polyunsaturated fatty acids significantly reduce HDL-cholesterol levels (Moreno, 2010).

Currently, products in the cosmetic industry made from pomegranate oil are being marketed. Pomegranate oil has been very useful in these products because, in addition to be a thick textured oil, it has properties for both topical and internal use. Some of these properties are regeneration of the epidermis, anti-inflammatory effect, anti-carcinogenic properties, protective chemo effect, reduction of fatty tissues in the body and normalization in lipid metabolism (Sánchez, 2009).

The raw fiber can be found in the edible part of the pomegranate (seeds); in Spanish pomegranates, the raw fiber content was between 5 and 22% of the total fruit, so the pomegranate can be considered as a source of natural fiber (Moreno, 2010). Tannins, meanwhile, are astringent flavor compounds, soluble in water, alcohol and acetone; they have the capacity to form complexes, as well as to precipitate proteins, metals and alkaloids (Arango, 2010). Tannins are astringent, antidiarrheal, antimicrobial, antifungal and useful for treating poisoning with alkaloids and heavy metals. Normally tannins can be found in the pericarp or the shell of the pomegranate (Moreno, 2010).

Punicalagin is the compound with the highest antioxidant capacity because it represents approximately 50% of this activity at least in pomegranate juice, followed by hydrolyzable tannins with 33% of the activity and finally there is 3% of the antioxidant activity from ellagic acid (Gil et al., 2000; García-Viguera and Pérez Vicente, 2004). Punicalagin is a polyphenol that can be hydrolyzed with ellagic acid and metabolized in the intestinal tract, thus generating urolithins which are compounds with high anti-inflammatory activity (Moreno, 2010). Sánchez (2009) mentioned some of the functions that punicalagin has, among which is its antioxidant capacity, its great support for the cardiovascular system and its anticancer capacity.

ACKNOWLEDGMENTS

Almanza-Tovanche thank SAGARPA-CONACYT for the financial support given to her studies at DIA-FCQ (UAdeC) within the project 2015-4-266936 "*Obtención, purificación y escalado de compuestos de extractos bioactivos con valor industrial, obtenidos usando tecnología avanzadas de extracción y a partir de cultivos, subproductos y recursos naturales poco valorados.*"

KEYWORDS

- **bioactive compounds**
- **polyphenols**
- **pomegranate**
- **shell**

REFERENCES

Aguilar, C. N., Aguilera-Carbó, A., Robledo, A., Ventura, J., Belmarés, R., Martínez, D., & Rodríguez-Herrera, R., (2008). Production of antioxidant nutraceuticals by solid-state cultures of pomegranate (*Punica granatum*) peel and creosote bush (*Larrea tridentata*) Leaves. *Food Technol. Biotechnol., 46*, 218–222.

Arango, G. J., (2010). Introduction to secondary metabolism, compounds derived from shikimic acid. Available at: https://documents.mx/documents/compuestos-derivados-de-shikimico.html (Accessed: 29 January 2018).

Aslam, M. N., Lansky, E. P., & Varani, J., (2006). Pomegranate as a cosmeceutical source: Pomegranate fractions promote proliferation and procollagen synthesis and inhibit matrix metalloproteinase-1 production in human skin cells. *Journal of Ethno Pharmacology, 103*(3), pp. 311–318. doi: 10.1016/j.jep.2005.07.027.

Ávalos, A., & Pérez, E., (2009). 'Metabolismo secundario de plantas,' *REDUCA (Biología) Serie Fisiología Vegetal*, 2(3), pp. 119–145. Available at: http://revistareduca.es/index.php/biologia/article/viewFile/798/814 (Accessed: 12 August 2020)

Ballard, T. S., Mallikarjunan, P., Zhou, K., & O'Keefe, S., (2010). Microwave-assisted extraction of phenolic antioxidant compounds from peanut skins. *Food Chemistry* (Vol. 120, No. 4 pp. 1185–1192). Elsevier Ltd,. doi: 10.1016/j.foodchem.2009.11.063.

Biblioteca Digital de la Mediciona Tradicional mexicana (2009). Available at: http://www.medicinatradicionalmexicana.unam.mx.

Biesalski, H. K., Dragsted, L. O., Elmadfa, I., Grossklaus, R., Müller, M., Schrenk, D., Walter, P., & Weber, P., (2009). Bioactive compounds: Definition and assessment of activity. *Nutrition, 25*(11, 12), 1202–1205. doi: 10.1016/j.nut.2009.04.023.

Caligiani, A., Bonzanini, F., Palla, G., Cirlini, M., & Bruni, R., (2010). 'Characterization of a potential nutraceutical Ingredient: Pomegranate (*Punica granatum* L.) seed oil unsaponifiable fraction. *Plant Foods for Human Nutrition* (Vol. 65, No. 3, pp. 277–283). Springer U.S. doi: 10.1007/s11130-010-0173-5.

de la Cruz Quiroz, R., Rodríguez Herrera, R., Contreras Esquivel, J. C., Aguilar, C. N., Aguilera-Carbó, A. & Prado Barragán, L. A., (2011). "The pomegranate: source of powerful bioactive agents,' *Revista de divulgación científica CIENCIACIERTA. 25*(7), 35–38. Available at: https://www.researchgate.net/publication/236135414_La_granada_fuente_de_potentes_agentes_bioactivos (Accessed: 12 August 2020)

Du, C. T., Wang, P. L., & Francis, F. J., (1975). Anthocyanins of pomegranate, *Punica granatum. Journal of Food Science* (Vol. 40, No. 2, pp. 417–418). Blackwell Publishing Ltd. doi: 10.1111/j.1365-2621.1975.tb02217.x.

Fresh Plaza (2016). *Mexico: Pomegranates Are Jalisco's Future*. Available at: https://www.freshplaza.com/article/164646/Mexico-Pomegranates-are-Jaliscos-future/ (Accessed: 12 August 2020).

García-Viguera, C., & Pérez Vicente, A., (2004). La granada. Alimento rico en polifenoles antioxidantes y bajo en calorías. *Alimentación nutrición y salud 11*(4), 113–120 Available at: https://digital.csic.es/bitstream/10261/17946/3/lecturaPDF.pdf (Accessed: 12 August 2020).

Gil, M. I., Tomás-Barberán, F. A., Hess-Pierce, B., Holcroft, D. M., & Kader, A. A., (2000). Antioxidant activity of pomegranate juice and its relationship with phenolic composition and processing. *Journal of Agricultural and Food Chemistry, 48*(10), 4581–4589.

Haidari, M., Ali, M., Ward, C. S., & Madjid, M., (2009). Pomegranate (*Punica granatum*) purified polyphenol extract inhibits the influenza virus and has a synergistic effect with oseltamivir. *Phytomedicine, 16*(12), 1127–1136. doi: 10.1016/j.phymed.2009.06.002.

Hao, J., Han, W., Huang, S., Xue, B., & Deng, X., (2002). Microwave-assisted extraction of artemisinin from *Artemisia annua* L. *Separation and Purification Technology, 28*(3), 191–196. doi: 10.1016/S1383-5866(02)00043-6.

Huang, T. H., Yang, Q., Harada, M., Li, G. Q., Yamahara, J., Roufogalis, B. D., & Li, Y., (2005). *Pomegranate Flower Extract Diminishes Cardiac Fibrosis in Zucker Diabetic Fatty Rats: Modulation of Cardiac Endothelin-1 and Nuclear Factor-kappaB pathways. Journal of Cardiovascular Pharmacology. 46*(6), 856–862. doi: 10.1097/01.fjc.0000190489.85058.7e.

Iqbal, S., Haleem, S., Akhtar, M., Zia-ul-Haq, M., & Akbar, J., (2008). Efficiency of pomegranate peel extracts in stabilization of sunflower oil under accelerated conditions. *Food Research International, 41*(2), 194–200. doi: 10.1016/j.foodres.2007.11.005.

Jaramillo, C. J., Espinoza, A. J., D'armas, H., Troccoli, L., & Rojas De Astudillo, L., (2016). 'Concentrations of alkaloids, cyanogenic glycosides, polyphenols and saponins in selected medicinal plants from Ecuador and their relationship with acute toxicity against *Artemia salina*,' *Revista de Biología Tropical. 64*(3), 1171–1184. doi: 10.15517/rbt.v64i3.19537.

Longtin, R., (2003). 'The pomegranate: nature's power fruit?,' *Journal of the National Cancer Institute, 95*(5), 346–348. doi: 10.1093/jnci/95.5.346.

López-Mejía, O. A., López-Malo, A., & Palou, E., (2010). 'La granada (*Punica granatum* L.) una fuente de antioxidantes de interés actual,' *Temas Selectos de Ingeniería de Alimentos, 4*(1), 64–73. Avilable at: https://tsia.udlap.mx/granada-punica-granatum-l-una-fuente-de-antioxidantes-de-interes-actual/ (Accessed: 12 August 2020).

Terpenes| Fundación CANNA: Scientific Research and Cannabis Testing. Available at: https://www.fundacion-canna.es/en/terpenes (Accessed: 12 August 2020).

Lu, J., Ding, K., & Yuan, Q., (2008) 'Determination of Punicalagin Isomers in Pomegranate Husk,' *Chromatographia*. Vieweg Verlag, 68(3–4), pp. 303–306. doi: 10.1365/s10337-008-0699-y.

Maharashtra Pomegranate Growers Research Association (2009). '*Pomegranate Facts*.' Available at: http://www.pomegranateindia.in/facts.htm (Accessed: 12 August 2020).

Martínez-Ramírez, A., Contreras-Esquivel, J. C., & Belares-Cerda, R., (2010). 'Microwave Assisted Polyphenol Extraction from Punica granatum L.,' *Acta Química Mexicana/Revista Científica de la Universidad Autónoma de Coahuila*, 2(4), pp. 1–5. Available at: https://docplayer.es/72568736-Extraccion-de-polifenoles-asistida-por-microondas-a-partir-de-punica-granatum-l.html (Accessed: 12 August 2020).

Melgarejo Moreno, P., & Agrónomo, I., (2008). 'Alternative Fruit Growing For Arid And Semi-Arid Areas. New Techniques For Saving Water And Energy.' Available at: https://cultivodelgranado.es/wp-content/uploads/2014/06/ponencia-granada.pdf (Accessed: 12 August 2020).

Milner, J. A., (2004). 'Molecular targets for bioactive food components,' *The Journal of Nutrition, 134*(3), p. 2492S–2498S. doi: 10.1093/jn/134.9.2492S.

Mohseni, A., (2009). 'The Situation of Pomegranate Orchards In Iran,' *Acta Horticulturae*, 818, 35–42. doi: 10.17660/ActaHortic.2009.818.3.

Moreno, P. M., (2010). 'El Granado,' *I Jornada nacional sobre el granado: producción, economía, industralización, alimentación y salud.*

Morton, J., (1987). *Pomegranate*. 352–355. In: Fruits of warm climates. Julia F. Morton, Miami, FL. Available at: https://www.hort.purdue.edu/newcrop/morton/pomegranate.html (Accessed: 12 August 2020).

Nijveldt, R. J., Van, N. E., Van, H. D. E., Boelens, P. G., Van, N. K., & Van, L. P. A., (2001). Flavonoids: A review of probable mechanisms of action and potential applications. *The American Journal of Clinical Nutrition, 74*(4), 418–425.

Özgüven, A. I., Yılmaz, M., & Yılmaz, C., (2009). 'The Situation Of Pomegranate And Minor Mediterranean Fruits In Turkey,' *Acta Horticulturae*, (818), pp. 43–48. doi: 10.17660/ActaHortic.2009.818.4.

Peris, J. B., Stübing, G. Y., & Figuerola, R., (1996). Guía de las plantas medicinales de la comunidad valenciana. Valencia: Las Provincias.

SAGARPA (2014a). *Capacita SAGARPA a productores de granada del Valle del Mezquital para exportación*. Available at: http://sagarpa.gob.mx/saladeprensa/2012/Paginas/2014B901.aspx (Accessed: 27 December 2017).

SAGARPA (2014b). Available at: http://www.sagarpa.gob.mx/saladeprensa/2012/Paginas/2014B602.aspx (Accessed: 26 June 2017).

Serrano, M. E. D., López, M. L., & Espuñes, T. D. R. S., (2006). 'Componentes bioactivos de alimentos funcionales de origen vegetal. *Revista Mexicana de Ciencias Farmaceuticas, 37*(4), pp. 58–68. doi: 10.1016/j.tifs.2016.09.006.

Snoei, G. (2016). *Granadas: aumentar la producción, el consumo y la competencia*. Available at: http://www.freshplaza.com/article/154127/Pomegranates-increasing-production%2C-consumption-and-competition (Accessed: 26 December 2017).

Sheets, M. D., Du, B. M. L., & Williamson, J. G., (2009). *La Granada en Florida*. Available at: http://edis.ifas.ufl.edu/hs294 (accessed on 24 July 2020).

The Daily Records, (2017). *Largest Pomegranate Producing Countries In The World 2017–2018, Top 10 Countries List*. Available at: http://www.thedailyrecords.com/2018-2019-2020-2021/world-famous-top-10-list/world/largest-pomegranate-producing-countries-world-statistics/6874/ (accessed on 24 July 2020).

Voravuthikunchai, S., Lortheeranuwat, A., Jeeju, W., Sririrak, T., Phongpaichit, S., & Supawita, T., (2004). 'Effective medicinal plants against enterohaemorrhagic *Escherichia coli* O157:H7,' *Journal of Ethnopharmacology, 94*(1), 49–54. doi: 10.1016/j.jep.2004.03.036.

Wong, P. J. E., Muñiz, M. D. B., Martínez, Á. G. C. G., Belmares, C. R. E., & Aguilar, C. N., (2015). Ultrasound-assisted extraction of polyphenols from native plants in the Mexican desert. *Ultrasonics Sonochemistry* (Vol. 22, pp. 474–481). Elsevier B.V. doi: 10.1016/j.ultsonch.2014.06.001.

Xia, E. Q., Ai, X. X., Zang, S. Y., Guan, T. T., Xu, X. R., & Li, H. B.,(2011). Ultrasound-assisted extraction of phillyrin from Forsythia suspense. *Ultrasonics Sonochemistry* (Vol. 18, No. 2, pp. 549–552). Elsevier B.V. doi: 10.1016/j.ultsonch.2010.09.015.

Yemiş, O., & Mazza, G., (2012). Optimization of furfural and 5-hydroxymethylfurfural production from wheat straw by a microwave-assisted process. *Bioresource Technology, 109*, 215–223. doi: 10.1016/j.biortech.2012.01.031.

CHAPTER 6

Lactic Acid Fermentation as a Tool for Obtaining Bioactive Compounds from Cruciferous

DANIELA IGA BUITRÓN,[1] MARGARITA DEL ROSARIO SALAZAR SÁNCHEZ,[2] EDGAR TORRES MARAVILLA,[3] LUIS BERMÚDEZ HUMARAN,[3] JUAN ALBERTO ASCACIO-VALDÉS,[1] JOSÉ FERNANDO SOLANILLA DUQUE,[2] and ADRIANA CAROLINA FLORES-GALLEGOS[1]

[1]*Food Research Department, School of Chemistry, Universidad Autónoma de Coahuila, Blvd. Venustiano Carranza S/N, Colonia República, 25280, Saltillo, Coahuila, México, Tel: +52 (844) 416 1238; E-mail: carolinaflores@uadec.edu.mx*

[2]*Universidad del Cauca, Popayán, Cauca, Colombia*

[3]*Micalis Institute, UMR INRA-AgroParisTech, Jouy en Josas, France*

ABSTRACT

Cruciferous such as brussels sprouts, cauliflower, broccoli, and cabbage are a rich source of health-promoting phytochemicals. Cruciferae family vegetables possess biological activities due to the presence of phenolics and glucosinolates (GLS). However, the health benefits of phenolic compounds and GLS are partially dependent on their microbial conversion. Post-harvest processing especially fermentation results in the extended shelf life of the product and could be a useful technological process since it favors the hydrolyzation of GLS to multiple potentially favorable breakdown products. In contrast with other bacteria, Lactobacilli are more resistant to phenolics and the fermentation with lactic acid bacteria (LAB) provides a natural way to modify nutritional and chemopreventive properties of food or food ingredients. This review aims to give an overview of the phytochemicals present in cruciferous vegetables and their metabolism through the lactic-acid fermentation process which could facilitate the possible application of selected bacteria to obtain beneficial breakdown products.

6.1 INTRODUCTION

Foods, especially fruits, and vegetables include numerous and diverse bioactive compounds. Fibers, vitamins, minerals, and phenolic compounds are molecules of nutritional interest (Septembre-Malaterre, Remize, and Poucheret, 2018). Recently, cruciferous vegetables including broccoli, cauliflower, cabbage, kale, brussels sprouts, and radish have drawn regard because of their potential protective attributes (Herr and Büchler, 2010). Epidemiological studies have shown that the regular intake of brassica vegetables can decrease de risk of various cancers, including lung, breast, and colorectal cancer (Palani et al., 2016) and cardiovascular diseases (Francisco et al., 2010). The main bioactive compounds found in these vegetables are phenolic acids, flavonoids, glucosinolates (GLS), and carotenoids (Lee, Boyce, and Breadmore, 2011). Also, antimicrobial, antioxidant, and anti-inflammatory activities of GLS and other sulfur compounds originating from Brassica vegetables have been reported. The GLS and phenolic fraction in brassica vegetables differ considerably with pre-harvest and post-harvest conditions. Post-harvest handling especially fermentation results in prolonged shelf-life of the product and produces numerous potentially advantageous breakdown products (Peñas et al., 2012). Lactic acid fermentation results in an entire degradation of GLS and augments the contents of health-promoting compounds. Also, fermentation enhances the antioxidant potential in vegetables (Palani et al., 2016). The skill to tolerate and metabolize phenolic compounds and GLS is strain- or species-dependent. The metabolism of phenolics in food may confer a selective advantage for microorganisms and thus impacts the selection of competing starter cultures for vegetable and fruit fermentations (Filannino et al., 2015).

6.2 PRODUCTION OF CRUCIFEROUS

Vegetables have long been cultivated as a food fount (Younho et al., 2017). Globally, Asia produces more than three-quarters of the vegetables. China, India, Turkey, Italy, Egypt, Spain, Brazil, Mexico, and the Russian Federation are the leading producers of fresh and processed vegetables with approximately 53 million hectares. The most-traded vegetables are tomatoes, onions, cabbages, cucumbers, and eggplants that account for about 40% of total world production. Mexico represents horticultural production in Latin America with 10.7 MMT (million metric tons), Brazil with 5.7 MMT, Argentina with 2.9 MMT, Chile with 2.2 MMT, Colombia 1.9 with MMT, and Peru 1.6 with MMT (FAOSTAT, 2018). Cruciferous vegetables, including cabbage, broccoli, and turnip are amongst the cultivation produced in the most abundant quantities. In accordance with the Ministry of Agriculture, Food, and Rural Affairs (MAFRA), in Korea, the annual production of cruciferous vegetables is about 2.6 million tons (Younho et al., 2017). The total output worldwide of cauliflower in 2016 was 25.21 MMT. Broccoli is cruciferous with the highest production worldwide, with a total volume of production of 50.8 MMT in 2016. China is the primary producer of cauliflower and broccoli (40.4% of the whole world production), followed by India (32.5%). Mexico ranks fifth in the world production of cauliflower and broccoli (2.3%) (SIAP, 2018). The merchandising of products of the *Brassica* genus, such as raw and frozen broccoli, in addition to by-products generated by agroindustry has increased considerably over time and in recent years presents significant advances and dynamics related to the promotion of new areas of agroindustry, increases in production and innovation in transformation processes, owing to the critical role of this vegetable in human nutrition for its nutritional value with functional properties (FAOSTAT, 2018).

6.3 NUTRIENTS PRESENT IN THE *BRASSICA* GENUS PLANTS

6.3.1 *PROTEINS*

Protein constitutes one of the major macronutrients of cruciferous vegetables. The protein fraction of cruciferous vegetables varies from 1.0% to 3.3% (w/w) of a fresh weight basis. The highest amount of protein is reported in kale (3.3%), while radish contains less than 1% protein on a fresh weight basis (Manchali, Chidambara, and Patil, 2011). The protein fraction of *Brassica* seeds are rich in lysine and has a notable content of sulfur amino acids (methionine and cysteine). Moreover, the quality of rapeseed protein has been compared with that of casein and considered for its balanced amino acid composition superior to that of other plants such as soybeans, peas, sunflower or wheat (Sarwar et al., 1984).

6.3.1.1 GLOBULINS

Cruciferins (12S globulins) are storage proteins formed by acidic (MW $\approx$ 40 kDa) and basic (MW $\approx$ 20 kDa) polypeptide chains. The oligomeric cruciferin structure is of type alpha (α) and beta (β) polypeptides chains where an acid chain is joined to a basic chain by disulfide bridges to form a subunit (Dalgalarrondo et al., 1986). It is a very small protein, since having a high molecular weight (300–350 kDa), it has a low Stoke radius value (5.7 - 5.5), which is reflected in its globular structure (Schwenke, 1990). They are rich in amino acids (Glutamine and Aspartate) and deficient in sulfur (Methionine and Cysteine). Likewise, the rigidity of its structure is not only due to the disulfide bridges between the subunits but also to hydrophobic interactions on account of the high hydrophobicity of the basic subunits. This strong hydrophobicity expresses its amino acid structure that has a marked impact not only on the protein structure but also on the functional properties of the global protein complex (Schwenke, 1990). Plietz et al., based on X-ray scatter tests, postulated a three-dimensional glomerular structure constituting a trigonal antiprism (Figure 6.1) (Plietz et al., 1983).

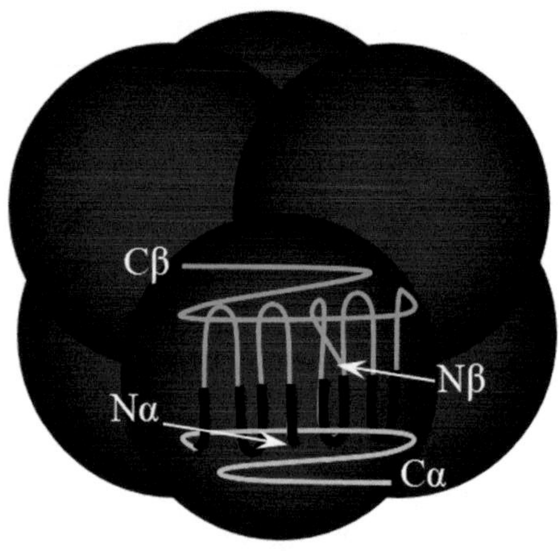

FIGURE 6.1 Hypothetical model of the structure of globulin 12 S.

Brassica globulin 12 S differs from other globulins in that its isoelectric point (pI) is slightly higher and the proportion of acidic amino acids (Glutamine + Aspartate)/basic amino acids (Lysine + Histidine + Arginine) close to 1 is lower than those seen in this type of globular proteins. This decrease in this ratio is mainly due to the high amidation levels of amino acids (61%) (Schwenke, 1990).

6.3.1.2 ALBUMINS

Albumins 2S or napins [in relation to *B. napus* where it was described for the first time (Crouch et al., 1983)] are characterized by being soluble in water, sediment by ultracentrifugation at a sedimentation rate of 1.7–2S and possess a high isoelectric point (pI) (pI > 10) being considered basic proteins (Schwenke et al., 1988). This strong basicity raised the possibility of grouping them within ribosomal or histonic proteins (Godon, 1996); however, they have chemical and structural characteristics that rule out this hypothesis. For example, their amino acid composition is rich in sulfur amino acids and deficient in lysine. They present two polypeptide chains, one small and the other larger, linked by disulfide bridges and mainly of helical conformation (Schwenke, 1990). This protein fraction also presents a high number of antigenic epitopes involved in the development of food allergies (González De La Peña et al., 1996).

6.3.2 CARBOHYDRATES

Cruciferous vegetables serve as a good source of carbohydrates, which ranges from 0.3% to 10% (w/w fresh weight basis) (Manchali, Chidambara, and Patil, 2011). Cruciferous further contains considerable amounts of dietary fiber and water-soluble simple sugars, including sucrose, d-glucose, and d-fructose (Younho et al., 2017). Sucrose and stachyose are the main oligosaccharides (Theyer et al., 1977), and its digestion requires the presence of two enzymes, invertase, and α-galactosidase, the latter being absent in the human gastrointestinal tract (Reddy et al., 1982). As a result, a microbiological fermentation process of the oligosaccharides takes place in the large intestine, resulting in the formation of methane, carbon dioxide, and hydrogen (Reddy et al., 1982). Branched polysaccharides such as cellulose, hemicellulose, and lignin present in the cell walls of cotyledons contain mainly glucose, galactose, arabinose, xylose, and galacturonic acid as constituent monomers (Slominski et al., 1994; Simbaya et al., 1995). The highest fiber content (4.6% (w/w fresh weight basis)) is found in Collard, followed by broccoli (30.4% w/w dry weight) and cauliflower (26.7% w/w dry weight basis) (Manchali, Chidambara, and Patil, 2011).

6.3.3 MICRONUTRIENTS

The major micronutrients of brassica's vegetables are vitamins and minerals. Cruciferous are a suitable fount of vitamin C (the ratio in broccoli is more than 50 mg/100 g of fresh weight), folic acid, tocopherols, and provitamin-A. The maximum mean b-carotene content of major cruciferous vegetables is 0.5–1.0 mg/ 100 g of fresh weight. Brassicas vegetables also contain α-tocopherol, and the maximum mean content of 0.47 mg/100 g has been found in broccoli. The main essential mineral elements in these vegetables are copper, calcium, iron, selenium, manganese, and zinc. Calcium, phosphorous, magnesium, sodium, and potassium constitute major macroelements (Manchali, Chidambara, and Patil, 2011).

6.4 PHYTOCHEMICALS IN CRUCIFEROUS VEGETABLES

In brassicas, nitrogen-containing compounds represent a major class of secondary metabolites, whereas glucosinolates are the major sulfur compounds. Flavonoids, anthocyanins, and carotenoids comprise biologically active colored compounds. Also, polyphenols and antioxidant enzymes constitute major health beneficial compounds (Manchali, Chidambara, and Patil, 2011).

6.4.1 GLUCOSINOLATES

GLS are sulfur-containing glycosides found in cruciferous vegetables primarily belonging to the *Brassicaceae* family. GLS is mainly present in the Brassicaceae, Capparaceae, and Caricaceae families (Fahey, Zalcmann and Talalay, 2001). These include *Brassica oleracea* var *Italica* (broccoli), *Raphanus sativus* (radish), *Brassica oleracea* var *capitata* (cabbage) or *Brassica oleracea* var *gemmifera* (Brussels sprouts), *Brassica oleracea* var *botrytis* (cauliflower) and *Brassica rapa* subsp rapa (turnip). Cabbage, broccoli and Brussels sprouts are the major source of glucosinolates for the human diet and they are often consumed vegetables (Verkerk and Dekker, 2008). The general structure of a GLS is shown in Figure 6.2. These phytochemicals are β-thioglucosides and (*Z*)-*cis*-*N*-hydroximinosulfate esters with a side chain R and sulfur-linked d-glucopyranose moiety (Verkerk and Dekker, 2008). The side chain defines the chemical and biological nature of GLS in the structure (Wittstock and Burow, 2010).

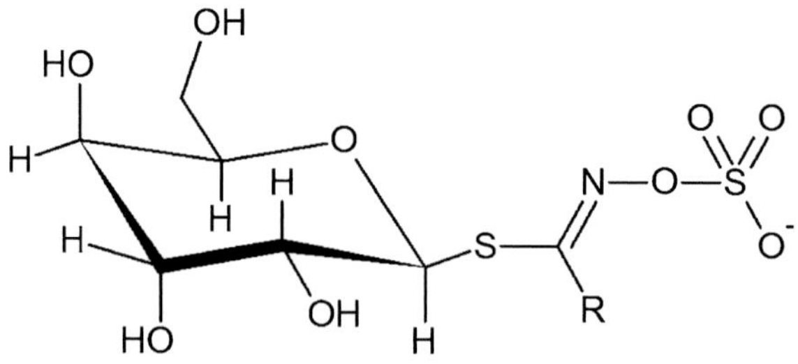

FIGURE 6.2 Structure of the glucosinolate molecule. R: lateral chain.

GLS are synthesized from amino acid precursors such as methionine, phenylalanine, tyrosine or tryptophan, by elongation of the main chain and in routes of oxidation or hydroxylation (Larsen, 1981; Giamoustaris and Mithe, 1995). There are more than 120 different GLS identified but only a few have been investigated deeply. They are known as sulfur and glucose-containing compounds and a few of the benzyl glucosinolates may have an additional sugar moiety such as rhamnose and arabinose bound to the aromatic ring (Fahey, Zalcmann and Talalay, 2001). Some of the GLS found in the plants are given in Table 6.1. Amino acids are the precursors of GLS and, depending on which amino acid acts as a precursor for their formation, they can be divided into three groups. Aliphatic GLS are derived from alanine, leucine, isoleucine, valine, and methionine. Aromatic GLS are derived from phenylalanine and tyrosine and indole GLS are derived from tryptophan (Bennett et al., 1995).

TABLE 6.1 Common Glucosinolates Found in *Brassica* (adapted from Cebeci, 2017)

Common Name	Structure of the Glucosinolate
Aliphatic Glucosinolates	
Sinigrin	2-propenyl
Glucoraphanin	4-methyl-sulfinyl butyl
Glucoraphanin	4-methyl-sulfinyl-3-butenyl
Glucoiberin	3-methyl-sulfinylpropyl
Glucoiberverin	3-methyl-thiopropyl
Glucoerucin	4-methyl-thiobutyl
Progoitrin	2-hydroxy-3-butenyl
Glucoalyssin	5-methylsulfinylpentyl
Glucocapparin	Methyl
Glucobrassicanapin	4-pentenyl
Gluconapin	3-Butenyl
Aromatic Glucosinolates	
Glucotropaeolin	Benzyl
Glucosinalbin	p-hydroxybenzyl
gluconasturtiin	2-phenethyl
Indole Glucosinolates	
Glucobrassicin	3-indolylmethyl
Neobrassicin	1-methoxy-3-indolylmethyl
4-Hydroxyglucobrassicin	4-hydroxy-3-indolylmethyl
4-Methoxyglucobrassicin	4-methoxy-3-indolylmethyl

The hydrolysis of GLS is mediated by the thioglucosidase enzyme myrosinase. Myrosinase is found detached from GLS in idioblasts or myrosin cells of the parenchymatous tissue. Cutting, chewing or food processing causes tissue damage, so the integrity of cells is disrupted and GLS and myrosinase come into contact. When GLS enzymatic hydrolysis is initiated, it results in the formation of GLS hydrolysis products such as isothiocyanates (ITCs), thiocyanates or nitriles (Holst and Williamson, 2004) and other less volatile or non-volatile compounds such as oxazolidinethiones and indoles (Peñas et al., 2012). A few glucosinolates exist in similar amounts in all cruciferous, while some Cruciferae contains an exceptionally content of specific glucosinolates and the corresponding isothiocyanate. Glucoraphanin, for example, accounts for 36%, 40%, 70%, and 0% of the total GLS of broccoli, cabbage, kale, and cabbage turnip respectively (Herr and Büchler, 2010). A few GLS derivates, especially isothiocyanates (ITCs), nitriles, and indoles have received attention as potential chemopreventive agents (Peñas et al., 2012).

6.4.2 PHYTIC ACID

Phytic acid or myo-inositol 1–6 hexaquis (dihydrogen phosphate) is a universal component in plant seeds (Morris and Ellis, 1981; Zhou et al., 1990). It occurs in plants as mono- or divalent cation salts of K^+, Ca^{2+}, and Mg^{2+}. Its main biological function is due to its high chelating power, the storage of cations and, due to its composition, the storage of phosphorus. Phytic acid is deemed an anti-nutritional factor as it interfaces with essential minerals and dietary proteins resulting in physiologically unavailable insoluble complexes (Erdman, 1979; Nwokodo and Bragg, 1987). However, some authors have linked dietary phytic acid intake to a reduction in the incidence of intestinal and colon cancer (Graf and Eaton, 1985; Graf and Eaton, 1990; Graf and Eaton, 1993; Midorikawa et al., 2001).

6.4.3 POLYPHENOLS

Phenolic compounds are secondary plant metabolites whose structure includes one or more aromatic rings substituted by one or more hydroxyl groups (Rodrìguez et al., 2009). The amplest classes of phenolics in vegetables are phenolic acids and flavonoids (Filannino et al., 2015). The most extended and diversified group of polyphenols in *Brassica* species are the flavonoids (mostly flavonols but also anthocyanins) and the hydroxycinnamic acids such as sinapic, caffeic, ferulic, and *p*-coumaric acids (Cartea, Francisco, Soengas, and Velasco, 2011; Avato and Argentieri, 2015).

Flavonoids constitute the most numerous of the phenolic compounds and are found across the plant kingdom. Their typical structure includes two benzene rings flanking 3-carbon ring. Multiple combinations of these molecules with hydroxyl groups, sugars, oxygen, and methyl give rise to a wide variety of flavonoids, such as flavonoids, flavanones, flavones, catechins, anthocyanins, and isoflavones. Flavonols are the broadest of the flavonoids (Cartea et al., 2011). The commonly isolated from brassica vegetables consist of quercetin, kaempferol, cyanidin, and isorhamnetin (Avato and Argentieri, 2015).

Hydroxycinnamic acids are a type of non-flavonoid phenolics that are characterized by having a C6-C3 structure. Hydroxycinnamic acids are plentiful in plants and are used in both structural and chemical plant defense strategies. They can appear freely or as a constituent of the plant cell wall (Cartea et al., 2011).

6.5 CRUCIFEROUS PHYTOCHEMICALS AND THEIR PROFIT IN HUMAN HEALTH

In recent years, brassicas have gained heed in cancer research because of their protective features. Nowadays, research has attested that phytochemicals from cruciferous induce detoxification enzymes, scavenge free radicals, reduce inflammation, stimulate the immune system, reduce the risk of cancers, inhibit tumorigenesis, and regulate the growth of cancer cells (Herr and Büchler, 2010).

6.5.1 BRASSICA PHYTOCHEMICALS AGAINST CANCER

Some of the clinical and pre-clinical research using cruciferous in different cancers were reported in the '70s. Studies suggested a relationship between the consumption of cruciferous and risk reduction of colorectal, pancreatic, lung, gastrointestinal, breast, and ovarian cancer (Manchali, Chidambara, and Patil, 2011). Some ITC has stronger properties than others; ITCs may exert their protective effects through multiple different processes (Herr and Büchler, 2010). SNF, as well as I3C have multi-faceted anticarcinogenic activities in cells, through the *in vivo* inhibition of the activation of the central factor of inflammation NF-κB, and also clogging carcinogenic stages *in vitro* and *in vivo* by initiation of apoptosis, cell cycle arrest and inhibition of histone deacetylases (Navarro et al., 2014). Also, SFN and the ITC derived from the GLS glucoraphanin are potent inductor of phase 2 detoxification enzymes (which help in the metabolism of xenobiotics to prevent carcinogenesis) such as quinone reductase, UDP-glucoronosyl transferases, and glutathione transferases, thereby blocking the action of potential carcinogens (Peñas et al., 2012). Phase 2 enzymes also inhibit phase 1 enzymes, which on the other hand, known to activate carcinogens. Chemoprevention of breast cancer through induction of phase 2 enzymes by I3C is reported in animal experiments (Manchali, Chidambara, and Patil, 2011). Also, SFN prevents carcinogen activation through inhibition of cytochrome P405, appears to inhibit angiogenesis, and exerts anti-inflammatory effects in cancer cells mediated through the inhibition of cyclooxygenase-2 expression. Allyl isothiocyanate, a breakdown product from sinigrin, has been studied since it induces cell death in colorectal and prostate cancer cells.

Furthermore, Allyl isothiocyanate is implicated in inflammation and cancer (Peñas et al., 2012). Regarding GLS and protection of gastric cancer, it has been demonstrated in experimental studies the block of Helicobacter pylori by the action of GLS and SFN. Bacterial infection with Helicobacter pylori is linked with an increase in the risk of gastric cancer (Herr and Büchler, 2010). Other phytochemicals also present in *Brassicaceae* are the phenolic compounds, main derivates of hydroxycinnamic acids (Baenas et al., 2017). The consumption of phenolic compounds on a diet is advantageous to health due to their chemopreventive effects against carcinogenesis and mutagenesis (Rodríguez et al., 2009).

6.5.2 INDUCTION OF APOPTOSIS IN CANCER CELLS

Among the bioactive compounds of cruciferous vegetables studied for cancer inhibition, ITCs, I3C, and phytoalexins are highly promising. SFN has demonstrated the inductance of apoptosis and cell cycle arrest in human colon adenocarcinoma (HT29) cells. ITCs are known to inhibit androgen-dependent and independent prostate cancer cells through the arrest of cells in the G2/M phase and induction of apoptosis (Manchali, Chidambara, and Patil, 2011). Otherwise, iberin, derived from glucoiberin decomposition, can inhibit the proliferation of human glioblastoma and neuroblastoma cells through the induction f cell apoptosis (Peñas et al., 2012). The major targets of isothiocyanates in inhibiting cancer cells include induction of caspase-dependent and independent apoptosis and cell cycle arrest (Manchali, Chidambara, and Patil, 2011).

6.5.3 ANTI-INFLAMMATORY ACTIVITY

Inflammation represents a crucial role in the evolution and progression of various diseases. In the inflammatory response, the macrophage population increases. Continuous pro-inflammatory macrophages and overflow of inflammatory mediators contribute to the development of chronic inflammatory diseases, such as atherosclerosis, type 2 diabetes, and cancer. Therefore, modulating macrophage-mediated inflammatory responses could reduce the incidence of several inflammatory diseases (Navarro et al., 2014). GLS and phenolic compounds have shown beneficial anti-inflammatory activity for human disease prevention (Baenas et al., 2017). NF-κB is a fundamental intermediary in the inflammatory process. NF-κB controls a battery of genes coding for cytokines, chemokines, adhesion molecules, and other elements implicated in the immune response. There is evidence of the role of ITCs in the modulation and inhibition of NF-κB (Navarro et al., 2014). Also, sinigrin, a thioglucoside of cruciferous vegetables, induce notable changes in the expression pattern of proinflammatory mediator genes (Lee et al., 2017).

6.5.4 ANTIOXIDANT ACTIVITY

Brassica species contain phenols, a group of more than 8000 different natural compounds. Phenols in plant foods are valuable because of their antioxidant activity and free radical-scavenging. Among phenols, the flavonoids exhibit the highest antioxidant activity. Furthermore, *Brassica* vegetables are widely considered to possess high levels of antioxidative activity, and their phenolic compounds are assumed to be the main dietary antioxidants (Radošević et al., 2017). Antioxidant activity of cruciferous is also owing to both bioactive compounds and enzymes such as catalase, superoxide dismutase and peroxidase, which are found in raw vegetables. Induction of oxidative stress in the cancer cell is one of the cytotoxicity mechanisms by which bioactive molecules derivate from cruciferous vegetables can kill cancer cells (Manchali, Chidambara, and Patil, 2011). Polyphenols act as iron and copper agents, thus preventing oxidation of ascorbic acid, inhibit the activity of enzymes involved in the formation of reactive oxygen species (such as xanthine oxidase) and block enzymatic and non-enzymatic lipid peroxidation (Scalbert et al., 2005).

6.5.5 HEART HEALTH

Bioactive compounds of cruciferous vegetables, particularly indolic compounds and phytoalexins (Maiyoh et al., 2007), help in heart health mainly through their ability to reduce low-density lipoprotein (LDL), and combat free radicals(Manchali, Chidambara, and Patil, 2011). It has been reported that I3C, a plant indole, reduce serum cholesterol levels in mice (Maiyoh et al., 2007).

6.6 LACTIC-ACID FERMENTATION AS A TOOL FOR OBTAINING BIOACTIVE COMPOUNDS

The bioaccessibility and bioavailability of phenolics and glucosinolate compounds differ significantly. The content of glucosinolates and other phytochemicals may be influenced not only by the species or variety but also by the postharvest treatments such as storage conditions and processing (Sosińska and Obiedziński, 2011; Francisco et al., 2010), it can also be modified by the traditional cooking techniques such as boiling, steaming, pressure cooking and microwaving (Francisco et al., 2010). All of those procedures could affect the health-promoting compounds of cruciferous (Oliviero, Verkerk and Dekker, 2012). Thermal process plays an essential role in the degradation of bioactive compounds and enzymes inactivation witch influenced the stability of plant phytochemicals (Dunja et al., 2018).

Fermentation seems to be ideal processing to improve the concentration of phytochemicals and their health benefits (Paramithiotis, Hondrodimou, and Drosinos, 2010). Fermentation has been associated with food safety, organoleptic modifications (taste, color, texture), simplicity of the preparation process and valorization of raw vegetable material. Hence, human interest in food fermentation lies in the advantages it has such as improve the nutrition health properties and the production of interest active compounds. (Septembre-Malaterre, Remize, and Poucheret, 2018). A fermented system is composed by the microorganisms, the substrate, and the environmental conditions. Microorganisms used for fermentation are very diverse; they include bacteria, yeast and molds and various types of fermentation can occur. Lactic acid fermentation has been found as one of the most economical methods in maintaining and improving the nutritional properties of vegetables (Kwaw et al., 2018). Amongst the most widespread fermented vegetables, cabbage is in the first place. Although in some cases the microbial species driving vegetable fermentation are not specifically known, in many cases lactic acid bacteria are involved and the most common species involved are *Lactobacillus plantarum, Lb. brevis, Lb. rhamnosus, Lb. acidophilus, Leuconostoc mesenteroides, Lc. Citreum, Lc. fallax* (Septembre-Malaterre, Remize, and Poucheret, 2018).

6.7 METABOLISM OF PHYTOCHEMICALS BY LACTIC ACID FERMENTATION

The beneficial effect of different compounds presents in plants (such as polyphenols) depends partly on their microbial conversion. This conversion can happen during the processing of the food or in the intestine after ingestion (Filannino et al., 2015). Post-harvest processing especially fermentation, produces several potentially beneficial breakdown products (Palani et al., 2016). LAB has shown great ability to metabolize these compounds; however, this ability depends on the strains or bacterial species. *Lactobacillus brevis, Lactobacillus fermentum,* and *Lactobacillus plantarum* are some of the Lactobacilli, which metabolize phenolic acids trough decarboxylation and/or reduction activities.

Further metabolic pathways identified in lactobacilli are based on glycosyl hydrolases, which convert flavonoid glucosides to the corresponding aglycones, and esterase, degrading methyl gallate, tannins, or phenolic acid esters (Filannino et al., 2015). Most studies have focused on changes in antioxidant activity of fermented vegetables, but a high diversity has been observed. Nevertheless, a common conclusion of many works is that the selection of a starter culture contributes to increasing the antioxidant activity during fermentation. Rodriguez et al. (2009) found that *Lactobacillus plantarum* possesses enzymes leading to the production of antioxidants that can impact in the human diet (Rodríguez et al., 2009). Besides, lactic acid fermentation results in complete degradation of glucosinolates and increased contents of SFN, ascorbigen and I3C (Palani et al., 2016).

In addition to releasing metabolites with biological activity, the fermentation of vegetables with these bacteria allows the production of new flavors and aromas (Selma, Espin, and Tomas-Barberan, 2009; Rodrìguez et al., 2009). The research indicates that enzymes of certain strains of lactic acid bacteria may be characterized by activity similar to myrosinase and can degrade glucosinolates, but it can also be degraded by endogenic myrosinase since it remains active during the fermentation process as a result of the lack of thermal inactivation. Sosińska and Obiedziński (2011), found significant amounts of ascorbigen in broccoli and cauliflower fermentations in comparison with the ones that undergo thermal proses (Sosińska and Obiedziński, 2011).

6.8 CONCLUSIONS

Cruciferous vegetables contain bioactive molecules associated with several health benefits, including anti-cancer, anti-inflammatory and antioxidant activities. Glucosinolate and their hydrolysis products and antioxidants such as phenolic compounds can influence in the initiation and progression of several diseases. Food fermentation plays an important role not only in preserving fruits and, but also in bioconversion and metabolism of phytochemicals. Lactic acid fermentation can enhance the content of glucosinolates and help in the metabolism of phenolics, increasing its bioavailability.

ACKNOWLEDGMENTS

The first author expresses recognition to National Council of Science and Technology of Mexico (CONACyT) for the economic support provided during her postgraduate studies. Authors want to thank Dirección de Investigación and Posgrado (DIP) from UAdeC for the support granted to the project "Desarrollo de cultivos iniciadores con propiedades probióticas para la fermentación de crucíferas" with code CGEPI-UADEC-C01-2019-66.

KEYWORDS

- **chemopreventive properties**
- **cruciferous**
- **glucosinolates**
- **isothiocyanates**
- **lactic acid bacteria**

REFERENCES

Avato, P., & Argentieri, M. P., (2015). Brassicaceae: A rich source of health-improving phytochemicals.' *Phytochem., 14,* 1019–1033.

Baenas, N., et al., (2017). Broccoli and radish sprouts are safe and rich in bioactive phytochemicals. *Postharvest Biology and Technology* (Vol. 127, pp. 60–67). Elsevier B.V. doi: 10.1016/j.postharvbio.2017.01.010.

Bennett, R. N., et al., (1995). Glucosinolate biosynthesis (Further characterization of the aldoxime-forming microsomal monooxygenases in oilseed rape Leaves). *Plant Physiol, 109*(1), 299–305.

Cartea, M. E., et al., (2011). Phenolic compounds in *Brassica* vegetables. *Molecules, 16*(1), 251–280. doi: 10.3390/molecules16010251.

Cebeci, F., (2017). *The Metabolism of Plant Glucosinolates by Gut Bacteria*. Doctoral thesis, University of East Anglia.

Dunja, Š., et al., (2018). *Comparative Analysis of Phytochemicals and Activity of Endogenous Enzymes Associated with their Stability, Bioavailability and Food Quality in Five Brassicaceae sprouts, 269*, 96–102. doi: 10.1016/j.foodchem.2018.06.133.

Fahey, J. W., Zalcmann, A. T., & Talalay, P., (2001). The chemical diversity and distribution of glucosinolates and isothiocyanates among plants. *Phytochemistry, 56*(1), 5–51.

Filannino, P., et al., (2015). Metabolism of phenolic compounds by *Lactobacillus* spp. during fermentation of cherry juice and broccoli puree. *Food Microbiology*, *46*, 272–279. doi: 10.1016/j.fm.2014.08.018.

Francisco, M., et al., (2010). Cooking methods of *Brassica rapa* affect the preservation of glucosinolates, phenolics and vitamin C. *Food Research International* (Vol. 43, No. 5, pp. 1455–1463). Elsevier Ltd. doi: 10.1016/j.foodres.2010.04.024.

Herr, I., & Büchler, M. W., (2010). Dietary constituents of broccoli and other cruciferous vegetables : Implications for prevention and therapy of cancer. *Cancer Treatment Reviews* (Vol. 36, No. 5, pp. 377–383). Elsevier Ltd. doi: 10.1016/j.ctrv.2010.01.002.

Holst, B., & Williamson, G., (2004). A critical review of the bioavailability of glucosinolates and related compounds. *Nat. Prod. Rep., 21*(3), 425–447.

Kwaw, E., et al., (2018). Effect of lactobacillus strains on phenolic profile, color attributes and antioxidant activities of lactic-acid-fermented mulberry juice. *Food Chemistry* (Vol. 250, pp. 148–154). Elsevier. doi: 10.1016/j.foodchem.2018.01.009.

Lee, H., et al., (2017). Sinigrin inhibits production of inflammatory mediators by suppressing NF- κ B/MAPK pathways or NLRP3 inflammasome activation in macrophages. *International Immunopharmacology* (Vol. 45, pp. 163–173). Elsevier B.V. doi: 10.1016/j.intimp.2017.01.032.

Lee, I. S. L., Boyce, M. C., & Breadmore, M. C., (2011). A rapid quantitative determination of phenolic acids in *Brassica* oleracea by capillary zone electrophoresis. *Food Chemistry* (Vol. 127, No. 2, pp. 797–801). Elsevier Ltd. doi: 10.1016/j.foodchem.2011.01.015.

Maiyoh, G. K., et al., (2007). Cruciferous indole-3-carbinol inhibits apolipoprotein B secretion in HepG2 cells. *The Journal of Nutrition, 137*(10), 2185–2189. doi: 137/10/2185 [pii].

Manchali, S., Chidambara, K. N., & Patil, B. S., (2011). Crucial facts about health benefits of popular cruciferous vegetables. *Journal of Functional Foods* (Vol. 4, No. 1, pp. 94–106). Elsevier Ltd. doi: 10.1016/j.jff.2011.08.004.

Navarro, S. L., et al., (2014). Cruciferous vegetables have variable effects on biomarkers of systemic inflammation in a randomized controlled trial in healthy young adults. *The Journal of Nutrition, Nutritional Immunology* (pp. 1850–1857). Elsevier Ltd. doi: 10.1016/j.foodres.2017.07.049.

Oliviero, T., Verkerk, R., & Dekker, M., (2012). Effect of water content and temperature on glucosinolate degradation kinetics in broccoli (Brassica oleracea var. italica). *Food Chemistry* (Vol. 132, No. 4, pp. 2037–2045). Elsevier Ltd. doi: 10.1016/j.foodchem.2011.12.045.

Palani, K., et al., (2016). Influence of fermentation on glucosinolates and glucobrassicin degradation products in sauerkraut. *Food Chemistry* (pp. 755–762). Elsevier Ltd, 190. doi: 10.1016/j.foodchem.2015.06.012.

Paramithiotis, S., Hondrodimou, O. L., & Drosinos, E. H., (2010). Development of the microbial community during spontaneous cauliflower fermentation. *Food Research International* (Vol. 43, No. 4, pp. 1098–1103). Elsevier Ltd. doi: 10.1016/j.foodres.2010.01.023.

Peñas, E., et al., (2012). Influence of fermentation conditions of *Brassica oleracea* L. var. capitata on the volatile glucosinolate hydrolysis compounds of sauerkrauts. *LWT-Food Science and Technology* (Vol. 48, No. 1, pp. 16–23). Elsevier Ltd. doi: 10.1016/j.lwt.2012.03.005.

Radošević, K., et al., (2017). Assessment of glucosinolates, antioxidative and antiproliferative activity of broccoli and collard extracts. *Journal of Food Composition and Analysis, 61*, 59–66. doi: 10.1016/j.jfca.2017.02.001.

Rodríguez, H., et al., (2009). Food phenolics and lactic acid bacteria. *International Journal of Food Microbiology* (Vol. 132, Nos. 2–3, pp. 79–90). Elsevier B.V. doi: 10.1016/j.ijfoodmicro.2009.03.025.

Rodrìguez, H., et al., (2009). Food phenolics and lactic acid bacteria. *Int. J. Food Microbiol, 132*, 79–90.

Scalbert, A., et al., (2005). Dietary polyphenols and the prevention of diseases. *Critical Reviews in Food Science and Nutrition, 45*(4), 287–306.

Selma, M. V., Espin, J. C., & Tomas-Barberan, F. A., (2009). Interaction between phenolics and gut micro biota: Role in human health. *Food Chemistry, 57*, 6485–6501.

Septembre-malaterre, A., Remize, F., & Poucheret, P., (2018). Fruits and vegetables, as a source of nutritional compounds and phytochemicals : Changes in bioactive compounds during lactic fermentation. *Food Research International* (Vol. 104, pp. 86–99). Elsevier. doi: 10.1016/j.foodres.2017.09.031.

SIAP (2016). '*Atlas 2016,*' 236. doi: 10.1016/j.tsep.2018.05.003.

Sosińska, E., & Obiedziński, M. W., (2011). Effect of processing on the content of glucobrassicin and its degradation products in broccoli and cauliflower. *Food Control, 22*(8), 1348–1356. doi: 10.1016/j.foodcont.2011.02.011.

Verkerk, R., & Dekker, M., (2008). Bioactive compounds in foods. In: Gilbert, J., & Senyuva, H. Z., (eds.), *Glucosinolates* (pp. 31–51). Blackwell Publishing: Oxford, UK.

Wittstock, U., & Burow, M., (2010). Glucosinolate breakdown in Arabidopsis: Mechanism, regulation, and biological significance. *Arabidopsis Book, 8*, e0134.

Younho, S., et al., (2017). Strategy for dual production of bioethanol and d-psicose as value-added products from cruciferous vegetable residue. *Bioresource Technology, 223*, 34–39. Available at: https://doi.org/10.1016/j.biortech.2016.10.021 (accessed on 24 July 2020).

CHAPTER 7

Potential of Agro-Food Residues to Produce Enzymes for Animal Nutrition

ERIKA NAVA-REYNA,[1] ANNA ILYINA,[2] GEORGINA MICHELENA-ALVAREZ,[3] and JOSÉ LUIS MARTÍNEZ-HERNÁNDEZ[2]

[1]*National Centers for Disciplinary Investigation in Water, Plant, Soil, and Atmosphere Relationship (CENID RASPA), Instituto Nacional de Investigaciones Forestales, Agrícolas y Pecuarias, Canal Sacramento Km. 6.5, Zona Industrial 4ta Etapa, ZC 35140, Gomez Palacio, Durango, Mexico*

[2]*Academic Staff of Nanobiosciences, School of Chemistry, Universidad Autónoma de Coahuila, Venustiano Carranza Blvd. w/o. República, ZC 25280, Saltillo, Coahuila, Mexico*

[3]*Instituto Cubano de Investigación de los Derivados de la Caña de Azúcar, Vía Blanca 804 y Central Road, San Miguel del Padrón, La Habana, Cuba, E-mail: nava.erika@inifap.gob.mx*

ABSTRACT

The supplement of animal diets with hydrolytic enzymes could also improve the nutritional factors by removing antinutrients from diet or being applied as a predigestion process of animal feed to degrade materials such as cellulose and hemicellulose and increasing its digestibility and quality. Several agro-food wastes have been reported to be efficient as substrates to produce enzymes in the animal feed industry, which not only contribute to their management but also transformed them into value-added products. This chapter presents an overview of agro-food waste potential to produce enzymes useful as a supplement in animal feed, such as cereals straw, whey, fruit shells, and bagasse, among others. A description of several enzymes used in animal feed, their production by fermentation of residues, purification processes, and their perspectives are also presented.

7.1 ENZYMES IN ANIMAL NUTRITION

Growing of world food demand has generated an intensive development of livestock and poultry, as well as increased competition in this area, has led producers to get higher efficiency and lower costs using new production technologies like the supplementation of animal diets with several enzymes to improve the nutrients uptake, accelerating their growing and increasing the performance of their products (meat, eggs, milk, etc.) (Adeoye et al., 2016; Chamorro et al., 2015; Salem et al., 2015).

Enzymes such as proteases, amylases, lipases, cellulases, xylanases, phosphatases, pectinases, among others, have been applied in several diet formulation, especially for monogastric animals, which do not produce all digestive enzymes to uptake all needed nutrients (Amerah, 2015; Castillo and Gatlin, 2015). This nutrient deficit uptake is due mainly to the chemical structure of plant cells, limited time for enzyme activity in some parts of the intestine, and the presence of anti-nutrient compounds in some feed (McDonald et al., 2010).

At first experiments with complementary enzymes in animal feed, results were variables, mainly due to their low grade of purity. However, biotechnological progress has let to obtain the biggest quantity and purity at a lower cost. Therefore, the supplementation of animal diets with enzymes to promote nutritional value is becoming an increasingly common practice.

There is a wide variety of enzymes used as feed additives depending on the composition of the diet (Table 7.1). For example, cereal-based diets contain soluble non-starch polysaccharides (NSP) in high levels, which cause an increase in digest viscosity in monogastric animals that affects the action of endogenous digestive enzymes on their substrates and influences the animal growth and productivity (Bedford and Classen, 1992; Choct and Annison, 1992; Yin et al., 2018). Thus, animal feed supplementation with several carbohydrases enzymes that hydrolyze NSP seems to be the best way to improve nutrient digestibility of plant-based feedstuffs, mainly in monogastric animals like poultry, pigs, or fishes. These kinds of enzymes include cellulases, xylanase, glucanase, α-amylase, β-mannanase, α-galactosidase, and pectinase, which hydrolyze carbohydrate polymers to oligo or polysaccharides with lower molecular weight (Adeola and Cowieson, 2011).

Another objective of animal feed supplementation is the increasing of metabolizable energy (ME) to improve animal performance, especially in poultry (McCracken and Quintin, 2000; Pirgozliev et al., 2015). Consequently, fats and oils are commonly added to animal diets, but their digestibility is influenced by factors like physicochemical characteristics of their constituent fatty acids, the source, and type of lipids, diet based material and

TABLE 7.1 Exogenous Feed Enzymes Used To Solve Main Digestive Problems in Animal Nutrition

Diet-based material	Digestive problem	Enzyme	Animal tested	Reference
Barley	1,3–1,4-β-glucans	1,3–1,1-β-Gucanases	Broilers chicks	(Fernandes et al., 2016)
Wheat	Arabinoxylans	Xylanase	Broiler chicks	(Zhang et al., 2014)
	Pentosans			
Corn, soybean meal and corn gluten meal	Low-energy diet	Lipase	Broiler chicks	(Hu et al., 2018)
Corn gluten meal and soybean meal	Starch	Amylopectase	Broiler chicks	(Yin et al., 2018)
	Xylans	Glucoamylase		
	Cellulose	Protease		
Kikuyu	Tannins	Protease	Mozambique tilapia	(Hlophe-Ginindza et al., 2016)
	Saponins	A-amylase		
	Phytate	Cellulase		
	Cellulose	Xylanase		
	Fiber	Phytase		
Cocoa pod husk	Phytate	Phytase	Hy-line silver brown laying hens	(Nortey et al., 2017)
	Lignin	Amylase		
	Hemicelluloses	Protease		
	Cellulose	Cellulase		
	Pectin	Xylanase		
		B-glucanase		
		Pectinase		

age of animals (Classen, 2017; Hu et al., 2018; Smink et al., 2010). Dietary fat is hydrolyzed in the small intestine by pancreatic lipases. Nevertheless, sometimes lipolytic enzymes secreted from the pancreas are insufficient, so the use of exogenous lipases could be an alternative way to offset that deficiency (Hedemann and Jensen, 2004; Zhang et al., 2018).

Proteins are also important nutrients in animal feed. Supplementation of animal diets with protease is the focus on the reduction of feed cost by the decrease in crude protein concentration and substitution of external protein sources in diet formulation with cheaper ones (Cowieson et al., 2017; Mahmood et al., 2017). Proteases such as pepsin, pancreatin, and papain can reduce feed conversion ratio (FCR) in pigs and broiler chicks (Freitas et al., 2011; Lewis et al., 1955). Proteases can also increase weight gain and apparent lean digestibility of energy when it is combined with other enzymes as xylanase (Barekatain et al., 2013; O'Shea et al., 2014).

Otherwise, anti-nutrient composition in animal feed is one of the main problems, which affects nutritional quality. These compounds include phytate, tannins, polyphenols, and oxalates that reduce protein digestibility and nutrient absorption. Phytic acid and their salts (phytates) are the main reserves of phosphorus in cereals, leguminous, seeds, dry fruits, etc., corresponding to more than 50% of total phosphorus in vegetables. However, there are not assailable by monogastric animals, which do not produce the hydrolytic enzymes necessary to process them. Phytic acid is the major anti-nutrient in monogastric animals feed, due to their chelating action during intestinal transit, binding minerals from the diet, and forming insoluble complexes that hinder the absorption of these minerals (Urbano et al., 2000). Phytates have also the ability to bind proteins in the diet and negatively affect their digestibility. This occurs with endogenous proteins involved in digestion too, such as the digestive enzymes trypsin and chymotrypsin, α-amylases, tyrosinases, and peptidases (Biehl and Baker, 1997). Therefore, the industry has been interested in isolation and improvement of phytases that can be applied as an animal diet supplement, especially from microbial origin. Phytases (Myo-inositol hexakisphosphate-phosphohydrolases; EC 3.1.3.8 and 3.1.3.26) are a subfamily of histidine acid phosphatases that catalyzes the phosphate monoester hydrolysis of phytic acid or phytates in inositol phosphate, myoinositol and inorganic phosphorus, which have the lower chelating capacity (Wyss et al., 1999). These enzymes are the most commonly used exogenous enzyme in the feed for monogastric animals, and their effect is influenced by factors like species and age of animals, Ca:Pratio and inorganic P content in the diet, the diet based material, particle size of limestone, and others (Amerah, 2015; Leske and Coon, 1999; Leytem et al., 2008; Manangi and Coon, 2007).

On the other hand, condensed tannins, a group of soluble phenolics, can form complexes with proteins, carbohydrates, minerals, and even with other plant secondary compounds (PSC) such as saponins, terpenoids, and alkaloids (MacAdam and Villalba, 2015), which causes their anti-nutritional effects. Some studies report that high-tannin inclusion decreases the activity of all digestive enzymes (Przywitowski et al., 2017; Singh, 1984). Besides, some condensed tannins may also injure the gut mucosa and organs, as well as interfere in physiological processes (Karimi et al., 2014; Mahgoub et al., 2008). These toxic compounds not only affect monogastric animals but also can disturb ruminants (Abd El Taw and Khattab, 2018). Tannases have a positive effect on animal growth, due to their ability to degrade part of the dietary tannins by hydrolysis of tannic acid to gallic acid and glucose (Khattab et al., 2017). This enzyme also hydrolyzes other compounds like methyl gallate, digallic acid, propyl gallate, ellagitannins, epicatechin gallate, and epigallocatechin gallate (Curiel et al., 2009; Lu and Chen, 2007). Furthermore, studies have demonstrated that supplementation of poultry diets with tannase improves feed efficiency, ME, and nutrient uptake (Abdulla et al., 2016, 2017).

Most of the commercial enzymes used in livestock are products from microbial fermentation. Once fermentation is over, enzymes are purified. The type and enzymatic activity vary depending on the substrate, microorganism used and culture conditions (Qasim et al., 2017; Reddy et al., 2015; Salgado et al., 2016; Shajitha et al., 2018).

Filamentous fungi are the main source of exogenous enzymes due to their stability. *Trichoderma longibrachiatum, Aspergillus niger, Aspergillus oryzae, Trichoderma viride*, and *Humicolainsolens*are some of the most commonly applied fungi. Bacteria are also used to produce enzymes for animal feed, such mainly from the genus *Bacillus* spp.

7.2 AGRO-FOOD RESIDUES TO PRODUCE ENZYMES

Agriculture, forest and food industries produce high quantities of residues worldwide (around 1.3 billion tons/ year), which could be a problem to manage them (Food and Agricultural Organization, 2013; Okello et al., 2013). The largest constituent of all plant materials is cellulose, representing 30 to 50% of its tissues, therefore, the agro-food waste contains mainly lignocellulose, formed by lignin, hemicellulose, and cellulose. These polymers have great nutritional potential and high biotechnological value, given by their chemical properties (Howard et al., 2003). Therefore, in recent years the awareness to reduce the amount of these residues have increased, thus the looking for alternative uses is a biotechnological priority due to their economic and environmental advantage. Fermentation of sugars into the residues by the adequate microorganisms lets to obtain high value and useful metabolites (Hodge et al., 2009). Agro-food residues such as sugarcane bagasse (Macrelli et al., 2012), pulp and citrus peel (Gómez et al., 2016), the coffee pulp (Machado et al., 2012), banana peel (Barman et al., 2015), husks from the cereal industry (da Silva Delabona et al., 2012), vinasses (Salgado et al., 2016), and so on, have been used to produce different bioproducts.

Interest in the study and commercialize microbial bioproducts such as enzymes, fuels, chemicals, organic acids, amino acids, polymers, etc., from renewable biomass-based feedstock, has been growing in last years, mainly by the limitation of fossil resources, climate change problems, sustainability and the growing consumer preference for natural and environmentally friendly products (Mondala, 2015). Moreover, using agro-food residues as raw materials permits not only the reduction of land disposal but also the production of marketable products.

Solid-state fermentation (SSF) consist of microbial growth on the surface and inside of a solid porous matrix, in the absence of almost absence of free water; however, the substrate must have enough moisture to support the growth and metabolism of the microorganism (Pandey, 2003). This process stimulates the natural growing of microorganisms in humid solids ant culture and SSF conditions favor the development of filamentous fungi (Hölker and Lenz, 2005). This process is widely used to process agro-food residues, using them as substrates for the production of industrial interest bioproducts.

Effectiveness of the SSF to process agro-food waste is largely due to the fact that the procedure has low energy requirements, produces less wastewater, and is environment-friendly when solving the problem of waste disposal (Pandey, 2003). Also, biotechnology advances have made that these processes produce high concentrations of the desired metabolite, product stability and the adaptability of microorganisms in the procedure, especially fungal systems with low free water content (Chen, 2013; Howard et al., 2003; Pandey, 2003).

Enzymes are one of the highest values bioproducts obtained from the fermentation process of agro-food residues. Submerged fermentation (SmF) has also useful to obtain enzymes, especially to obtain bacterial proteins due to the high requirements of water potential (Subramaniyam and Vimala, 2012). Nonetheless, most of these bioproducts are from the fungal source, since they are commonly extracellular metabolites, while bacterial ones are typically intracellular and therefore, they are more economically profitable (Subramaniyam and Vimala, 2012), and they are mostly obtained by SSF.

Industrial enzymes market was valued in 2016 for 5 billion dollars and is projected to reach 6.2 billion dollars by 2020, with a compound annual rate of 7% from 2015–2020 (Global Market Insights, 2017). Enzymes are regularly used in multiple areas such as the food, animal feed, tanning, textiles, laundry, pharmaceutical, cosmetics, and fine chemical industries. These industrial applications mean more than 80% of the global enzyme market (van Oort, 2009). In the market segmentation corresponding to animal feed enzymes market, it is estimated to grow with a CAGR of 7.8% to 2026, especially by the demand for ingredients that improve the livestock quality and keep the metabolism rate intact (Transparency Market Research, 2017). Thus, more efficient and lower cost fermentation processes are needed to respond to the growing demand of these products, so the use of agro-food residues as raw materials is an opportunity to decrease costs and also the subsequent use of the fermented substrate for animal feed.

7.2.1 PHYTASES

Attempts have been made to get phytases from cereals and vegetables (Esmaeilipour et al. 2012; Guo et al. 2015). Nevertheless, the most common source of these enzymes is microorganisms, because of their higher physicochemical stability. This has driven to the development of several biotechnological processes to produce phytases, mainly from fungi, due to their optimal pH is between 5.5 and 2.5, pH values similar to those in the digestive tract of monogastric animals, in contrast with bacterial phytases which have an optimal pH close to neutral (Tran et al., 2011). Purification processes of fungal phytases are also more economically feasible than bacterial ones(Subramaniyam and Vimala, 2012). Over 200 fungi have been reported as potential producers of phytases, mainly strains of genera *Mucor, Penicillium, Rhizopus,* and *Aspergillus. Aspergillus niger* has been reported as the best phytases producers (Costa et al., 2009; Gupta et al., 2014; Liu et al., 2010).

Several studies have been made to produce fungal phytases by solid-state fermentation (SSF) of different agro-food residues as wheat bran, rice flour, canola meal, and cranberry pomace (Costa et al., 2009; Gupta et al., 2014). Phytase production is influenced by glucose and dextrin content, humidity, and sulfate concentration (Buddhiwant et al., 2015).

Agro-food by-products and oil cakes are low-cost energy-rich substrates that can be applied for enzyme production. Groundnut oil cake, for example, has been tested as a substrate for animal feed enzymes production by *Aspergillus niger*, due to its low cost and worldwide availability. Moreover, the subproduct of the SSF (koji) could be used directly as a supplement for animal nutrition without any exogenous enzyme (Buddhiwant et al., 2015).

Gunashree and Venkateswaran (2008) compared the effect of media ingredients on phytase production in SSF and SmF. They found that sucrose increases enzyme production in both systems, while peptone is a favorable nitrogen source for SmF. Moreover, they report that none of the surfactants nor the metal salts stimulate phytase production by SSF, but they increase the production in SmF.

Phytases used for animal feed should have activity through the digestive tract, so they have to be pH stables, and also resistant to high temperature during the feed processing and storage. Therefore, there is a big interest to produce phytases with major stability to temperature by fermentation with thermos-tolerant microorganisms, such as *A. flavus* and *R. microspores* which presented an optimal temperature of 55 and 45°C, respectively, when they are growing in the de-oiled mustard cake (*Brassica juncea*) and sugarcane bagasse (Gaind and Singh, 2015; Sato et al., 2016).

Other mechanisms to reduce pH and temperature sensibility of most of the native microbial phytases, increase productivity, and reduce manufacturing costs, is the exploitation of genomic techniques. Ranjan and Satyanarayana (2016) produced recombinant phytase from the codon-optimized phytase gene of the thermophilic mold *Sporotricum thermophile* expressed in *Pichia pastoris.* They reported an enzyme activity over 40 fold higher than the native fungal strain. Furthermore, the enzyme demonstrated high efficiency in dephytinization of broiler feeds in *in vitro* experiments. Elseways, Singh et al. (2018) inserted a thermostable and protease-resistant HAP-phytase gene of *Sporotrichum thermophile* in *Pichia pastoris* X-33. The resulting protein showed wide substrate specificity and similar biochemical parameters to the native enzyme. The recombinant phytase also exhibited peroxidase activity when metavanadate was added, inhibiting phytase activity.

7.2.2 CARBOHYDROTASES

Carbohydratases production by fermentation of agro-food residues let to the obtaining of a complex of enzymes that include cellulases, xylanases, amylases, pectinases, among others (Awasthi et al., 2018).

NSP in animal feed represents from 8.3 to 9.8% in cereals (Slominski et al., 2000). The main NSP in wheat and corn, the most common diet-base materials, are arabinoxylans, thus xylanases have been widely used in cereal diets, especially for poultry (Amorim et al., 2017). Amorim et al. (2017) used SSF of the cocoa meal by *A. awamori* to remove undesirable compounds such as theobromine and caffeine to increase the use and value of the residue and produce xylanase with thermal stability. Winery and olive oil residues can also be applied as substrates for carbohydrases production. Salgado et al. (2015) reported the use of Winery and mixtures of olive mill waste as a substrate for cellulose and xylanases production by SSF with *A. uvarum* MUM 08.01. They also optimize the production process in respect of substrate composition, aeration, temperature and initial moisture level in a packed-bed bioreactor. Palm kernel cake and palm pressed fiber also can be used as substrates for xylanase and lipase production by SSF of *Aspergillus* strains, at the time that the fermentation residue can be used as an ingredient in animal feed (Oliveira et al., 2018). Other studies are focused on produce more resistant xylanases to environmental conditions. In this line, Sydenham et al. (2013) cloned in *A. niger*strain CBS 513.88 four novel xylanases: *GtXyn10A* and GtXyn10B from *Gloeophyllum trabeum*, OpXyn11A from *Ophiostoma piliferum*, and *CcXyn11C* from *Coprinopsis cinérea*. They applied bioinformatics analysis for enzyme selection and 3D modeling of the obtained proteins. All xylanases had a broad spectrum of pH (from pH 3.4 to 6.5) and optima temperature from 40 to 70°C, necessary to resist animal digestive tract and feed pretreatments; consequently, they may be applied in animal feed.

Cellulose is the most important compound in the plant cell wall. Then, plant biomass contains significant amounts of this polymer and could be applied as raw material for cellulases obtaining. These enzymes have been produced using several agro-food wastes as a substrate in SSF. For example, potato starchy wastes, olive pomace, sugar beet pulp, sugarcane molasses and fodder yeast, olive cake, cornflour, and wheat bran (Fang and Xia, 2017; Kahil and Hassan, 2015; Leite et al., 2016; Xia et al., 2017). Diversity and proportion of produced enzymes are influenced by media and incubation time (Tirado-González et al., 2016). Additionally, cellulases production by agro-food residues fermentation favors filamentous fungi, which morphology let them penetrate the support surface because of the presence of turgid pressure at the tip of their mycelium (Kahil and Hassan, 2015). According to with Bhargav et al., (2008), factors such as moisture, particle size, pH, incubation temperature and time, medium enrichment with carbon or nitrogen, and inoculum size, are important in the optimization of the fermentation procedure. Recombinant DNA technology allows the identification of novel cellulose genes, increasing the enzyme production, and improving their activity. *A. nidulans, A. niger, A. oryzae,* and *T. reesei*, are the most common fungal strains used for the production of recombinant cellulases (Kaur et al., 2014; Lima et al., 2016; Zhao et al., 2018). The improvement of their expression is reached by four mechanisms: gene fusion strategy, over-expression of chaperones, screening for multi-copy strains and the use of protease deficient host strains (Juturu and Wu, 2014; Sharma et al., 2009; Ward, 2012).

On the other hand, endo-β-(1,4)-mannanase (3.2.1.78) hydrolyzes glucomannan, which could stimulate gastrointestinal beneficial bacteria to improve the host's health (Albrecht et al., 2011). Corn cob, palm kernel cake, soluble coffee wastes, apple pomace, copra paste, and wheat bran, have been applied for mannanase production by SSF with filamentous fungi (Abdeshahian et al., 2010; Wu et al., 2011; Zhang and Sang, 2015). In spite of attempts to look for novel β-mannanases, increase their activity by optimizing fermentation conditions or creating genetically modified strains (Regalado et al., 2000; Regmi et al., 2016), the application of this enzyme is still limited by its high cost and low production (Albrecht et al., 2011). Genetic engineering can be applied to resolve these restrictions. For instance, the genome mining of microbial sequences has been applied to identify novel β-mannanases from several organisms (Kumar and Wyman, 2009). To cite an instance, Tang et al. (2016) identified two glycosyl hydrolase family 5 β-mannanases (AoMan5A and AoMan5B) from *A. oryzae*(CBM-1) and clone them in *P. pastoris* GS115. Produced enzymes showed high stability under simulative gastric fluid and prilling process, in addition to their ability to degrade the pretreated konjac flour and produce prebiotics.

Protein engineering has been also applied to obtain more capable β-glucanases, enzymes that degrade β-glucans. These linear polymers of β-D-glycosyl residues are the major constituent of endosperm cell walls in cereals and the other main type of NSP besides arabinoxylans (Limberger-Bayer et al., 2014; Zielke et al., 2017). Typically, endoglucanases are produced in different agro-food residues, including wheat bran, soy bran, corn cob, corn straw, rice peel, or sugarcane bagasse (Garcia et al., 2015; Ong and Straw, 2015). Due to the fact that most of these enzymes present instability during high-temperature processes, several attempts have been made to improve their thermostability, since the looking for new genetic resources of thermophiles to genetic engineering (Mao et al., 2016; Niu et al., 2018; Wang et al., 2015), but these methods often decrease enzyme activity. According to You et al. (2016), optimization of residual charge-charge interactions could be a suitable technique for thermostability improvement of β-glucanase using the enzyme redesign algorithm Enzyme Thermal Stability System (ETSS). They mined an endo-1,3–1,4-β-glucanase from *Talaromyces leycettanus* JCM12802 and cloned the gene in *P. pastoris.* The protein showed a wide pH range, resistance to most metal ions and chemical reagents, a great enzymatic activity.

7.3 FORECAST ENZYMES PRODUCTION WITH AGRO-FOOD RESIDUES FOR ANIMAL NUTRITION

Animal feed enzymes market will continue growing in the next years. Moreover, it is estimated to exceed USD 2 billion by 2024 with a CAGR over 7.4% during 2017–2022. This increase in the market will be driven mainly by growing usage of feed enzymes to reach better feed conversion rate that permits increase productivity and quality of animal products. Poultry will continue as the leading segment in the market. Phytase will also be the major animal feed enzyme, and its market will grow around 6.5% up to 2024 (Global Market Insights, 2017).

Due to the expected market growing, more efficient enzymes and conditions to getting bigger production yields will be a biotechnological priority. Some desire characteristics for enzymes used in animal nutrition are unable to reach by natural resources, while others have great characteristics for commercial use but with insufficient yields. Therefore, modern molecular tools and genetic engineering are key options to discovery novel enzymes genes or create genetically modified microorganisms to get higher production yields and more efficient catalytic proteins (Haitjema et al., 2017; Ranjan and Satyanarayana, 2016; Ushasree et al., 2017; Youssef et al., 2013). The genome mining of genomic and metagenomics DNA databases represents a suitable, faster, simpler and more economically efficient tool to enzyme screening than classical methods (Tang et al., 2016; Ushasree et al., 2017; Wang et al., 2015; You et al., 2016).

cDNA cloning is another potential method for isolation of several eukaryotic enzymes (Ranjan and Satyanarayana, 2016), that can be used with known genes or novel enzyme codifying regions found by data mining (Albrecht et al., 2011; Singh et al., 2018; Wang et al., 2015).

KEYWORDS

- **agro-food residues**
- **animal feed**
- **fermentation**
- **hydrolytic enzymes**
- **solid-state fermentation**
- **submerged fermentation**

REFERENCES

Abd, E. T. A. M., & Khattab, M. S. A., (2018). Utilization of polyethylene glycol and tannase enzyme to reduce the negative effect of tannins on digestibility, milk production, and animal performance. *Asian J. Anim. Vet. Adv., 13*, 201–209.

Abdeshahian, P., Samat, N., Hamid, A. A., & Yusoff, W. M. W., (2010). Utilization of palm kernel cake for production of β-mannanase by *Aspergillus niger* FTCC 5003 in solid substrate fermentation using an aerated column bioreactor. *J. Ind. Microbiol. Biotechnol., 37*, 103–109.

Abdulla, J., Rose, S. P., Mackenzie, A. M., Mirza, W., & Pirgozliev, V., (2016). Exogenous tannase improves feeding value of a diet containing field beans (*Viciafaba*) when fed to broilers. *Br. Poult. Sci., 57*, 246–250.

Abdulla, J. M., Rose, S. P., Mackenzie, A. M., & Pirgozliev, V. R., (2017). Feeding value of field beans (*Viciafaba* L. var. minor) with and without enzyme containing tannase, pectinase and xylanase activities for broilers. *Arch. Anim. Nutr. 71*, 150–164.

Adeola, O., & Cowieson, A. J., (2011). Opportunities and challenges in using exogenous enzymes to improve nonruminant animal production. *J. Anim. Sci., 89*, 3189–3218.

Adeoye, A. A., Jaramillo-Torres, A., Fox, S. W., Merrifield, D. L., & Davies, S. J., (2016). Supplementation of formulated diets for tilapia (*Oreochromisniloticus*) with selected exogenous enzymes: Overall performance and effects on intestinal histology and micro biota. *Anim. Feed Sci. Technol., 215*, 133–143.

Albrecht, S., Van, M. G. C. J., Xu, J., Schols, H. A., Voragen, A. G. J., & Gruppen, H., (2011). Enzymatic production and characterization of konjac glucomannan oligosaccharides. *J. Agric. Food Chem., 59*, 12658–12666.

Amerah, A. M., (2015). Interactions between wheat characteristics and feed enzyme supplementation in broiler diets. *Anim. Feed Sci. Technol., 199*, 1–9.

Amorim, G. M., Oliveira, A. C., Gutarra, M. L. E., Godoy, M. G., & Freire, D. M. G., (2017). Solid-state fermentation as a tool for methylxanthine reduction and simultaneous xylanase production in cocoa meal. *Biocatal. Agric. Biotechnol., 11*, 34–41.

Awasthi, M. K., Wong, J. W. C., Kumar, S., Awasthi, S. K., Wang, Q., Wang, M., Ren, X., et al., (2018). Biodegradation of food waste using microbial cultures producing thermo stable A-amylase and cellulase under different pH and temperature. *Bioresour. Technol., 248*, 160–170.

Barekatain, M. R., Antipatis, C., Choct, M., & Iji, P. A., (2013). Interaction between protease and xylanase in broiler chicken diets containing sorghum distillers' dried grains with solubles. *Anim. Feed Sci. Technol., 182*, 71–81.

Barman, S., Sit, N., Badwaik, L. S., & Deka, S. C., (2015). Pectinase production by *Aspergillus niger* using banana (*Musa balbisiana*) peel as substrate and its effect on clarification of banana juice. *J. Food Sci. Technol., 52*, 3579–3589.

Bedford, M. R., & Classen, H. L., (1992). Reduction of intestinal viscosity through manipulation of dietary rye and pentosanase concentration is effected through changes in the carbohydrate composition of the intestinal aqueous phase and results in improved growth rate and food conversion efficiency. *J. Nutr., 122*, 560–569.

Bhargav, S., Panda, B. P., Ali, M., & Javed, S., (2008). Solid-state fermentation: An overview. *Chem. Biochem. Eng. Q, 22*, 49–70.

Biehl, R. R., & Baker, D. H., (1997). Microbial phytase improves amino acid utilization in young chicks fed diets based on soybean meal but not diets based on peanut meal. *Poult. Sci., 76*, 355–360.

Buddhiwant, P., Bhavsar, K., Ravi, K. V., & Khire, J. M., (2015). Phytase production by solid-state fermentation of groundnut oil cake by *Aspergillus niger*: A bioprocess optimization study for animal feedstock applications. *Prep. Biochem. Biotechnol., 46*, 531–538.

Castillo, S., & Gatlin, D. M., (2015). Dietary supplementation of exogenous carbohydrase enzymes in fish nutrition: A review. *Aquaculture., 435*, 286–292.

Chamorro, S., Viveros, A., Rebolé, A., Rica, B. D., Arija, I., & Brenes, A., (2015). Influence of dietary enzyme addition on polyphenol utilization and meat lipid oxidation of chicks fed grape pomace. *Food Res. Int., 73*, 197–203.

Chen, H., (2013). Modern solid-state fermentation. In: Chen, H., (ed.), *Modern Solid State Fermentation* (pp. 243–305). Springer: Dordrecht.

Choct, M., & Annison, G., (1992). Anti-nutritive effect of wheat pentosans in broiler chickens: Roles of viscosity and gut microflora. *Br. Poult. Sci., 33*, 821–834.

Classen, H. L., (2017). Diet energy and feed intake in chickens. *Anim. Feed Sci. Technol., 233*, 13–21.

Costa, M., Lerchundi, G., Villarroel, F., Torres, M., & Schöbitz, R., (2009). Phytase production by *Aspergillus ficuum* in submerged and solid state fermentation using agroindustrial waste as support. *Rev. Colomb. Biotecnol, 11*, 73–83.

Cowieson, A. J., Lu, H., Ajuwon, K. M., Knap, I., & Adeola, O., (2017). Interactive effects of dietary protein source and exogenous protease on growth performance, immune competence and jejunal health of broiler chickens. *Anim. Prod. Sci., 57*, 252–261.

Curiel, J. A., Rodríguez, H., Acebrón, I., Mancheño, J. M., De Las, R. B., & Muñoz, R., (2009). Production and physicochemical properties of recombinant *Lactobacillus plantarum* Tannase. *J. Agric. Food Chem., 57*, 6224–6230.

Da Silva, D. P., Buzon, P. R. D. P., Codima, C. A., Tremacoldi, C. R., Rodrigues, A., & Sanchez, F. C., (2012). Using Amazon forest fungi and agricultural residues as a strategy to produce cellulolytic enzymes. *Biomass Bioenergy, 37*, 243–250.

Fang, H., & Xia, L., (2015). Cellulase production by recombinant *Trichoderma reesei* and its application in enzymatic hydrolysis of agricultural residues. *Fuel, 143*, 211–2016.

Fernandes, V. O., Costa, M., Ribeiro, T., Serrano, L., Cardoso, V., Santos, H., Lordelo, M., Ferreira, L. M. A., & Fontes, C. M. G. A., (2016). 1, 3–1, 4-β-Glucanases and not 1, 4-β-glucanases improve the nutritive value of barley-based diets for broilers. *Anim. Feed. Sci. Technol., 211*, 153–163.

Food and Agricultural Organization, (2013). *Food Waste Harms Climate, Water, Land and Biodiversity-New FAO Report*. http://www.fao.org/news/story/en/item/196220/icode/ (accessed on 24 July 2020).

Freitas, D. M., Vieira, S. L., Angel, C. R., Favero, A., & Maiorka, A., (2011). Performance and nutrient utilization of broilers fed diets supplemented with a novel mono-component protease. *J. Appl. Poult. Res., 20*, 322–334.

Gaind, S., & Singh, S., (2015). Production, purification and characterization of neutral phytase from thermo tolerant *Aspergillus flavus* ITCC 6720. *Int. Biodeterior. Biodegradation*, 99, 15–22.

Garcia, N. F. L., Da Silva, S. F. R., Gonçalves, F. A., Da Paz, M. F., Fonseca, G. G., & Leite, R. S. R., (2015). Production of β-glucosidase on solid-state fermentation by *Lichtheimia ramosa* in agro-industrial residues: Characterization and catalytic properties of the enzymatic extract. *Electron. J. Biotechnol., 18*, 314–319.

Global Market Insights, (2017). *Enzymes Market Size-Industry SHARE, Analysis Report 2024*. https://www.gminsights.com/industry-analysis/enzymes-market (accessed on 24 July 2020).

Gómez, B., Yáñez, R., Parajó, J. C., & Alonso, J. L., (2016). Production of pectin-derived oligosaccharides from lemon peels by extraction, enzymatic hydrolysis and membrane filtration. *J. Chem. Technol. Biotechnol., 91*, 234–247.

Gunashree, B. S., & Venkateswaran, G., (2008). Effect of different cultural conditions for phytase production by *Aspergillus niger* CFR 335 in submerged and solid-state fermentations. *J. Ind. Microbiol. Biotechnol., 35*, 1587–1596.

Gupta, R. K., Gangoliya, S. S., & Singh, N. K., (2014). Isolation of thermo tolerant phytase producing fungi and optimization of phytase production by *Aspergillus niger* NRF9 in solid state fermentation using response surface methodology. *Biotechnol. Bioprocess Eng. 19*, 996–1004.

Haitjema, C. H., Gilmore, S. P., Henske, J. K., Solomon, K. V., De Groot, R., Kuo, A., Mondo, S. J., et al., (2017). A parts list for fungal cellulosomes revealed by comparative genomics. *Nat. Microbiol., 2*, 17087.

Hedemann, M., & Jensen, B., (2004). Variations in enzyme activity in stomach and pancreatic tissue and digesta in piglets around weaning. *Arch. Anim. Nutr., 58*, 47–59.

Hlophe-Ginindza, S. N., Moyo, N. A. G., Ngambi, J. W., & Ncube, I., (2016). The effect of exogenous enzyme supplementation on growth performance and digestive enzyme activities in *Oreochromis mossambicus* fed Kikuyu-based diets. *Aquac. Res., 47*, 3777–3787.

Hodge, D. B., Andersson, C., Berglund, K. A., & Rova, U., (2009). Detoxification requirements for bioconversion of softwood dilute acid hydrolyzates to succinic acid. *Enzyme Microb. Technol., 44*, 309–316.

Hölker, U., & Lenz, J., (2005). Solid-state fermentation-are there any biotechnological advantages? *Curr. Opin. Microbiol., 8*, 301–306.

Howard, R. L., Abotsi, E., Van, J. R. E. L., & Howard, S., (2003). Lignocellulose biotechnology: Issues of bioconversion and enzyme production. *African J. Biotechnol. 2*, 602–619.

Hu, Y. D., Lan, D., Zhu, Y., Pang, H. Z., Mu, X. P., & Hu, X. F., (2018). Effect of diets with different energy and lipase levels on performance, digestibility, and carcass trait in broilers. *Asian-Australasian J. Anim. Sci., 31*, 1275–1284.

Juturu, V., & Wu, J. C., (2014). Microbial cellulases: Engineering, production, and applications. *Renew. Sustain. Energy Rev., 33*, 188–203.

Kahil, T., & Hassan, H. M., (2015). Economic co-production of cellulase and α-amylase by fungi grown on agro-industrial wastes using solid-state fermentation conditions. *Middle-East J. Appl. Sci., 195*, 184–195.

Karimi, H., Kia, H. D., & Hosseinkhani, A., (2014). Histological effects of different levels of sorghum grain on the liver and kidney of Ghezel×Arkhar-merino crossbred lambs. *Anim. Vet. Sci., 2*, 130–134.

Kaur, B., Oberoi, H. S., & Chadha, B. S., (2014). Enhanced cellulase producing mutants developed from heterokaryotic *Aspergillus* strain. *Bioresour. Technol., 156*, 100–107.

Khattab, M. S. A., El Tawab, A. M. A., & Fouad, M. T., (2017). Isolation and characterization of anaerobic bacteria from frozen rumen liquid and its potential characterizations. *Int. J. Dairy Sci., 12*, 47–51.

Kumar, R., & Wyman, C. E., (2009). Effects of cellulase and xylanase enzymes on the deconstruction of solids from pretreatment of poplar by leading technologies. *Biotechnol. Prog., 25*, 302–314.

Leite, P., Salgado, J. M., Venâncio, A., Domínguez, J. M., & Belo, I., (2016). Ultrasounds pretreatment of olive pomace to improve xylanase and cellulase production by solid-state fermentation. *Bioresour. Technol., 214*, 737–746.

Leske, K., & Coon, C., (1999). A bioassay to determine the effect of phytase on phytate phosphorus hydrolysis and total phosphorus retention of feed ingredients as determined with broilers and laying hens. *Poult. Sci., 78*, 1151–1157.

Lewis, C. J., Catron, D. V., Liu, C. H., Speer, V. C., & Ashton, G. C., (1955). Swine nutrition, enzyme supplementation of baby pig diets. *J. Agric. Food Chem., 3*, 1047–1050.

Leytem, A. B., Willing, B. P., & Thacker, P. A., (2008). Phytate utilization and phosphorus excretion by broiler chickens fed diets containing cereal grains varying in phytate and phytase content. *Anim. Feed Sci. Technol.. 146*, 160–168.

Lima, M. S., Damasio, A. R. D. L., Crnkovic, P. M., Pinto, M. R., Da Silva, A. M., Da Silva, J. C. R., Segato, F., De Lucas, R. C., Jorge, J. A., & Polizeli, M. D. L. T. D. M., (2016). Co-cultivation of *Aspergillus nidulans* recombinant strains produces an enzymatic cocktail as alternative to alkaline sugarcane bagasse pretreatment. *Front. Microbiol., 7*, 583.

Limberger-Bayer, V. M., De Francisco, A., Chan, A., Oro, T., Ogliari, P. J., & Barreto, P. L. M., (2014). Barley β-glucans extraction and partial characterization. *Food Chem., 154*, 84–89.

Liu, N., Ru, Y., Wang, J., & Xu, T., (2010). Effect of dietary sodium phytate and microbial phytase on the lipase activity and lipid metabolism of broiler chickens. *Br. J. Nutr., 103*, 862–868.

Lu, M. J., & Chen, C., (2007). Enzymatic tannase treatment of green tea increases *in vitro* inhibitory activity against N-nitrosation of dimethylamine. *Process Biochem., 42*, 1285–1290.

MacAdam, J., & Villalba, J., (2015). Beneficial effects of temperate forage legumes that contain condensed tannins. *Agriculture, 5*, 475–491.

Machado, E. M. S., Rodriguez-Jasso, R. M., Teixeira, J. A., & Mussatto, S. I., (2012). Growth of fungal strains on coffee industry residues with removal of polyphenolic compounds. *Biochem. Eng. J., 60*, 87–90. https://doi.org/10.1016/j.bej.2011.10.007 (accessed on 24 July 2020).

Macrelli, S., Mogensen, J., & Zacchi, G., (2012). Techno-economic evaluation of 2nd generation bioethanol production from sugar cane bagasse and leaves integrated with the sugar-based ethanol process. *Biotechnol. Biofuels, 5*, 22.

Mahgoub, O., Kadim, I. T., Tageldin, M. H., Al-Marzooqi, W. S., Khalaf, S., Ali, A. A., & Al-Amri, I., (2008). Pathological features in sheep fed rations containing phenols and condensed tannins. *J. Anim. Vet. Adv., 7*, 1105–1109.

Mahmood, T., Mirza, M. A., Nawaz, H., Shahid, M., Athar, M., & Hussain, M., (2017). Effect of supplementing exogenous protease in low protein poultry by-product meal-based diets on growth performance and nutrient digestibility in broilers. *Anim. Feed Sci. Technol., 228*, 23–31.

Manangi, M. K., & Coon, C. N., (2007). The effect of calcium carbonate particle size and solubility on the utilization of phosphorus from phytase for broilers. *Int. J. Poult. Sci., 6*, 85–90.

Mao, S., Gao, P., Lu, Z., Lu, F., Zhang, C., Zhao, H., & Bie, X., (2016). Engineering of a thermostable β-1, 3–1, 4-glucanase from *Bacillus altitudinis* YC-9 to improve its catalytic efficiency. *J. Sci. Food Agric., 96*, 109–115.

McCracken, K. J., & Quintin, G., (2000). Metabolisable energy content of diets and broiler performance as affected by wheat specific weight and enzyme supplementation. *Br. Poult. Sci., 41*, 332–342.

McDonald, P., Edwards, R., Greenhalgh, J. F. D., Morgan, C., Sinclair, L., & Wilkinson, R. G., (2010). Evaluation of foods: Energy content of foods and energy partition within the animal. *Anim. Nutr.*, 254–280.

Mondala, A. H., (2015). Direct fungal fermentation of lignocellulosic biomass into itaconic, fumaric, and malic acids: Current and future prospects. *J. Ind. Microbiol. Biotechnol., 42*, 487–506.

Niu, C., Liu, C., Li, Y., Zheng, F., Wang, J., & Li, Q., (2018). Production of a thermostable 1,3–1,4-β-glucanase mutant in *Bacillus subtilis* WB600 at a high fermentation capacity and its potential application in the brewing industry. *Int. J. Biol. Macromol., 107*, 28–34.

Nortey, T. N., Kpogo, D. V., Naazie, A., & Oddoye, E. O. K., (2017). Cocoa pod husk plus enzymes is a potential feed ingredient for Hy-line silver brown laying hens. *Sci. Dev., 1*, 19–30.

O'Shea, C. J., Mc Alpine, P. O., Solan, P., Curran, T., Varley, P. F., Walsh, A. M., & Doherty, J. V. O., (2014). The effect of protease and xylanase enzymes on growth performance, nutrient digestibility, and manure odor in grower–finisher pigs. *Anim. Feed Sci. Technol., 189*, 88–97.

Okello, C., Pindozzi, S., Faugno, S., & Boccia, L., (2013). Bioenergy potential of agricultural and forest residues in Uganda. *Biomass Bioenergy, 56*, 515–525.

Oliveira, A. C., Amorim, G. M., Azevêdo, J. A. G., Godoy, M. G., & Freire, D. M. G., (2018). Solid-state fermentation of co-products from palm oil processing: Production of lipase and xylanase and effects on chemical composition. *Biocatal. Biotransformation,* 1–8.

Ong, L. G. A., & Straw, A. R., (2015). Solid state fermentation for glucanase production using acid/heat treated rice straw. *IJPBS, 4*, 106–109.

Pandey, A., (2003). Solid-state fermentation. *Biochem. Eng. J., 13*, 81–84.

Pirgozliev, V., Rose, S. P., Pellny, T., Amerah, A. M., Wickramasinghe, M., Ulker, M., Rakszegi, M., et al., (2015). Energy utilization and growth performance of chickens fed novel wheat inbred lines selected for different pentosan levels with and without xylanase supplementation. *Poult. Sci., 94*, 232–239.

Przywitowski, M., Mikulski, D., Jankowski, J., Juśkiewicz, J., Mikulska, M., & Zdunczyk, Z., (2017). The effect of varying levels of high- and low-tannin faba bean (*Viciafaba* L.) seeds on gastrointestinal function and growth performance in turkeys. *J. Anim. Feed Sci.*, 257–265.

Qasim, S. S., Shakir, K. A., Al-Shaibani, A. B., & Walsh, M. K., (2017). Optimization of culture conditions to produce phytase from *Aspergillus tubingensis* SKA. *Food Nutr. Sci., 8*, 733–745.

Ranjan, B., & Satyanarayana, T., (2016). Recombinant HAP phytase of the thermophilic mold *Sporotrichum thermophile*: Expression of the codon-optimized phytase gene in *Pichia pastoris* and applications. *Mol. Biotechnol., 58*, 137–147.

Reddy, G. P. K., Narasimha, G., Kumar, K. D., Ramanjaneyulu, G., Ramya, A., Shanti, B. S., & Rajasekhar, B., (2015). Cellulase production by *Aspergillus niger* on different natural lignocellulosic substrates. *Int. J. Curr. Microbiol. Appl. Sci., 4*, 835–845.

Regalado, C., García-Almendarez, B. E., Venegas-Barrera, L. M., Téllez-Jurado, A., Rodríguez-Serrano, G., Huerta-Ochoa, S., & Whitaker, J. R., (2000). Production, partial purification and properties of β-mannanases obtained by solid substrate fermentation of spent soluble coffee wastes and copra paste using *Aspergillus oryzae*and *Aspergillus niger*. *J. Sci. Food Agric., 80*, 1343–1350.

Regmi, S., G. C. P., Choi, Y. H., Choi, Y. S., Choi, J. E., Cho, S. S., & Yoo, J. C., (2016). A multi-tolerant low molecular weight mannanase from *Bacillus* sp. CSB39 and its compatibility as an industrial biocatalyst. *Enzyme Microb. Technol. 92*, 76–85.

Salem, A. Z. M., Alsersy, H., Camacho, L. M., El-Adawy, M. M., Elghandour, M. M. Y., Kholif, A. E., Rivero, N., et al., (2015). Feed intake, nutrient digestibility, nitrogen utilization, and ruminal fermentation activities in sheep fed *Atriplex halimus* ensiled with three developed enzyme cocktails. *Czech J. Anim. Sci., 60*, 185–194.

Salgado, J. M., Abrunhosa, L., Venâncio, A., Domínguez, J. M., & Belo, I., (2016). Combined bioremediation and enzyme production by *Aspergillus* sp. in olive mill and winery wastewaters. *Int. Biodeterior. Biodegrad., 110*, 16–23.

Salgado, J. M., Abrunhosa, L., Venâncio, A., Domínguez, J. M., & Belo, I., (2015). Enhancing the bioconversion of winery and olive mill waste mixtures into lignocellulolytic enzymes and animal feed by *Aspergillus uvarum* using a packed-bed bioreactor. *J. Agric. Food Chem., 63*, 9306–9314.

Sato, V. S., Jorge, J. A., & Guimarães, L. H. S., (2016). Characterization of a thermotolerant phytase produced by Rhizopusmicrosporus var. microsporus biofilm on an inert support using sugarcane bagasse as carbon source. *Appl. Biochem. Biotechnol., 179*, 610–624.

Shajitha, G., & Nisha, M. K., (2018). Tannase production from agro-wastes as substrate by *Trichoderma viride. IJCRLS, 7*, 1994–1997.

Sharma, R., Katoch, M., Srivastava, P. S., & Qazi, G. N., (2009). Approaches for refining heterologous protein production in filamentous fungi. *World J. Microbiol. Biotechnol., 25*, 2083–2094.

Singh, B., Sharma, K. K., Kumari, A., Kumar, A., & Gakhar, S. K., (2018). Molecular modeling and docking of recombinant HAP-phytase of a thermophilimould *Sporotrichum thermophile* reveals insights into molecular catalysis and biochemical properties. *Int. J. Biol. Macromol., 115*, 501–508.

Singh, U., (1984). The inhibition of digestive enzymes by polyphenols of chickpea (*Cicer Arietinum* L.)and pi geonpea (*Cajanus cajun* (L.) Millsp.). *Nutr. Rep. Int., 29*, 745–753.

Smink, W., Gerrits, W. J. J., Hovenier, R., Geelen, M. J. H., Verstegen, M. W. A., & Beynen, A. C., (2010). Effect of dietary fat sources on fatty acid deposition and lipid metabolism in broiler chickens. *Poult. Sci., 89*, 2432–2440.

Subramaniyam, R., & Vimala, R., (2012). Solid state and submerged fermentation for the production of bioactive substances: A comparative study. *Int. J. Sci. Nat., 3*, 480–486.

Sydenham, R., Zheng, Y., Riemens, A., Tsang, A., Powlowski, J., & Storms, R., (2014). Cloning and enzymatic characterization of four thermostable fungal endo-1, 4-β-xylanases. *Artic. Appl. Microbiol. Biotechnol., 98*, 3613–3628.

Tang, C. D., Shi, H. L., Tang, Q. H., Zhou, J. S., Yao, L. G., Jiao, Z. J., & Kan, Y. C., (2016). Genome mining and motif truncation of glycoside hydrolase family 5 endo-β-1,4-mannanase encoded by *Aspergillus oryzae* RIB40 for potential konjac flour hydrolysis or feed additive. *Enzyme Microb. Technol., 93, 94*, 99–104.

Tirado-González, D. N., Jáuregui-Rincón, J., Tirado-Estrada, G. G., Martínez-Hernández, P. A., Guevara-Lara, F., & Miranda-Romero, L. A., (2016). Production of cellulases and xylanases by white-rot fungi cultured in corn Stover media for ruminant feed applications. *Anim. Feed Sci. Technol., 221*, 147–156.

Tran, T. T., Hatti-Kaul, R., Dalsgaard, S., & Yu, S., (2011). A simple and fast kinetic assay for phytases using phytic acid-protein complex as substrate. *Anal. Biochem., 410*, 177–184.

Transparency Market Research, (2017). *Animal Feed Enzymes Market-Global Industry Analysis, Size, Share, Growth, Trends, Forecast 2026*. https://www.transparencymarketresearch.com/animal-feed-enzymes.html (accessed on 24 July 2020).

Urbano, G., López-Jurado, M., Aranda, P., Vidal-Valverde, C., Tenorio, E., & Porres, J., (2000). The role of phytic acid in legumes: Antinutrient or beneficial function? *J. Physiol. Biochem., 56*, 283–294.

Ushasree, M. V., Shyam, K., Vidya, J., & Pandey, A., (2017). Microbial phytase: Impact of advances in genetic engineering in revolutionizing its properties and applications. *Bioresour. Technol., 245*, 1790–1799.

Van, O. M., (2009). Enzymes in food technology-introduction. In: Whitehurst, R. J., & Van, O. M., (eds.), *Enzymes in Food Technology* (pp. 1–17). Wiley-Blackwell: Oxford.

Wang, C., Luo, H., Niu, C., Shi, P., Huang, H., Meng, K., Bai, Y., Wang, K., Hua, H., & Yao, B., (2015). Biochemical characterization of a thermophilic β-mannanase from *Talaromyces leycettanus* JCM12802 with high specific activity. *Appl. Microbiol. Biotechnol., 99*, 1217–1228.

Ward, O. P., (2012). Production of recombinant proteins by filamentous fungi. *Biotechnol. Adv., 30*, 1119–1139.

Wu, M., Tang, C., Li, J., Zhang, H., & Guo, J., (2011). Bimutation breeding of *Aspergillus niger* strain for enhancing β-mannanase production by solid-state fermentation. *Carbohydr. Res., 346*, 2149–2155.

Wyss, M., Brugger, R., Kronenberger, A., Rémy, R., Fimbel, R., Oesterhelt, G., Loon, A. P. G. M. V., Re, R., & Lehmann, M., (1999). Biochemical characterization of fungal phytases (myo-Inositol Hexakisphosphate Phosphohydrolases): Catalytic properties. *Appl. Environ. Microbiol. 65*, 367–373.

Xia, D. Q. H., Yee, Y. H., Illias, R. M., Mahadi, N. M., Bakar, F. D. A., & Murad, A. M. A., (2017). Characterization of recombinant *Trichoderma reesei* cellobiohydrolase and the potential of cellulase mixture in hydrolyzing oil palm empty fruit bunches. *Malaysian Appl. Biol. 46*, 11–19.

Yin, D., Yin, X., Wang, X., Lei, Z., Wang, M., Guo, Y., Aggrey, S. E., et al., (2018). Supplementation of amylase combined with glucoamylase or protease changes intestinal micro biota diversity and benefits for broilers fed a diet of newly harvested corn. *J. Anim. Sci. Biotechnol. 9*, 24.

You, S., Tu, T., Zhang, L., Wang, Y., Huang, H., Ma, R., Shi, P., et al., (2016). Improvement of the thermo stability and catalytic efficiency of a highly active β-glucanase from *Talaromyces leycettanus* JCM12802 by optimizing residual charge–charge interactions. *Biotechnol. Biofuels. 9*, 124.

Youssef, N. H., Couger, M. B., Struchtemeyer, C. G., Liggenstoffer, A. S., Prade, R. A., Najar, F. Z., Atiyeh, H. K., et al., (2013). The genome of the anaerobic fungus *Orpinomyces* sp. strain c1a reveals the unique evolutionary history of a remarkable plant biomass degrader. *Appl. Environ. Microbiol. 79*, 4620–4634.

Zhang, H., & Sang, Q., (2015). Production and extraction optimization of xylanase and β-mannanase by *Penicillium chrysogenum* QML-2 and primary application in saccharification of corn cob. *Biochem. Eng. J., 97*, 101–110.

Zhang, L., Xu, J., Lei, L., Jiang, Y., Gao, F., & Zhou, G. H., (2014). Effects of xylanase supplementation on growth performance, nutrient digestibility and non-starch polysaccharide degradation in different sections of the gastrointestinal tract of broilers fed wheat-based diets. *Asian-Australas. J. Anim. Sci., 27*, 855–861.

Zhang, S., Zhang, X., Qiao, H., Chen, J., Fang, C., Deng, Z., & Guan, W., (2018). Effect of timing of post-weaning supplementation of soybean oil and exogenous lipase on growth performance, blood biochemical profiles, intestinal morphology and caecal microbial composition in weaning pigs. *Ital. J. Anim. Sci.,* 1–9.

Zhao, C., Deng, L., & Fang, H., (2018). Mixed culture of recombinant *Trichoderma reesei* and *Aspergillus niger* for cellulase production to increase the cellulose degrading capability. *Biomass Bioenergy, 112*, 93–98.

Zielke, C., Kosik, O., Ainalem, M. L., Lovegrove, A., Stradner, A., & Nilsson, L., (2017). Characterization of cereal β-glucan extracts from oat and barley and quantification of proteinaceous matter. *PLoS One, 12*, e0172034.

CHAPTER 8

Biotechnological Valorization of Whey: A By-Product from the Dairy Industry

HILDA KARINA SÁENZ-HIDALGO,[1] ALEXANDRO GUEVARA-AGUILAR,[1] JOSÉ JUAN BUENROSTRO-FIGUEROA,[1] RAMIRO BAEZA-JIMÉNEZ,[1] ADRIANA C. FLORES-GALLEGOS,[2] and MÓNICA ALVARADO-GONZÁLEZ[1]

1Centro de Investigación en Alimentación y Desarrollo A.C. Unidad Delicias, Chihuahua, México, E-mail: salvarado@ciad.mx (Mónica Alvarado-González)

[2]Food Research Department, School of Chemistry, Universidad Autónoma de Coahuila, Blvd. Venustiano Carranza S/N, Colonia República, 25280, Saltillo, Coahuila, México

ABSTRACT

Whey is a by-product generated from the dairy industry and contains about 55% of milk nutrients; however, for years, it has been neglected and underutilized. The typical average composition consisted of lactose (72%), protein (10%), and minerals (12%) for both sweet and acid wheys. About 9 L of whey are generated per kilogram of cheese manufactured. Worldwide production of whey is estimated at 190 million metric tons and growing. Due to the organic matter content, whey is considered highly pollutant (biochemical oxygen demand: 40,000–60,000 ppm and chemical oxygen demand: 50,000–80,000 ppm). Traditionally, whey has been used as animal feed, fertilizer, and beverage additive. Outstanding scientific efforts have been conducted to develop viable alternatives to recover the valuable components of whey. Biotechnological processes are an alternative to valorize whey through lactose bioconversion acting as a carbon source for the preparation of novel bioproducts such as biofuels, biopolymers, bacteriocins, enzymes, organic acids, single-cell proteins or even bioelectricity, using bacteria, yeasts, and algae. In this chapter, biotechnological alternatives are detailed for whey valorization, to reduce the environmental impact by discharge untreated whey into the drain and obtain high-value-added compounds.

8.1 INTRODUCTION

The dairy industry is one of the fastest and diverse industries. A wide variety of products are already prepared and new and novel ones are under study or development. As in other industries, during the manufacture of dairy products, some by-products are also formed and attention must be paid to diminish the environmental impact. Whey is one of those by-products and its worldwide production is about 190 million metric tons and growing. Whey contains about 55% of milk nutrients, including lactose, proteins, fat, and minerals, and then, it is highly pollutant but on the same hand, it is a potent mixture for the production of value-added compounds.

Different biotechnological processes will be described as the use of whey as raw materials to obtain protein concentrates, beverages, biofuels, biopolymers, bacteriocins, enzymes, and organic acids, according to technical literature.

8.2 MILK WHEY

Whey is a by-product generated from dairy industries, mainly in cheese making or casein manufacture, is formed by coagulation of milk casein proteins. It exhibits a yellowish color due to the content of riboflavin, called lactochrome (Powers, 2003). Depending on the method employed for casein coagulation, sweet or acid whey can be produced. Sweet whey, with a pH of approximately 6.4, is obtained through enzymatic coagulation process using rennet containing the protease chymosin.

On the other hand, acid whey, with a pH of 4.8, the casein coagulation results from direct acidification of milk by activity of lactobacilli or addition of organic acid (Marwaha and Kennedy, 1998; Smithers, 2008; Zadow, 1994). Due to whey is obtained after removing fat and casein from milk (1 kg of cheese generates about 9 L of whey), this contains 85–95% of the milk volume and retains about 55% of total milk nutrients (Kosikowski, 1979; Siso, 1996) including approximately 20% of total proteins of milk (Walsh, 2002). Water is the main component of both sweet and acid wheys with ~95%, followed by lactose (72%), proteins (10%), and minerals (12%), namely, calcium, potassium, and sodium (Table 8.1). Protein is one of the main components which confers value to whey because of high nutritional value, health benefits, and therapeutic potential (Beaulieu, Dupont, and Lemieux, 2006; Smithers, 2008; Yalcin, 2006) which have several applications in food, cosmetics and pharmaceutical industries (Table 8.2) (Audic, Chaufer, and Daufin, 2003; Gallardo-Escamilla, Kelly, and Delahunty, 2007; Goudarzi et al., 2014; Zall, 1984).

TABLE 8.1 General Chemical Composition of Sweet Whey and Acid Whey

Component	Sweet whey (g/L)	Acid whey (g/L)	Reference
Total solids	63–70	63–70	(Chatzipaschali, and Stamatis, 2012 Panesar et al., 2007; Yadav et al., 2015)
Lactose	46–52	44–46	(Chatzipaschali, and Stamatis, 2012 Panesar et al., 2007; Yadav et al., 2015)
Proteins	6–10	6–8	(Chatzipaschali, and Stamatis, 2012 Panesar et al., 2007; Yadav et al., 2015)
pH	6.2–6.4	4.6–5.0	(Jinjarak et al., 2006; Zadow, 1994)
Minerals	2.5–4.7	4.3 – 7.2	(Yadav et al., 2015; Zadow, 1994)
Fat	5.0	0.4	(Yadav et al., 2015))
Ash	5.0	8.0	(Chatzipaschali, and Stamatis, 2012 Panesar et al., 2007; Yadav et al., 2015)
Calcium	0.4–0.6	1.2–1.6	(Chatzipaschali, and Stamatis, 2012 Panesar et al., 2007; Yadav et al., 2015)
Phosphate	1.0–3.0	2–4.5	(Chatzipaschali, and Stamatis, 2012 Panesar et al., 2007; Yadav et al., 2015)
Lactate	2.0	6.4	(Chatzipaschali, and Stamatis, 2012 Panesar et al., 2007; Yadav et al., 2015)
Chloride	1.1	1.1	(Chatzipaschali, and Stamatis, 2012 Panesar et al., 2007; Yadav et al., 2015)

TABLE 8.2 Protein Composition of the Whey Adapted from Yadav et al., (2015) and Spălăţelu (Vicol) (2012)

Protein	Concentration (g/L)
β-Lactoglobulin (β-LG)	3–2
α-Lactoalbumin (α-LA)	1.2 –1.5
Glycomacropeptides (GMP)	1.2
Immunoglobulin G	0.7–0.8
Bovine Serum Albumin (BSA)	0.4–0.5
Lactoferrin (LF)	0.1
Lactoperoxidase (LP)	0.03–0.05

Whey also contains some metallic ions such as zinc and copper (Venetsaneas, 2009) and other compounds, including citric acid, lactic acid, uric acid, urea (non-protein nitrogen) and B group vitamins (Chandan, Kilara and Shah, 2009; Kosikowski, 1979; Marwaha and Kennedy, 1998; Siso, 1996). Mineral content, pH, and fraction of proteins are the main compositional difference between sweet and acid wheys. Acid whey presents lower levels of proteins and higher quantities of ash, compared to sweet whey, generating an unpleasant acid and salty flavor for food applications (Shiby and Mishra, 2013).

Global production of whey is estimated at around 180 to 190 million metric tons and this volume is increasing directly to milk production (Spălăţelu, 2012). The main whey producer countries are the European Union and the US, with about 70% of the total world production. For years, whey has been neglected and underutilized, employing direct disposal to the environment. Due to whey is a by-product rich in lactose and protein, it is a very biodegradable substrate (~99%) with high strength organic pollutants, exhibiting a biochemical oxygen demand (BOD) of 40,000–60,000 ppm and a chemical oxygen demand (COD) of 50,000–80,000 ppm (Chatzipaschali and Stamatis, 2012). Lactose, the most abundant constituent of whey (70–72%), is the main responsible for both elevated BOD and COD values (Domingues et al., 1999; Patel and Murthy, 2011). Approximately 50% of worldwide whey production is used in direct feed, spread as fertilizer, or dispose into municipal sewers. The remanent is treated and industrially transformed in whey powder, lactose, and whey protein concentrates (WPC).

However, the direct spraying of whey on fields for extended periods turns out in salts deposition with the consequent reduction of crop yield (Ghaly and El-Taweel, 1997). Lactose and proteins from whey are degraded in the soil, leading to rapid consumption of oxygen causing crop death, reduction of redox potential, which conducts to Fe and Mn solubilization and the further contamination of water.

The main environmental problems related to the dairy industry affect water, air, and biodiversity. Whey effluent decomposes rapidly and depletes the dissolved oxygen level of water and release of odors. Also, a higher concentration of whey wastes is toxic to certain varieties of fishes because when casein precipitates generate an odorous black sludge. Soluble organics of the whey promote the release of gases, causing the odor, turbidity, and promote eutrophication (Deshpande, Patil, and Anekar, 2012). Eutrophication occurs when in a body of water high amounts of N and P derivatives are dissolved, inducing the excessive growth of plants, algae, and bacteria that consume oxygen in the water, within consequence blocks sunlight path and photosynthesis rate decreases, affecting water quality and fish diet (Schindler, 2006).

8.3 BIOTECHNOLOGICAL PROCESSES

Whey is a good source of nutrients that could be exploited by biotechnology, mainly to convert them into value-added products (Smithers, 2008). So far, whey has been submitted to traditional physicochemical methods to precipitate proteins and reduce fat content; however, lactose is still present in the remaining whey. Thus, the use of biotechnological processes such as fermentation and enzymatic modification, are employed to transform whey into value-added products with important industrial applications.

8.3.1 TRADITIONAL PROCESSES

Typically, whey is not used for human consumption, but as animal feed, sprayed as fertilizer or simply discharged. For many years whey was a supplement for animal feed (pigs, cattle, sheep) and formula for weaned in lactating provides good quality proteins and lactose as energy sources, minerals, and vitamins (Sienkiewicz and Riedel, 1986; Tunick, 2008). Some studies have indicated that spraying whey on the land increases groundwater pollution and salt deposition, affecting fertility (Marwaha and Kennedy, 1998; Kosikowski, 1979). On the other hand, the direct disposal of whey into water bodies represents a severe environmental problem for both elevated COD and BOD values. Thus, any of the uses mentioned before, cause environmental damage, and those also imply transportation cost. Both uses turn out in environmental problems and transportation costs. With the recent advances in biotechnology, whey can be revalorized due to its nutrient content (lactose, minerals, and proteins), with the consequent reduction of environmental impact as well as a sustainable economy (Panesar, 2007).

The first efforts to concentrate whey consists of dry it by heating methods. However, these efforts were economically and commercially not feasible, due to changes in color (yellow-brownish) and composition (protein denature) of the product, as well as elevated costs.

After that, with the use of new and novel processes, namely, spray drying, operational aspects such as thermal degradation and costs are reduced. It was not until 1970 with the application of membrane filtration, when it was possible to perform a sufficient concentration and separation of whey components without modifications. Technologies traditionally available for the treatment of industrial whey are discussed below.

8.3.1.1 CONCENTRATED WHOLE WHEY (CW)

Whey is concentrated to be used directly. Initially, whey is drained by centrifugation or vibration to separate fine particles of curd. After the separation of curd fines, fat is recovered and then converted into whey butter, which is widely applied in the food industry (Jinjarak, 2006). Whey is then clarified after those processes with 5% total solids, ready for the concentration process. Concentration is a combined procedure, consisted of water evaporation followed by reverse osmosis (RO) and the final concentration of total solids is about 45–65%. The CW is stored in refrigeration with constant agitation, and due to lactose is in the form of a saturated solution, that favors its crystallization. The spray drying process achieves the drying of whey concentrate. Concentrated whey is applied in the food industry as an emulsifier, gelling, and whipping agents to soups, salad dressing, dairy and bakery products (Walsh, 2002).

Separation and fractionation of whey by ultrafiltration (UF), is a method of filtration by membrane that combined with other such as diafiltration (DF), microfiltration (MF), nanofiltration (NF), and reverse osmosis (RO) followed by spray drying allows the separation and fractionation of whey proteins, obtaining dry (<5% moisture) products. The principle of whey ultrafiltration consists of passing it through a membrane to retain the whey proteins (retentate) where lactose and minerals flow through a membrane (permeate whey).

8.3.1.2 WHEY PERMEATE PROCESSING

8.3.1.2.1 Demineralization

Due to the high mineral content of whey and to be employed in food applications, it needs to be reduced. Demineralization process is performed by ion-exchange, electrodialysis (ED), or nanofiltration followed by spray drying (Houldsworth, 1980; Pepper and Orchard, 1982; Ryder, 1980). Different degrees of demineralization can be obtained (25, 50, or 90%); however, the ion exchange process is used when complete demineralization is necessary (28). Ion exchange is a discontinuous process due to the massive amounts of chemical compounds for resin regeneration and requires enormous volumes of water and generated higher costs. Demineralized whey is used in the manufacture of baby food, protein-based infant food, pharmaceutical, and chemical industries (Figure 8.1).

8.3.1.2.2 Lactose

Lactose, named milk sugar (4-θ-β-D-galactopyranosyl-D-glucose), is one of the components mostly used from whey. It can be removed from whey and permeate whey by two methods. Recovery lactose process consists of chemical precipitation of proteins and minerals and mechanical separation, obtaining pure lactose serum with a concentration higher than 65%. The lactose serum is refined and crystallized or alternative process where

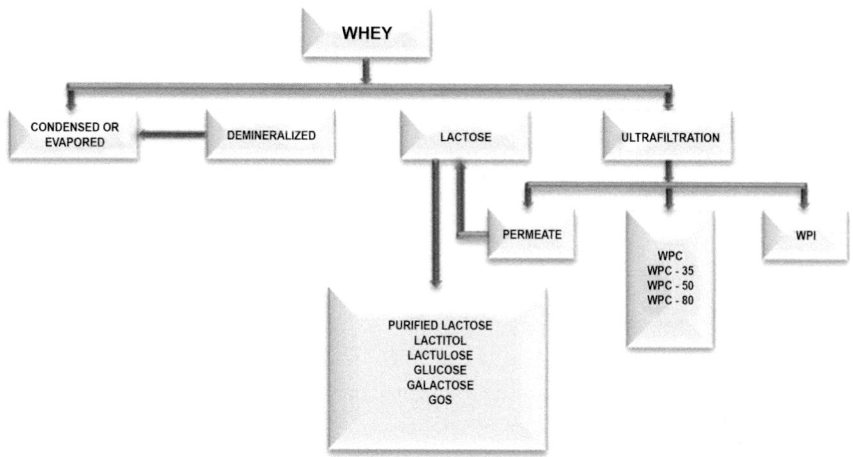

FIGURE 8.1 Scheme of traditional processing of whey.

Source: Modified from Mollea et al., (2013).

lactose can be hydrolyzed into its constituent sugars (glucose and galactose) by the enzyme galactosidase (lactase). The low sweetness of lactose (16% less sweet than sucrose) allows it to be used in the confectionery industries (Joesten, Hogg, and Castellion, 2006), baking industries to promote crust browning, pharmaceutical as an excipient, preparation of infant formula and chemical industry (McSweeney and Fox, 2009).

8.3.1.2.3 Lactulose

Other compounds can be derived from lactose (Figure 8.2). Lactulose (4-O-b-D-galactopyranosyl-D-fructose), is derived from an alkaline isomerization of lactose [33]. It is used as a sweeter, is more soluble than lactose, sweetness value is higher than sucrose (48–62%). It can be used as a sweetener for diabetics in food preparation, ideal for the elderly (Mayer, 2004). The main application of lactulose has been as a laxative in the treatment of constipation, chronic hepatic encephalopathy and hyperammonemia [35] and most recently prebiotic properties have been attributed (Oliveira, 2011).

8.3.1.2.4 Lactitol

Lactitol (4-O-(b-galactopyranosyl)-D-sorbitol) is a sugar alcohol, formed by catalytic hydrogenation of lactose (O'Donnell and Kearsley, 2012). This

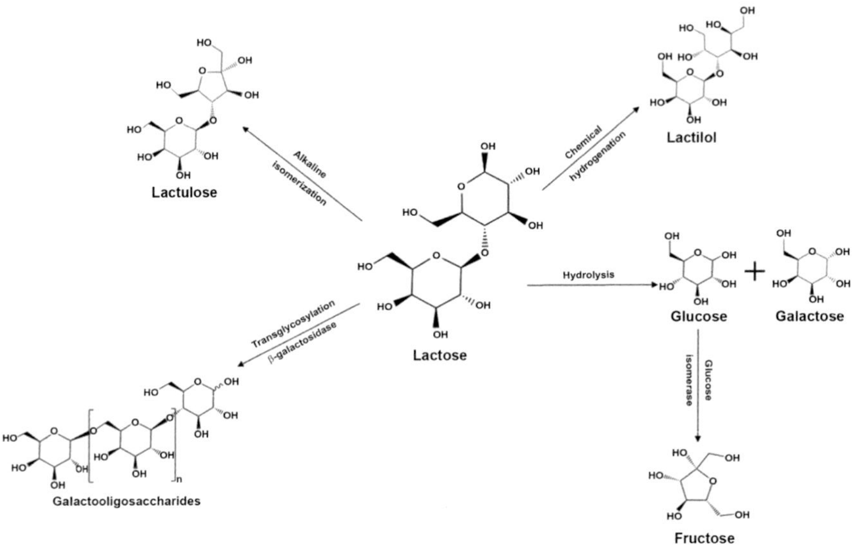

FIGURE 8.2 Compounds derived from lactose.

sugar is a low-calorie sweetener approximately 40% of the sweetness of sucrose. Lactitol is used in diets to lose weight or for diabetics. Lactitol is also used in chronic hepatic encephalopathy and constipation (Faruqui and Joshi, 2012). Prebiotic effect of lactitol has been studied due to it can be metabolized in the colon (Dills, 1989).

8.3.1.2.5 *Glucose and Galactose*

Another utilization of lactose from whey is the obtaining of glucose and galactose by hydrolysis. After this hydrolysis, the obtained solution has a sweetening power of glucose 80% and galactose 60% compared to sucrose. However, the sweetness can be increased to 110% by conversion of glucose to fructose by glucose isomerase (Joesten, Hogg, and Castellion, 2006; Kosaric and Asher, 1985; Moulin and Galzy, 1984), replacing corn starch syrup and saccharose. Glucose, galactose, and fructose have applications for the production of ice-cream, confectionery and drink soft (Gänzle, Haase, and Jelen, 2008).

8.3.1.2.6 *Galactooligosaccharides (GOS)*

Galactooligosaccharides (GOS) produced from lactose present in the whey through transglycosylation reaction catalyzed by β-galactosidase (Golowczyc et al., 2013; Jovanovic-Malinovska et al., 2012; Torres et al., 2010).

GOS are made from 2 to 8 galactose molecules linked to glucose. They are no digestible molecules with activity prebiotic with a beneficial effect on the host, stimulating the growth and or the activity of probiotic bacteria in the colon, as a consequence human health is improved.

8.3.1.3 WHEY RETENTATE

Liquid whey retentate consists of protein, fat, and insoluble salts, allowing its concentration to increase the protein: whey protein concentrate (WPC) (Figure 8.3). The different concentrates are known as: WPC-35 (with 35% protein), WPC-50 (with 50% protein) and WPC-80 (with 80% protein)

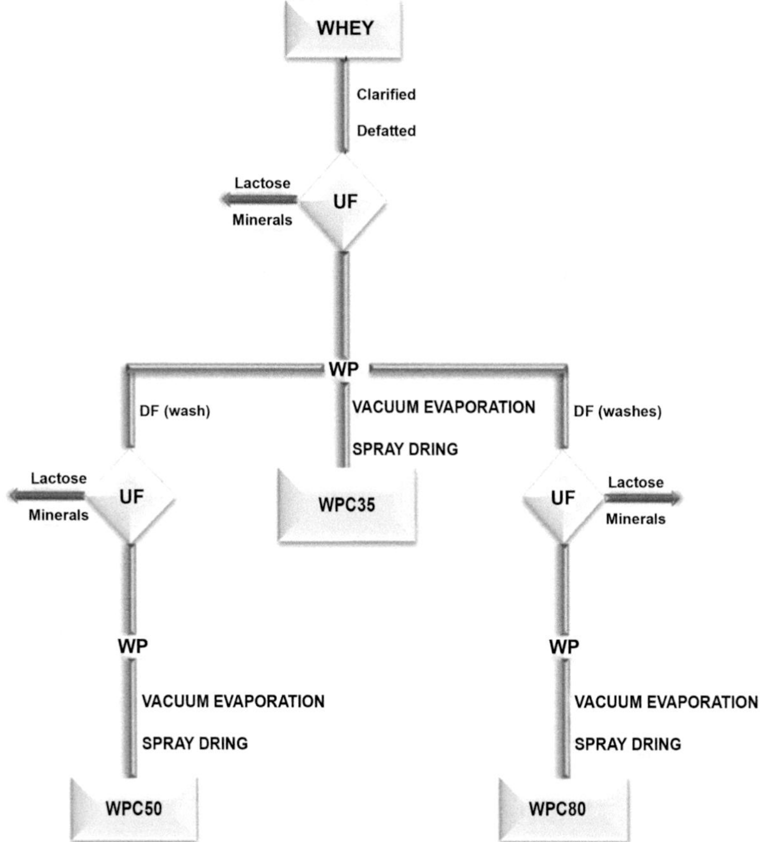

FIGURE 8.3 Diagram of the process to obtain why protein concentrate.
Source: Modified from Yadav et al., (2015).

(Chandan, Kilara, and Shah, 2009). Whey protein isolate (WPI) contains 90% protein and none lactose (Figure 8.4). Protein is the most important component of whey followed by lactose. Also is a key factor in meeting the nutritional needs of the global population. An inadequate protein intake induces stunted growth, decreased muscle capacity, susceptibility to infection, and decreased mental performance. Nowadays, products with high protein content are very well positioned in the market, and they are used as additives, nutraceuticals, and therapeutics.

8.3.1.3.1 Whey Protein Concentrate 35 (WPC 35)

First, whey is clarified and defatted. Secondly, whey is ultrafiltered, and after that, retentate contains lactose and proteins in equal parts and some minerals. The permeate is concentrated by vacuum evaporation and spray dried to obtain WPC 35.

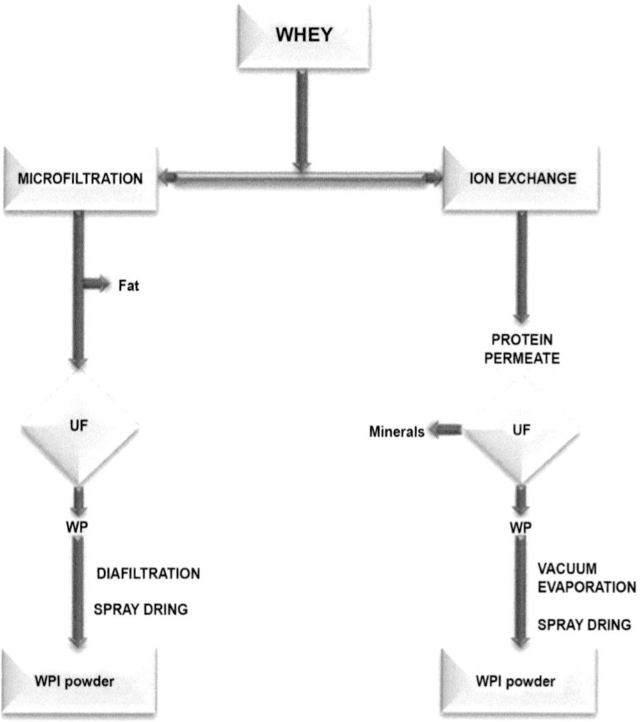

FIGURE 8.4 Diagram of the process to obtain why protein isolated.

Source: Modified from Yadav et al., (2015).

Source: Modified from Yadav et al., (2015).

8.3.1.3.2 Whey Protein Concentrate 50 (WPC 50)

The initial procedure is similar to the manufacture of WPC 30. The retentate is washed (diafiltration) and ultrafiltered again to be further concentrated by vacuum evaporation and spray drying. After that, WPC 50 is obtained.

8.3.1.3.3 Whey Protein Concentrate 80 (WPC 80)

As can be seen in Figure 8.3, the manufacture of WPC 80 consists of an increased number of washes (two or three). The extra washes contribute to removing the residual lactose with the aim of concentrate proteins.

8.3.1.3.4 Whey Protein Isolate (WPI)

Whey protein isolate is a protein product that contains at least 97% protein, and practically without lactose and moisture-free. The two processes commonly used to obtain WPI are ion exchange and microfiltration by membranes.

Ion exchange is a method based on the amphoteric properties of the proteins, that is, net charge proteins are dependent on environment composition, pH, and amino composition. Acid media, net charge proteins are positive, and neutral or slightly alkaline conditions net charge proteins are negative. In addition, taking into account the isoelectric point that is the pH at which the number of positive and negative charges is equal. Protein adsorption on an ion exchange process can be conducted in a resin packed column with carboxymethyl cellulose. Whey is adjusted to pH 3.2 before flowing through the ion exchange matrix. For desorption, pH is increased to 7–7.5 and the solution recovered is filtered and concentrated by UF. To achieve a higher protein content, the ultrafiltered solution is further submitted to vacuum evaporation followed by spray drying. This process permits to obtain a 97% protein content (dry weight basis), 3% ash, lactose traces, and 0.2% fat.

Another method employed for obtaining WPI is microfiltration. Here, the purpose is to remove fat because it alters the functional properties of whey proteins, mainly foaming properties. Defatted whey is subjected to UF, then DF to obtain WPI in solution and concentrated by spray drying.

8.3.2 BEVERAGES

All over the world dairy industries generate sufficient amounts of whey per liter of milk processed, according to the process employed and products manufactured.

Approximately 50% of total world cheese whey production is transformed and treated into a variety of food products (S. M. S and F. K. J, 1988).

Depending on the processing techniques used for the casein separation from liquid milk, the type and composition of whey are dependent. Frequently, the type of whey encountered comes from cheese manufacture or certain casein cheese products, where processing is based on coagulating the casein by rennet. In sweet whey, the coagulation of casein occurs approximately at pH 6.5 (Panesar et al., 2007).

One of the products prepared from whey is beverages, including alcoholic, fermented, carbonated, or made from deproteinized whey (Jarc and Hadžiosmanović, 1994). Holsinger et al., (1974) mentioned four basic types of whey beverages:

- Beverages from whole whey.
- Non-alcoholic beverages from deproteinized whey (Fermented and Unfermented beverages).
- Alcoholic beverages (Beverages containing less than 1% alcohol, whey beer, whey wine, or alcoholic beverages containing protein).
- Protein beverages.

However, in this chapter, we will group them into three types: fruit juices, fermented beverages, and alcoholic beverages.

8.3.2.1 WHEY BEVERAGES (FRUIT JUICES MIXED WITH WHEY)

Due to the properties of whey proteins, beverages made from this, have increased their interest due to their nutritional properties in products such as whey protein supplement bars, whey protein concentrates, whey protein isolates, and beverages (P. J, 1992).

Several difficulties occur during the processes of whey beverage production (Chavan and Nalawade, 2015; Ryan et al., 2012):

a. Fruit beverages based on sweet or acid whey are susceptible to precipitation of denatured proteins, because of the thermal treatment applied.
b. High water contents allow bacterial deterioration.
c. Crystallization of lactose during storage at refrigerated temperature.
d. The undesired salty-sour flavor of whey by its high content of minerals.
e. The efficacy of thermal treatments can be affected by high viscosity.

Source: Modified from Yadav et al., (2015).

In spite of the disadvantages of using whey for the preparation of beverages, significant advantages confer its use, for instance:

a. Buffering capacity of whey can be assayed for the survival of probiotic bacteria in the gastrointestinal tract.
b. Whey has a bland flavor and on many occasions, they can act as a carrier for the aroma compounds.
c. Addition of whey improves the 'mouthfeel' of the drink by increasing the viscosity of the beverage.
d. Whey is having a broad range of solubility from pH 3–8.
e. Whey can also be used to solve the problems associated with the cloudiness of tropical fruit juices and produce a stable cloud juice (Chavan and Nalawade, 2015).

Whey beverages have been a challenge for the acceptance of consumers due to the sensory profile of these products, which turns out in the development of various formulations to find out the optimal mixture (Castro et al., 2013; Sakhale and Ranveer, 2012; Djurić et al., 2004; Goudarzi et al., 2014). The functional properties of the drinks can be regulated depending on the fruit used in the formulation since each fruit constitutes a rich source of bioactive molecules such as polyphenols, carotenoids, and phytosterols. For example, in strawberries and blueberries, the presence of phenolic compounds have been linked to several functional properties, including the elimination of free radicals and reactive oxygen species (Brambilla et al., 2008; Zafra-Stone, 2007).

Researchers at Faculty of Technology of the University of Novi Sad suggested that citrus fruits, pear, strawberry, mango, raspberry, apple, and passion fruit have been efficient disguising the sour taste, the salty and undesirable odor of whey. For this reason, they prepared whey beverages using different fruits (orange, pear, peach, and apple), as well as, sucrose and citric acid, and they suggested the optimal blend formulation (Djurić et al., 2004).

A ready-to-drink beverage was formulated by Chatterjee et al., (2015), with concentrated whey and orange juice, mixed with an adequate amount of sugar, stabilizer, citric acid, and flavor. All of those parameters were statistically evaluated as well as stability during storage. Authors proposed that in the functional food market, beverages based on whey with orange juice can be an excellent option due to its storage, nutritional and sensory properties.

In other studies, it has been reported that it is possible to obtain a new functional beverage that does not need a cold chain using only Ricotta cheese whey and fruit juice because they observed that this formulation with pasteurization prevented the formation of precipitate at the bottom of the bottles (Rizzolo and Cortellino, 2017). In 2005, Beucler et al., (Beucler, Drake, and Allen Foegeding, 2005) designed a beverage made of whey protein to meet the requirement of thirst-quenching beverages and the comparison of perception and acceptance against a commercial were that whey contents of 25 to 50% could be successfully added for their dairy sour flavor and salty taste.

8.3.2.2 FERMENTED WHEY BEVERAGES

Due to the nutritional properties of whey, it can be used for the preparation of alcoholic and lactic beverages by means of fermentation, employing yeasts, and selective bacteria producing lactic acid (Kumar and Vandna, 2015). Cultured milk beverages require some additional ingredients to ensure the stability of the coagulated casein, for such purpose alginate, pectin, xanthan gum, and carboxymethyl cellulose can be used to improve the mouthfeel of the end product. On the other hand, fermented whey beverages do not have this problem of thin sedimentation (Gallardo-Escamilla, Kelly, and Delahunty, 2007).

The production of fermented beverages can be an interesting alternative for the use of whey since the fermentation process does not affect the nutritional value of unfermented whey; however, it has a lower level of lactose. The fermented whey beverages can be an option with good sensory properties (Barukčić, Božanić , and Tratnik, 2008).

The difference between whey yogurt and fermented whey beverages is how whey is incorporated. In yogurt, it is usually added as whey proteins resulting in a cost reduction whereas for fermented beverages are used as reconstituted or liquid (Castro et al., 2013).

In some recent studies, whey was fermented by the following strains: *Lactobacillus acidophilus*, *Lactobacillus delbrueckiis* sp. *bulgaricus*, *Streptococcus thermophillus*, *Lactobacillus rhamnosus,* and *Bifidobacterium animalis* sp. *Lactis* (Almeida, Tamime, and Oliveira, 2008; Pescuma et al., 2008).

Two of the bacteria strains widely used in the manufacture of probiotics are*Bifidobacterium* and *Lactobacillus acidophilus*, with well-documented clinical effects and desirable properties which are used in the fermentation of beverages to obtain a product with functional properties (Shah, 2007; Kailasapathy, and Chin, 2000).

According to Skryplonek et al., (2015) the manufacture fermented probiotic beverages with acid whey can be performed by *Bifidobacterium animalis* or *Lactobacillus acidophilus*. Many researches have included different studies related to sensory, physical, and chemical characterizations of fermented whey beverages: Gallardo-Escamilla (2007) evaluated mouthfeel and flavor of fermented whey add with hydrocolloids. On the same hand, a probiotic beverage was developed from whey and orange and pineapple juices (Shukla and Kushwaha, 2017; Shukla, 2012). On the other hand, Pogon et al., (2014) characterized orange whey fermented beverages.

In 2016, Janiaski et al., (2016) established the attributes, tastes and sensory profiles of Brazilian consumers regarding their preference of whey beverages (fermented or not) and yogurts with different contents (fat, with or without sugar). They also found the ideal product that was characterized by the sweet taste, acrid aroma, artificial strawberry aroma and taste, viscosity, low acid taste, highest levels of brightness, and particulate matter.

8.3.2.3 ALCOHOLIC DRINKS (WHEY BEER, WHEY WINE, AND WHEY CHAMPAGNE)

Alcoholic beverages can be divided as follows: beverages with low alcohol content (≤1.5%), whey wine, and "whey beer." They are obtained from whey permeate and characterized by a low alcohol content with the addition of yeast strains (*Kluyveromyces fragilis* or *Saccharomyces lactis*). Sparkling wine is called whey champagne (Barukčić, Božanić, and Tratnik, 2008).

Dragone et al., (2009) produced an alcoholic beverage by whey fermentation and characterized its volatile profile. They identified forty volatile compounds in this beverage. For the production of beer, malt may or may not be used, and it may be enriched with starch hydrolysates, vitamins, and minerals. The presence of milk fat can be a problem because it can cause undesirable taste and smell as well as foam loss in beer (Garg, 2017).

Process of whey wine includes clearing, deproteinization, lactose hydrolysis by β-galactosidase, decanting and cooling, the addition of yeasts and fermentation, decanting, aging, filtering and bottling. Whey wine is mostly flavored with fruit aromas and its alcohol content ranges 10 and 11% (Palmer and M. R. F, 1978).

Kosikowski and Wzorek (1977) developed a table whey-based on fermenting whey permeate with high lactose and lactose fermenting yeast. They demonstrated that an acceptable table whey wine could be derived from whey permeates. On the other hand, Friend et al., (1982) compared industrial alcohol production using whey and grain fermentation. They examined six strains of a trained lactose fermenting *Kluyveromyces* yeast for their ability to utilize lactose in sweet-whey permeate.

8.3.3 OTHER DAIRY PRODUCTS

8.3.3.1 WHEY BUTTERMILK

Recently, whey butter has been widely studied, including the evaluation of their sensorial characteristics, physicochemical properties and compared to sweet cream, whey, and cultured butters (Jinjarak et al., 2006). The authors concluded that there is not significant difference for the acceptability of the different products. Functional properties and compositional among sweet, sour, and whey buttermilk were compared, evaluating the emulsifying, viscosity, foaming properties and solubility, finding that whey buttermilk exhibited better emulsifying capacity, stable levels of protein solubility and viscosity between pH 4 to 6 than sweet or cultured buttermilk (Sodini et al., 2006). Thus, they suggested whey buttermilk could be used in the formulation of low pH food. Morin et al., (2006) reported that whey buttermilk has similar amounts of phospholipids concerning regular buttermilk.

8.3.3.2 WHEY CHEESES

Whey cheeses are made all over the world according to traditional techniques, and thus, chemical composition differs in each of them (Table 8.3).

TABLE 8.3 Composition of Whey Cheeses Adapted From Pintado et al., (2001)

Cheese	Fat (%w/w)	Lactose (%w/w)	Protein (%w/w)	Solid total (%w/w)	Reference
Manouri	36.67	2.49	10.86	51.93	Veinoglou et al., 1984
Anthotyro	18.5	n.a.	n.a.	33.82	Tzanetaski et al., 1977
Requijào	29.5	3.5	8.5	41	Pintado and Malcata, 1999
Ricotta	10.2	4.1	6.1	20/30	Zino et al., 1993
Ricotone	n.a.	1.5	11.3	17.5	Kosikowski, 1982b

The effect of goat breed on the chemical, sensory and nutritional characteristics of ricotta cheese were evaluated (Pizzillo et al., 2005). Obtained results indicated that lipid composition and sensory properties of ricotta cheese is affected by breed of goat. Brunost, a typical Norwegian dairy product, is a type of cheese produced by whey concentration up to 80% total solids and crystallization of the lactose. This group of cheeses includes Mysost, Gjetost, Primost, Flmemyost, Gudbrandsdalost, and Niesost (McSweeney, Ottogalli, and Fox, 2004).

Source: Modified from Yadav et al., (2015).

Whey cheeses are made all over the world (manouri, anthotyro, requijao, ricotta, ricotone cheese) according to traditional techniques, therefore, the chemical and nutritional composition differs in each of them depending on the origin and type of whey ((Pintado, Macedo, and Malcata, 2001)).

8.3.4 BIOFUELS

All the concerns derived from petroleum depletion as well as the environmental impact and price fluctuations have conducted to the development of new and novel energy resources. Biofuels are potential alternatives for petroleum derivatives; however, some valuable technical aspects need to be further studied: raw materials, catalysts, and reaction system (Marchetti, 2012).

In technical literature has been mentioned that raw material impacts on 70% of the total price of the biofuel obtained; thus, the different raw material must be evaluated. In this sense, whey has an important role to play. Whey is mainly obtained during cheese-making, in dairy and cheese industries. Whey constitutes a very relevant environmental issue due to the high concentration of organic compounds turning out in elevated BOD and COD (Ergüder et al., 2001; Mockaitis, 2006; Das, Raychaudhuri, Ghosh, 2016).

And then? Is really whey a potential alternative for biofuels? The answer is yes! Even when it is a pollutant, its chemical composition makes of whey and interesting platform to obtain several value-added compounds (Boura, 2017). Apart from lactose, whey also contains proteins, lipids and mineral salts, which can be submitted to the biotechnological process to obtain different products, as it is shown in Figure 8.5.

According to Figure 8.5, whey can be suitable for practical approaches in both food and energy fields. However, the valorization described in this section is related to biofuels (energy).

8.3.4.1 ETHANOL

This is the most explored biofuel produced from whey. Due to its negative effect on bacteria metabolism, ethanol is mainly synthesized by yeast. For whey, various strains of yeasts, including *Kluyveromyces, Saccharomyces* and *Candida*, have been tested. It is known that species of *Kluyveromyces* and *Candida* can metabolize lactose, whereas *Saccharomyces cerevisiae* is not. However, some outstanding efforts have been conducted for the genetical modification of *S. cerevisiae* due to its high ethanol productivity and conversion rates.

In their review, Das et al. (2016), mentioned a technical bottleneck for *Kluyveromyces* species: 10% of lactose is the highest content tolerated by

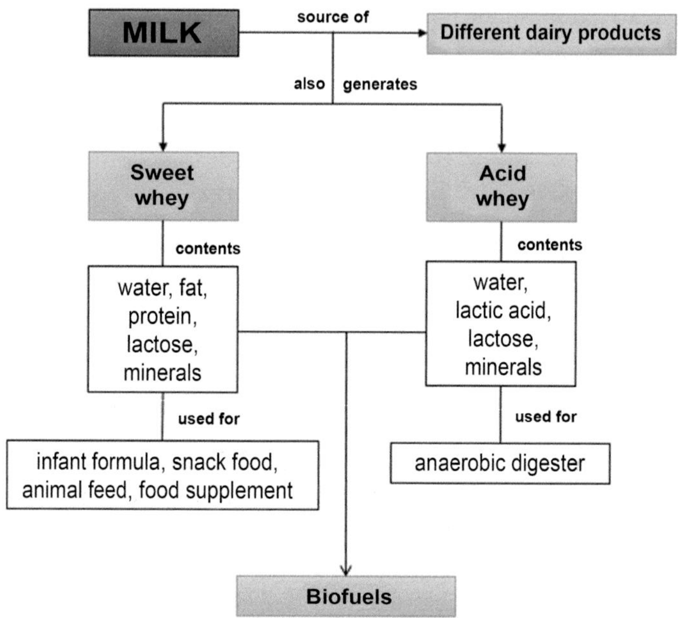

FIGURE 8.5 Products obtained from milk and whey.

this yeast. They also indicated some operational parameters explored in order to increase the resistance of this yeast using metabolic engineering, whey permeate solutions, temperature, and pH, immobilization of the yeast, as well as batch and continuous reaction systems.

Concerning *Candida* species, Das et al. (2016) cited some studies where this yeast can utilize higher levels of lactose compared to *Kluyveromyces* species.

In a recent study, Boura et al. (2017) refer that cheese whey was submitted to anaerobic acidogenesis by a UASB mixed anaerobic culture and alcoholic fermentation by kefir. Both cultures were immobilized on γ-alumina to obtain organic acids and ethanol. From their results, they indicate that during acidogenesis 12 g/L of organic acids and 0.2 g/L of ethanol can be obtained whereas for a continuous process consisted in two bioreactors connected in series, turned out in both increased organic acid (15 g/L) and ethanol (0.3–0.4 g/L) contents.

8.3.4.2 BIODIESEL

Biodiesel is another important biofuel which has been widely studied from different raw materials. Several reviews have been reported indicating the economic and technological feasibility for the industrial production of biodiesel.

Source: Modified from Yadav et al., (2015).

For whey, biodiesel can also be obtained and not for the content of fat found in whey, but, for the utilization of whey as a broth for harvesting some oleaginous yeasts. In this sense, Carota et al. (2017) described a sustainable use of ricotta cheese whey for microbial biodiesel production. In their research, these authors refer that 18 strains of oleaginous yeasts were investigated for their growth and lipid-producing capabilities on this substrate. Among them, *Cryptococcus curvatus* NRRL Y-1511 and *Cryptococcus laurentii* UCD 68–201 adequately grew therein producing substantial amounts of lipids (6.8 and 5.1 g/L, respectively). A high similarity between the percentage of fatty acid methyl esters composition of lipids from the former and the latter strain was found with a predominance of oleic acid (52.8 vs. 48.7%) and total saturated fatty acids (37.9 vs. 40.8%).

In another study conducted by Pirozzi et al. (2013), the oleaginous yeasts *Lipomyces starkey* were grown in the presence of dairy industry wastewaters. They reported that yeasts could degrade the organic components and to produce triglycerides. When using wastewaters from the Ricotta cheese production or residual whey as a growth medium, *L. starkey* was cultured without dilution nor external organic supplement. On the other hand, yeasts could only partially degrade wastewaters from the Mozzarella cheese production, due to the accumulation of a metabolic product beyond the threshold of toxicity. In this case, a dilution of the waste was required to obtain a more efficient degradation of the carbon compounds and a higher yield in oleaginous biomass. According to their findings, the fatty acid distribution of the microbial oils obtained showed a relevant content of oleic acid.

Even when fewer studies have been devoted for biodiesel, it is important to highlight that even when whey is defined as residue and pollutant, it is an important platform for other compounds and its valorization help to the sustainability of dairy industries.

8.3.4.3 HYDROGEN

This gas is considered as the fuel for the future and even when different metabolic pathways lead to hydrogen production, fermentation has quite potential. Several reports have been documented related to the batch and continuous production from whey, as well as some operational variables during fermentation.

For instance, Davila-Vazquez et al. (2008), studied the effects of initial pH and initial substrate concentration on both hydrogen molar yield and volumetric hydrogen production rate. Lactose, cheese whey powder, and glucose were used as substrates and heat-treated anaerobic granular sludge as inoculum. With lactose, 3.6 mol H_2/mol lactose and 5.6 mmol H_2/L/h were measured at pH 7.5 and 5 g lactose/L. Powder yielded 3.1 mol H_2/mol lactose at pH 6 and 15 g powder/L while 8.1 mmol H_2/L/h was attained at pH 7.5 and 25 g powder/L. Finally, glucose led to 1.46 mol H_2/mol substrate (pH 7.5, 5 g glucose/L), with a 8.9 mmol H_2/L/h, at pH 8.12 and 15 g glucose/L. These authors mentioned that their findings could be of significance when alkaline pre-treatments are performed on organic feedstock.

In a further study, Davila-Vazquez et al. (2009) investigated the continuous hydrogen production using cheese whey in a continuous stirred tank reactor. Three hydraulic retention times were tested to reach the highest volumetric hydrogen production rate and then assay four organic loading rates. From their results, the authors indicated that the highest volumetric hydrogen production rate (46.61 mmol H_2/L/h) and hydrogen molar yield of 2.8 mol H_2/mol lactose were found at an organic loading rates of 138.6 g lactose/L/d, and also, that volumetric hydrogen production rates are crucial for the full-scale practical application of fermentation technologies.

In a recent batch hydrogen production with cheese whey as substrate, Muñoz-Páez et al. (2014) explored the effect of temperature and addition of buffer. According to their observations, when H_2 production reached a plateau, the headspace of the reactors was flushed with N_2 and reactors were re-incubated. Afterward, only the reactors with phosphate buffer showed a second cycle of H_2 production and 48% more H_2 was obtained. The absence of a second cycle in non-buffered reactors could be related to a lower final pH than in the buffered reactors; the low pH could drive the fermentation to solvents production.

Some other studies related to H_2, also refer to the formation of methane (CH_4). Venetsaneas et al. (2009) used cheese whey for H_2 and CH_4 generation in a two-stage continuous process with alternative pH controlling approaches. Their study focused on mesophilic fermentative H_2 production from undiluted cheese whey at a hydraulic retention time of 24 h. Addition of $NaHCO_3$or an automatic pH controller were used, to maintain the pH culture at 5.2. The H_2 production rate was 2.9 ± 0.2 L/L reactor/d, while the yield of H_2 produced was approximately 0.78 ± 0.05 mol H_2/mol glucose consumed, with alkalinity addition, while the respective values when using pH control were 1.9 ± 0.1 L/L reactor/d and 0.61 ± 0.04 mol H_2/mol glucose consumed. The corresponding yields of H_2 produced were 2.9 L of H_2/L cheese whey and 1.9 L of H_2/L cheese whey, respectively. The effluent from the hydrogenogenic reactor was further digested to biogas in a continuous mesophilic anaerobic bioreactor. The anaerobic digester was operated at a

Source: Modified from Yadav et al., (2015).

hydraulic retention time of 20 d and produced approximately 1 L CH_4/d, corresponding to a yield of 6.7 L CH_4/L of influent.

On the other hand, Cota-Navarro et al. (2011) assayed the continuous H_2 and CH_4 production in a two-stage process using mixed cultures and cheese whey powder as a substrate. The effect of operational parameters such as hydraulic retention time and organic loading rate on the volumetric H_2 and CH_4 production rates were evaluated. In the case of H_2, a volumetric production of 28 L H_2/L/d was achieved during stable operation in a CSTR at hydraulic retention time and organic loading rate of 6 h and 142 g lactose/L/d, respectively. Concerning CH_4 production in an UASB reactor, the acidified effluent from the hydrogen-producing bioreactor was efficiently treated obtaining COD removals above 90% at organic loading rate and hydraulic retention time of 20 g COD/L/d and 6 h, respectively. The two-stage process for continuous production of H_2 and CH_4 recovered over 70% of the energy present in the substrate.

According to the aforementioned, whey can be useful for the production of several biofuels. Depending on the amount available and the facilities at any particular research center, it will be the potential application for this residue.

8.3.5 BIOPOLYMERS

Biopolymers are defined as compounds formed by biological monomers, fully or partially recyclable and that no generate toxic substances during their production, mainly obtained from plants and animals (Castilho, Mitchell, and Freire, 2009). However, in the last years have been gained importance the production of microbial biopolymers, due to the high commercial value. The microbial biopolymers are produced by the fermentation process, and there are three main groups according to their industrial interest: polysaccharides, proteins, and polyhydroxyalkanoates (PHA).

PHA are biopolymers accumulated and stored by several microorganisms under limited growth conditions. These accumulations are carbon and energy reserves, originated when carbon source is in excess and another nutrient such as nitrogen, phosphate, sulfur, iron, magnesium, potassium or oxygen in the medium are limited. Once that limitation is supplemented, PHA is polymerized and metabolized as carbon source and energy (Ahn, Park, and Lee, 2001).

PHA is composed of linear 3-hydroxy fatty acids monomers, generally produced as a polymer of 103 and 104 monomers, accumulated as inclusions of 0.2–0.5 μ of diameter, composed principally of four monomers: 3-hydroxybutyrate (3HB), 3-hydroxyvalerate (3HV), 3-hydroxy-2-metilvalerate

(3HMV) and 3-hydroxy-2-metylbutirate (3HMB). The importance of PHA is due to their properties are similar than those found in plastics of chemical origin, which constitutes a practical approach for environmental problems derived from common plastics (Koller et al., 2008). According to Wong and Lee (1998), up to 50% of total production cost is related to substrate or carbon source. Additionally, several factors such as PHA productivity, content and yield, type of substrate or carbon source, and the recovery method used, they all affect the PHA production cost (Choi and Lee, 1999).

The main of whey valorization is to recover it for the obtention of sub-products with a high market value. According to Panesar et al. (2007), about 50% of total produced whey worldwide is transformed into different foodstuff (mainly human and animal feed) and the rest remains as waste. Because of its composition, several biotechnological processes (aerobic and anaerobic fermentation, lactose hydrolysis, microbial fuel cells, aerobic digestion) can be applied to use whey as a substrate for the production of value-added compounds (Prazeres, Carvalho, and Rivas, 2012).

One of such compounds is PHA. Conversion of this waste into biopolymers decreases the production costs of PHA and simultaneously saves money for waste treatment (Koller et al., 2011). Nevertheless, whey contains a high amount of lactose, and most of the microorganisms cannot employ it as carbon source (Yellore and Desai, 1998); thus, whey should be submitted to the activity of enzymes (β-galactosidases) previously. There are some microorganisms able to produce those enzymes and improve the conversion bioprocess, such as *Aspergillus niger*, *A. oryzae*, *Kluyveromyces lactis*, *K. marxianus*, *Lactobacillus bulgaricus*, *Lactococcus lactis, Streptococcus thermophilus* and some species of *Bacillus* genus.

Research studies are focused on the search of microorganisms that ferment fast and completely cheap substrates, with minimal addition of nitrogen nutrients and high stereospecificity at reduced values of pH and high temperatures, that produce a little amount of biomass and the generation of by-products are depreciable. PHA production from whey has been focused on the development of bioprocess by submerged fermentation (Table 8.4), found a range yield of 12–81% depending on the microorganism. Maximum yields have been obtained using recombinant strains of *Escherichia coli.* Recently, alternative strategies as the use of solid-state fermentation have been proposed for PHA production (Castilho et al., 2009). However, there are no reports about the use of whey in the solid-state fermentation process for PHA production, an option that has demonstrated good results in other metabolites production, increasing yield, which impacts directly on production cost.

TABLE 8.4 PHA Production Bioprocesses Using Cheese Whey as a Carbon Source

Strain	Process	Type of PHA	Yield (%)	Reference
Azotobacter chrococcum	SmF	P(3HB)	75	Khanafari et al. (2006)
Azohydro monaslata DSM123	SmF	P(3HB)	NR (1.67 g/L)	Baei et al. (2010)
Bacillus megaterium CCM 2073	SmF	P(3HB)	51.57	Obruca et al. (2011)
Recombinant *Escherichia coli* CGSC4401	SmF	P(3HB)	80.5	Ahn et al. (2001)
Recombinant *E. coli (C. necator* genes)	SmF	P(3HB)	81.3	Lee et al. (1997)
Recombinant *E. coli (C. necator* genes)	SmF	P(3HB)	80	Wong and Lee (1998)
Recombinant *E. coli (Azotobacter* sp. genes)	SmF	P(3HB)	72.9	Kim (2000)
Haloferax mediterranei	SmF	PHA	50	Koller et al., (2007)
Haloferax mediterranei ATCC 33500	SmF	P(3HV)	65	Pais et al. (2016)
Methylobacterium sp.	SmF	P(3HB)	67	Nath et al. (2008)
Pseudomonas hydrogenovora	SmF	3HV, 3HB	12	Koller et al. (2008)
Thermus thermophilus HB8	SmF	3HV	35	Pantazaki et al. (2009)

Source: Modified from Yadav et al., (2015).

Once produced, PHA can be easily degraded in the environment by the presence of soil microorganisms, which can produce PHB depolymerases, enzymes that break the ester bond of a polymer into monomers and oligomer soluble in water. Final products of PHA degradation are water and carbon dioxide (Jiang et al., 2012).

The selection of a microorganism strain depends on several factors including the ability of the strain to use cheap carbon sources, cost of culture media, growth rate, polymers synthesis rate, quality and quantity of PHA, and total process costs (Chanprateep, 2010). Other factor that impacts on PHA production costs is the downstream processing (disruption of the cell membrane, polymers separation from cell debris, polymer purification), which can be reduced by optimization of PHA production conditions, such as C/N ratio, temperature, strain, pH, medium composition, and other stress factors that enhance PHA content into the cells, improving both the efficiency and the costs of downstream processing, The latter can be up to 40% of the total production costs (Choi and Lee, 1999).

The valorization of agroindustrial by-products allows obtaining a wide range of biomolecules with great industrial interest, such as biopolymers. The development of bioprocess to generate biopolymers using different agro-industrial by-products has increased. It is possible to use whey as a carbon source to obtain PHB as an alternative to synthetic materials. However, more research is necessary to reduce costs and an alternative is the implementation of new production lines in dairy plants that already operate with the further evaluation of scaling and comparison to other processes such as solid-state fermentation.

8.3.6 BACTERIOCINS, ENZYMES, AND ORGANIC ACIDS

Different products can be obtained from whey using the biotransformation process. Some of those products are bacteriocins, enzymes, and organic acids, which have proven useful and interesting applications for food, medicine, cosmetics, etc.

8.3.6.1 BACTERIOCINS

These compounds have a large potential for the inhibition of some microorganisms, and they do not alter or deteriorate foods and extend their shelf life. The use of bacteriocins has been approved and studied for dairy products, meat products, and some vegetables (Marcos et al., 2013). It is known that bacteriocins can be obtained in different ways, but those that are produced from the lactic acid bacteria isolated from whey, have potential applications. A very important bacteriocin, which has been approved for by the FDA, is Nisin (Cleveland, 2001). That compound is produced by the lactic acid bacteria *Lactococcus lactis* (Table 8.5)and activity of 6,400 U/mL has been measured, which is an elevated activity compared to that produced by *Enterococcus faecalis* (360 U/mL). These bacteriocins are employed for the conservation of food, their production is easy and their production costs are low.

TABLE 8.5 Bacteriocins Produced from Whey (W) Fermentation

Bacteriocin	Strain	Culture media	Activity (U/mL)	References
Enterocin	*Enterococcus faecalis*	Whey +glucose	360	Ananou et al., (2008)
Nisin	*Lactococcus lactis*	Whey + soybean peptone and $MgSO_4/MnSO_4$ mixture	575; 6,400	González-Toledo et al., (2010)
		Whey permeate + casein hydrolysate		Liu et al., (2005)
Pediocin	*Pediococcus acidilactici*	Deproteinized whey +yeast extract	720	Guerra et al., (2007)

8.3.6.2 ENZYMES

It has been reported that some bacteria and yeasts, when added to whey and permeated whey, are good enzyme producers. However, when whey is supplemented with nitrogen sources and other nutrients, yields are enhanced.

The enzyme that has been most used by the food industry is β-D-galactosidase, better known as lactase. This enzyme is obtained from yeast and its function is to catalyze the hydrolysis of lactose to glucose and galactose (Panesar Parmjit et al., 2006). For the manufacture of beer and bioprocessing of the starch the enzyme used is amylase, which using the microorganism *Aspergillus niger* and with proper supplementation can obtain a good activity of this enzyme (Akpan et al., 1999). The maximum activity for the enzymes of amylase and galactosidase are reported in Table 8.6, obtained an activity of 4,360 and 500 U/mL, this activity can be observed due to the carbon and nitrogen source with which they are supplemented, obtaining a good response.

TABLE 8.6 Enzymes Produced by Whey

Enzymes	Microorganism	Culture media	Activity	References
α-amylase	*Bacillus sp.*	Whey + defatted soya, nutrients salts and corn starch	1,295 U/mL	Bajpai et al., (1991)
		Whey + corn starch, corn gluten, and nutrients salts	4,360 U/mL	Bajpai et al., (1992)
β-D-Galactosidase	*Kluyveromyces marxianus*	Whey + lactose	291 U/L	Rech and Ayub (2007)
	Kluyveromyces fragilis	Lactose and nutrients	18,300 U/L	Mahoney et al., (1975)
	Kluyveromyces lactis	Hydrolysis of lactose	500 U/mL	Siso (1994)
	Streptococcus thermophiles	Lactose protease peptone	18,200 U/L	Rao and Dutta (1977)
Protease	*Aspergillus terreus*	Whey	2.25 U/mL	Ali (2008)
		Whey + starch and gelatin	15 U/mL	El-Shora and Metwally (2008)
	Aspergillus niger	Whey + starch and yeast extract	16 U/mL	El-Shora and Metwally (2008)
	Serratia marcescens	Whey + glucose	3800 U/mL	Romero et al., (2001)

Source: Modified from Yadav et al., (2015).

3.6.3 ORGANIC ACIDS

Due to the high contents of nutrients contained in whey, it is considered a very good option for the production of organic acids, via a fermentation process. Some acids, such as propionic, lactic, citric and acetic, can be produced by fermenting the whey and permeate whey. From the economic point of view, lactic acid is the one that represents the highest production, due to the diverse use that can be given to it, from dairy products to food additive and plastic degrader (Bogaert, 1997). Another acid that is very useful, especially in the food industry is citric acid, which is used as an antioxidant, acidifier and ingredient that improves flavor (Kumar, 2003). These two acids by the *Lactobacillus bulgaricus* and *Aspergillus niger* microorganisms are responsible for giving a higher production yield according to the shown in Table 8.7, where the lactic acid shows 20.89 g/L and the citric acid gives a yield of 106.5 g/L, being supplementals with lactose and sucrose respectively.

8.3.7 SINGLE CELL PROTEIN (SCP)

Cheese whey is a waste by-product of the dairy industry that contains a low level of non-coagulable proteins (0.6–0.7 % w/v), which precludes it from being accepted as a high-grade material (Willetts and Unai, 1987). In contrast, it is rich in lactose (4–5 % w/v) which corresponds to a chemical oxygen demand (COD) between 40,000 to 60,000 ppm, which may disrupt the biological process of sewage-disposal plants (Ghaly and Singh, 1989). For this reason, whey can be more profitable if lactose is converted into single-cell protein (SCP) by microorganisms. SCP is the protein extracted from cultivated microbial biomass and can be used as supplementation of a staple diet by replacing conventional sources like fishmeal or soymeal, which are more expensive. Its production responds to increasing demand for protein sources of high nutrition value in human foods and animal feeds (Emtiazi et al., 2003).

For the production of SCP, fungi, bacteria and algae can be used nut yeasts are more suitable due to the higher nutritional quality of yeasts protein (Young and Scrimshaw, 1975); fewer amino acids containing sulfur have been reported in bacterial than in yeasts proteins but the higher protein content of yeasts make them a natural protein concentrate (Peppler, 1970). Microbial biomass has been produced commercially from whey since the 1940s, but its production for use as food started in France at Fromageries Le Bel around 1958 (Yves, 1979).

Within the production of SCP, different microorganisms have been evaluated, even yeasts as *Kluyveromyces*, *Candida,* and *Trichosporon* species are

TABLE 8.7 Organic Acids Produced from Whey and Its Derivates by Fermentation Processes

Organic Acid	Strain	Substrate	Bioconversion	Yield (g/L)	References
Acetic acid	*Kluyveromyces marxianus* and *Acetobacter pasteurans* or *Gluconoaceto bacterliquefaciens*	Whey permeate	Lactose to ethanol and ethanol to acetic acid	55 and 52	Parrondo et al., (2009)
Citric acid	*Aspergillus niger*	Whey	Whey to citric acid Supplemented whit sucrose	106.5	El-Holi and Al-Delaimy (2003)
	Candidalipolytica	Whey	Conversion of whey to citric acid Supplemented whit date seed ash, sources of nitrogen and glucose	42	Abou-Zeid et al., (1983)
	Metschnikowia pulcherrima	Whey	Conversion of whey to citric acid Supplemented whit unclarified whey and nutrients	2.5	Singh et al., (2004)
	Yarrowia lipolytica	Whey	Conversion of whey to citric acid Supplemented whit fructose and glucose	49.23	Yalcin et al., (2009)
Gibberellic acid	*Fusarium moniliforme*	Whey permeate	Conversion of lactose to gibberellic acid	0.68	Kahlon and Malhotra (1986)
Gluconic acid	*Aspergillus niger*	Deproteinized whey	Conversion of lactose and glucose	92	Mukhopadhyay et al., (2005)
Lactic acid	*Lactobacillus helveticus*	Whey ultrafiltrate Hydrolyzed whey	Conversion of whey supplemented whit yeast extract and lactose Conversion of whey supplemented with yeast autolysate and transferring into whey permeate with corn-steep	9.7 5.5	Roy et al., (1987) Amrane and Prigent (1993)

usually employed as they naturally metabolize whey lactose (Mansour et al., 1993). Sandhu and Waraich (Sandhu and Waraich, 1983) evaluated the production of ß-galactosidase among thirteen yeasts including *Brettanomyces anomalus*, *Kluyveromyces fragilis*, *Wingea robertsii,* and *Trichosporon cutaneum* for single-cell production from cheese whey supplemented with minerals and yeast extract, which demonstrated to be a suitable media for yield, biomass production, conversion efficiency and yield, while *W. robertsii* proved to be the best strains regarding yield and shorter incubation period.

The production of microbial biomass is done mainly by continuous fermentation systems (Ghaly et al., 1992). In the case of production of SCP requires both aeration and agitation to maintain a homogeneous medium, to provide intimate contact between the microorganisms and their substrate, to provide a uniform temperature distribution and better efficiency of heat removal and to ensure that adequate oxygen is obtained by all yeasts (Ghaly, Kamal, and Correia, 2005). For this purpose, different production systems have been applied for SCP production including continuous stirred tank and bubble-column reactors. The latter is simple in construction, lack of moving parts, low shear forces and high energy efficiency for mass transfer (Deckwer and Wolf-Dieter, 1992; Parasu, 1999). Hosseini et al (2003) achieved the fermentation of deproteinized cheese whey resulting in 17.3 g/L biomass under optimum conditions (aeration, 7.5 vvm; L/D ratio, 3.5; pH, 3.5) using *Trichosporon* in a bubble column reactor. This system also offers a more stable physical environment for growth while in stirred tank reactors, high shear forces can arise near the impeller causing cell stress/damage thus lowering productivity.

The production process of SCP faces some challenges such as low efficiency of COD removal, fate and recovery of soluble whey protein after fermentation and economy of the process and contamination problem. Yadav et al., (2014) demonstrate that the use of high-cell-density inoculum resulted in higher COD removal (80 %) with a biomass productivity of 0.25 g/L h in batch system for *Kluyveromyces marxianus*, but applying this finding in a continuous process they achieved a COD reduction of up to 78.5 % with biomass productivity of 0.26 g/L h. Ghaly et al., (1992) obtained a total biomass output (viable and dead cells) of 37 g/L also using *K. marxianus* for cheese whey fermentation through the development of a mathematical model describing the kinetics of continuous production of SCP. This model took into account the effect of substrate concentration and cell death rate, substrate utilization during the fermentation process and its use for growth and maintenance, demonstrating that the use of models may lead to the development of better strategies for the optimization of SCP production which ensures its economic viability. Wei et al., (2013) also analyzed a mathematical model of cheese whey fermentation for SCP production with impulsive state feedback control. This was introduced in order to obtain the optimal conditions for microorganism growth and decrease the inhibition of the microorganism concentration. They conclude that researchers in this area should give suitable feedback state, appropriate initial concentration of microorganism and substrate and control parameters in order to obtain a steady and optimal production.

8.3.8 BIOELECTRICITY: MICROBIAL FUEL CELL (MFC)

Microbial fuel cells can convert the energy available in organic substrates into electricity through the following reactions described by Antonopoulou et al. [168]:

$$\text{Anodic reaction: } C_6H_{12}0_6 + H_20 \rightarrow 6CO_2 + 24e^- + 24H^+$$

$$\text{Cathodic reaction: } O_2 + 4e^- + 4H^+ \rightarrow 2H_2O$$

Electrons produced by the bacteria are conveyed to the anode electrode and flow to the cathode ones linked by a conductive material containing a resistor or operated under a load. Electrons are able to be transferred to the anode by mediators, shuttles and nanowires. Moreover, some chemical mediators have been used to increase the efficiency of power generation in MFCs. Usually, MFC two-chamber are used where reached electrons in the cathode compartment combine with the diffused protons from the anode one via a membrane and oxygen originated from atmosphere, resulting in water formation (Nasirahmadi and Safekordi, 2011).

Electricity can be generated using glucose, lactate or acetate as fuel pure compounds. However, MFCs can be also used in the wastewater treatment such as domestic water, swine water and food wastewater. Within the latter, cheese whey can be considered due to its richness in lactose. The aerobic digestion of this wastewater offers an excellent alternative for energy conservation and for the control of a serious environmental problem. However, this procedure involves some challenges for anaerobic digestion of raw cheese whey due to low alkalinity, its tendency to get easily acidified and the high COD concentration. For this reason, Antonopoulou et al., proposed the use of diluted cheese whey "feta" in a two-chamber mediator-less MFC by indigenous microbial consortium contained in wastewater. By this strategy, they obtained a maximum power density of 18.4 mW/m^2 (corresponding to a current density of 80 mA/m2 and a MFC voltage of 0.23 V. For comparison reasons, the experiment was also carried out using either glucose or lactose, obtaining a maximum power density of 15.2 mW/m^2 and 17.2 mW/m^2, respectively. Thus, the power generation using this wastewater as energy source was demonstrated.

Nasirahmadi and Safekordi (Nasirahmadi and Safekordi, 2011) investigated the production of bioelectricity from treated whey using *E. coli* in the anode chamber of MFC using two mediators: riboflavin and humic acid. When in absence of mediator, *E. coli* was able to use whey as substrate to generate bioelectricity with an open-circuit voltage of 751.1 mV at room temperature; the voltage was stable for more than 24 h. Regarding mediators, humic acid was a few times more effective than riboflavin and enhanced the electrical energy; the power ad current production was increased to 324.8 μW and 1194.6 μA, respectively. Kassongo and Togo (Kassongo and Togo, 2010) also investigated whey remediation and MFC for fueling. They used three different anodic setups: raw whey inoculated with *Enterobacter cloacae*, sterile whey also inoculated with *E. cloacae* and raw whey alone. In the latter experiment, the native whey microbes achieved a 44.7 % COD removal efficiency and 0.04 % coulombic efficiency with a maximum power density of 0.4 W/m^2. The use of an exogenous electricigenic bacteria (*E. cloacae*) resulted in a dropping of COD removal to 5 % while the coulombic efficiency was the highest (3.7 %) and the power density of 16.7 W/m^2. The experiment with a combination of *E. cloacae* and raw whey gave 1.1 W/m^2, 0.5% of coulombic efficiency and 22.1% COD removal. Thus, it was demonstrated that whey can be used as a fuel in the anodic chamber for electricity generation, with its partial remediation (COD removal), but that there is no synergism between *E. cloacae* and the indigenous bacteria in whey.

8.4 FUTURE TRENDS/PERSPECTIVES

Along with this chapter, we have reviewed the current applications of whey. However, the progressive advances in food science and technology lead to new and novel uses of raw materials as well as some by-products. The scientific evidence described is a clear attempt regarding the more efficient technologies for processing whey instead of discharge it into waterways or sprayed onto farmland.

For its chemical composition, whey-based ingredients can be employed for the preparation of functional foods and nutraceuticals, which could be helpful in weight loss, aging and nutrition (through improving flavor and texture). Some examples of these new products can include bakery, beverages, ice-cream, mayonnaise, confectionery, soup, sauces, and others. On the other hand, whey can also be used for the production of biofuels. In this sense, strain´s improvement can conduct higher conversions of lactose in order to increase yields and productivities and also new reactors designs as well as their operation modes, can be explored.

Source: Modified from Yadav et al., (2015).

8.5 CONCLUSIONS

Whey is constituted of several nutrients such as protein, lactose, fat, and minerals. This composition is responsible for both high COD and BOD levels, which adversely affect the environment when whey is discarded. At the same time, such an organic load is convenient from nutritional and functional values. First, physical and chemical methods were employed to separate proteins and fat from whey showing important pollution load reduction, and on the same hand, it was possible to recover such products for food applications. Secondly, biological treatment was used due to their higher efficiency and for their eco-friendly impact, turning out in sustainable process for the dairy industry. Nowadays, new and novel advances in biotechnology allowed to transform residues or by-products into value-added compounds. In the case of whey, fermentation technology as well as biocatalysis made possible the utilization of the organic matter content of whey for the production of a diverse products, including biofuels, biopolymers, bacteriocins, enzymes, organic acids, single-cell proteins or even bioelectricity, by means of low-cost alternatives bacteria, yeast and algae cells. Even when several bioprocesses were described, it is still possible to design a new one to increase the applications of whey, to improve the technologies used, to reduce the environmental impact, and to reduce production costs.

KEYWORDS

- **whey**
- **valorization**
- **bioconversion**
- **chemical oxygen demand**
- **value-added**

REFERENCES

Abou-Zeid, A. Z. A., Baghlaf, A. O., Khan, J. A., & Makhashin, S. S., (1983). Utilization of date seeds and cheese whey in production of citric acid by *Candida lipolytica*. *Agricultural Wastes*, *8*(3), 131–142.

Ahn, W. S., Park, S. J., & Lee, S. Y., (2001). Production of poly(3-hydroxybutyrate) from whey by cell recycle fed-batch culture of recombinant *Escherichia coli*. *Biotechnology Letters*, *23*(3), 235–240.

Aider, M., & De Halleux, D., (2007). Isomerization of lactose and lactulose production. *Trends in Food Science and Technology*, *18*(7), 356–364.

Akpan, I., Bankole, M. O., Adesemowo, A. M., & Latunde-Dada, G. O., (1999). Production of amylase by *A. niger* in a cheap solid medium using rice bran and agricultural materials. *Tropical Science*, *39*(2), 77–79.

Ali, U. F., (2008). Utilization of whey amended with some agro-industrial by-products for the improvement of protease production by *Aspergillus terreus* and its compatibility with commercial detergents. *Research Journal of Agriculture and Biological Sciences*, *4*(6), 886–891.

Almeida, K. E., Tamime, A. Y., & Oliveira, M. N., (2008). Acidification rates of probiotic bacteria in minasfrescal cheese whey. *LWT-Food Science and Technology*, *41*(2), 311–316.

Amrane, A., & Prigent, Y., (1993). Influence of media composition on lactic acid production rate from whey by *Lactobacillus helveticus*. *Biotechnology Letters*, *15*(3), 239–244.

Ananou, S., Muñoz, A., Gálvez, A., Martínez-Bueno, M., Maqueda, M., & Valdivia, E., (2008). Optimization of enterocin AS-48 production on a whey-based substrate. *International Dairy Journal*, *18*(9), 923–927.

Antonopoulou, G., Stamatelatou, K., Bebelis, S., & Lyberatos, G., (2008). *Electricity Generation from Cheese Whey Using a Microbial Fuel Cell.* Conference paper.

Arasaratnam, V., Senthuran, A., & Balasubramaniam, K., (1996). Supplementation of whey with glucose and different nitrogen sources for lactic acid production by *Lactobacillus delbrueckii*. *Enzyme and Microbial Technology*, *19*(7), 482–486.

Audic, J. L., Chaufer, B., & Daufin, G., (2003). Non-food applications of milk components and dairy co-products: A review. *Le Lait*, *83*(6), 417–438.

Baei, S. M., Najafpour, G. D., Lasemi, Z., Tab, F., Younesi, H., Issazadeh, H., & Khodab, M., (2010). Optimization PHAs production from dairy industry wastewater (Cheese Whey) by *Azohydromonaslata* DSMZ 1123. *Iranica Journal of Energy and Environment*, *1*(2).

Bajpai, P., Gera, R. K., & Bajpai, P. K., (1992). Optimization studies for the production of α-amylase using cheese whey medium. *Enzyme and Microbial Technology*, *14*(8), 679–683.

Bajpai, P., Verma, N., Neer, J., & Bajpai, P. K., (1991). Utilization of cheese whey for production of α-amylase enzyme. *Journal of Biotechnology*, *18*(3), 265–270.

Barukčić, I., Božanić, R., & Tratnik, L., (2008). *Whey-Based Beverages-a New Generation of Diary Products*, 58, 257–274.

Beaulieu, J., Dupont, C., & Lemieux, P., (2006). Whey proteins and peptides: Beneficial effects on immune health. *Clinical Practice*, *3*(1), 69.

Beucler, J., Drake, M., Allen Foegeding, E., (2005). *Design of a Beverage from Whey Permeate*, *70*, S277–S285.

Bogaert, J. C., (1997). Production and novel applications of natural L(+) lactic acid: Food, pharmaceutics and biodegradable polymers. *Cerevisa*, *22*, 46–50.

Boura, K., Kandylis, P., Bekatorou, A., Kolliopoulos, D., Vasileiou, D., Panas, P., Kanellaki, M., & Koutinas, A. A., (2017). New generation biofuel from whey: Successive acidogenesis and alcoholic fermentation using immobilized cultures on *γ*-alumina. *Energy Conversion and Management*, *135*, 256–260.

Brambilla, A., Lo Scalzo, R., Bertolo, G., & Torreggiani, D., (2008). *Steam-Blanched High Bush Blueberry (Vaccinium corymbosum L.) Juice: Phenolic Profile and Antioxidant Capacity in Relation to Cultivar Selection*, *56*, 2643–2648.

Carota, E., Crognale, S., D'Annibale, A., Gallo, A. M., Stazi, S. R., & Petruccioli, M., (2017). A sustainable use of ricotta cheese whey for microbial biodiesel production. *Science of the Total Environment*, *584–585*, 554–560.

Castilho, L. R., Mitchell, D. A., & Freire, D. M. G., (2009). Production of polyhydroxyalkanoates (PHAs) from waste materials and by-products by submerged and solid-state fermentation. *Bioresource Technology*, *100*(23), 5996–6009.

Castro, W. F., Cruz, A. G., Bisinotto, M. S., Guerreiro, L. M. R., Faria, J. A. F., Bolini, H. M. A., Cunha, R. L., & Deliza, R., (2013). Development of probiotic dairy beverages: Rheological properties and application of mathematical models in sensory evaluation. *Journal of Dairy Science*, *96*(1), 16–25.

Chandan, R. C., Kilara, A., & Shah, N. P., (2009). Dairy Processing and Quality Assurance: An overview. *Dairy Process. Qual. Assur.* John Wiley, London.

Chanprateep, S., (2010). Current trends in biodegradable polyhydroxyalkanoates. *Journal of Bioscience and Bioengineering*, *110*(6), 621–632.

Chatterjee, G., De Neve, J., Dutta, A., & Das, S., (2015). *Formulation and Statistical Evaluation of a Ready-to-Drink Whey Based Orange Beverage and its Storage Stability*, *14*, 253–264.

Chatzipaschali, A. A., & Stamatis, A. G., (2012). Biotechnological utilization with a focus on anaerobic treatment of cheese whey: Current status and prospects. *Energies*, *5*(9), 3492.

Chavan, R., A Nalawade, T., K., (2015). *Whey Based Beverage: Its Functionality, Formulations, Health Benefits and Applications* (Vol. 6, pp. 1–8).

Choi, J., & Lee, S. Y., (1999). Factors affecting the economics of polyhydroxyalkanoate production by bacterial fermentation. *Applied Microbiology and Biotechnology*, *51*(1), 13–21.

Cleveland, J., Montville, T. J., Nes, I. F., & Chikindas, M. L., (2001). Bacteriocins: Safe, natural antimicrobials for food preservation. *International Journal of Food Microbiology*, *71* (1), 1–20.

Colomban, A., Roger, L., & Boyaval, P., (1993). Production of propionic acid from whey permeate by sequential fermentation, ultra filtration, and cell recycling. *Biotechnology and Bioengineering*, *42*(9), 1091–1098.

Cota-Navarro, C. B., Carrillo-Reyes, J., Davila-Vazquez, G., Alatriste-Mondragón, F., & Razo-Flores, E., (2011). Continuous hydrogen and methane production in a two-stage cheese whey fermentation system. *Water Science and Technology*, *64*(2), 367–374.

Das, M., Raychaudhuri, A., & Ghosh, S. K., (2016). Supply chain of bioethanol production from whey: A review. *Procedia Environmental Sciences*, *35*, 833–846.

Davila-Vazquez, G., Alatriste-Mondragón, F., De León-Rodríguez, A., & Razo-Flores, E., (2008). Fermentative hydrogen production in batch experiments using lactose, cheese whey and glucose: Influence of initial substrate concentration and pH. *International Journal of Hydrogen Energy*, *33*(19), 4989–4997.

Davila-Vazquez, G., Cota-Navarro, C. B., Rosales-Colunga, L. M., De León-Rodríguez, A., & Razo-Flores, E., (2009). Continuous biohydrogen production using cheese whey: Improving the hydrogen production rate. *International Journal of Hydrogen Energy*, *34*(10), 4296–4304.

Deckwer & Wolf-Dieter, (1992). *Bubble Column Reactors*. Wiley: New York.

Deshpande, D., Patil, P., & Anekar, S., (2012). Biomethanation of dairy waste. *Research Journal of Chemical Sciences*, *2*(4), 35–39.

Dills, Jr. W. L., (1989). Sugar alcohols as bulk sweeteners. *Annual Review of Nutrition*, *9*(1), 161–186.

Djurić, M., Carić, M., Milanović, S., Tekić, M., & Panić, M., (2004). *Development of Whey-Based Beverages*, *219*, 321–328.

Domingues, L., Dantas, M. M., Lima, N., & Teixeira, J. A., (1999). Continuous ethanol fermentation of lactose by a recombinant flocculating *Saccharomyces cerevisiae* strain. *Biotechnology and Bioengineering*, *64*(6), 692–697.

Dragone, G., Mussatto, S. I., Oliveira, J. M., & Teixeira, J. A., (2009). Characterization of volatile compounds in an alcoholic beverage produced by whey fermentation. *Food Chemistry*, *112*(4), 929–935.

El-Holi, M. A., & Al-Delaimy, S., (2003). Citric acid production from whey with sugars and additives by *Aspergillus niger*. *African Journal of Biotechnology*, *2*(10), 356–359.

El-Shora, H., & Metwally, M., (2008). Production, purification and characterization of proteases from whey by some fungi. *Annals of Microbiology*, *58*(3), 495–502.

Emtiazi, G., Etemadifar, Z., & Tavassoli, M., (2003). A novel nitrogen fixing cellulytic bacterium associated with root of corn is a candidate for production of single cell protein, *Biomass and Bioenergy, 25*(4), 423–426.

Ergüder, T. H., Tezel, U., Güven, E., & Demirer, G. N., (2001). Anaerobic biotransformation and methane generation potential of cheese whey in batch and UASB reactors. *Waste Management*, *21*(7), 643–650.

Faruqui, A. A., & Joshi, C., (2012). Lactitol: A review of its use in the treatment of constipation. *Int. J. Recent Adv. Pharm. Res.*, *2*(1), 1–5.

Friend, B. A., Cunningham, M. L., & Shahani, K. M., (1982). Industrial alcohol production via whey and grain fermentation. *Agricultural Wastes*, *4*(1), 55–63.

Gallardo-Escamilla, F. J., Kelly, A. L., & Delahunty, C. M., (2007). Mouthfeel and flavor of fermented whey with added hydrocolloids. *International Dairy Journal*, *17*(4), 308–315.

Gänzle, M. G., Haase, G., & Jelen, P., (2008). Lactose: Crystallization, hydrolysis and value-added derivatives. *International Dairy Journal*, *18*(7), 685–694.

Garg, N., (2017). Chapter 8-Technology for the Production of Agricultural Wines. In: Kosseva, M. R., Joshi, V. K., & Panesar, P. S., (eds.), *Science and Technology of Fruit Wine Production* (pp. 463–486). Academic Press: San Diego.

Ghaly, A. E., & El-Taweel, A. A., (1997). Continuous ethanol production from cheese whey fermentation by Candida pseudotropicalis. *Energy Sources*, *19*(10), 1043–1063.

Ghaly, A. E., & Singh, R. K., (1989). Pollution potential reduction of cheese whey through yeast fermentation. *Appl. Biochem. Biotechnol., 22*, 181–203.

Ghaly, A. E., Ben-Hassan, R. M., & Ben-Abdallah, N., (1992). Measurement of heat generated by mixing during batch production of single cell protein from cheese whey. *Biotechnol. Prog., 8*, 404–409.

Ghaly, A. E., Kamal, & Correia, L. R., (2005). Kinetic modeling of continuous submerged fermentation of cheese whey for single cell protein production. *Bioresource Technology*, 96, 1143–1152.

Ghasemi, M., Najafpour, G., Rahimnejad, M., Beigi, P. A., Sedighi, M., & B. H., (2009). Effect of different media on production of lactic acid from whey by *Lactobacillus bulgaricus*. *African Journal of Biotechnology*, *8*(1), 81–84.

Golowczyc, M., Vera, C., Santos, M., Guerrero, C., Carasi, P., Illanes, A., Gómez-Zavaglia, A., & Tymczyszyn, E., (2013). Use of whey permeate containing in situ synthesized galacto-oligosaccharides for the growth and preservation of Lactobacillus plantarum. *Journal of Dairy Research*, *80*(3), 374–381.

González-Toledo, S. Y., Domínguez-Domínguez, J., García-Almendárez, B. E., Prado-Barragán, L. A., & Regalado-González, C., (2010). Optimization of nisin production by *Lactococcus lactis* UQ2 using supplemented whey as alternative culture medium. *Journal of Food Science*, *75*(6), M347–M353.

Source: Modified from Yadav et al., (2015).

Goudarzi, M., Madadlou, A., Mousavi, M., & Emam-Djomeh, Z., (2014). *Formulation of apple Juice Beverages Containing Whey Protein Isolate or Whey Protein Hydrolysate Based on Sensory and Physicochemical Analysis, 68.*

Guerra, N. P., Bernárdez, P. F., & Castro, L. P., (2007). Fed-batch pediocin production on whey using different feeding media. *Enzyme and Microbial Technology*, *41*(3), 397–406.

Holsinger, V. H., Posati, L. P., & DeVilbiss, E. D., (1974). Whey beverages: A review. *Journal of Dairy Science*, *57*(8), 849–859.

Hosseini, M., Seyed, A. S., & Jafar, T., (2003). Application of a Bubble-Column Reactor for the Production of a Single-Cell Protein from Cheese Whey. *Ind. Eng. Chem. Res.*, *42*, 764–766.

Houldsworth, D., (1980). Demineralization of whey by means of ion exchange and electro dialysis. *International Journal of Dairy Technology*, *33*(2), 45–51.

Janiaski, D. R., Pimentel, T. C., Cruz, A. G., & Prudencio, S. H., (2016). Strawberry-flavored yogurts and whey beverages: What is the sensory profile of the ideal product? *Journal of Dairy Science*, *99*(7), 5273–5283.

Jarc, S., P. K., & Hadžiosmanović, M., (1994). Chemical, bacteriological and sensory quality indices of whey-fruit drinks. *Mljekarstvo*, *44*(3), 186–196.

Jiang, Y., Marang, L., Tamis, J., Van, L. M. C. M., Dijkman, H., & Kleerebezem, R., (2012). Waste to resource: Converting paper mill wastewater to bioplastic. *Water Research*, *46*(17), 5517–5530.

Jinjarak, S., Olabi, A., Jiménez-Flores, R., & Walker, J. H., (2006). Sensory, Functional, and Analytical Comparisons of Whey Butter with Other Butters. *Journal of Dairy Science*, *89*(7), 2428–2440.

Jinjarak, S., Olabi, A., Jiménez-Flores, R., & Walker, J., (2006). Sensory, functional, and analytical comparisons of whey butter with other butters1. *Journal of Dairy Science*, *89*(7), 2428–2440.

Joesten, M. D., Hogg, J. L., & Castellion, M. E., (2006). *The World of Chemistry: Essentials: Essentials*. Cengage Learning.

Jovanovic-Malinovska, R., Fernandes, P., Winkelhausen, E., & Fonseca, L., (2012). Galacto-oligosaccharides synthesis from lactose and whey by β-galactosidase immobilized in PVA. *Applied Biochemistry and Biotechnology*, *168*(5), 1197–1211.

Kahlon, S. S., & Malhotra, S., (1986). Production of gibberellic acid by fungal mycelium immobilized in sodium alginate. *Enzyme and Microbial Technology*, *8*(10), 613–616.

Kailasapathy, K., & Chin, J., (2000). Survival and therapeutic potential of probiotic organisms with reference to *Lactobacillus acidophilus* and *Bifidobacterium* spp. *Immunology and Cell Biology*, *78*(1), 80–88.

Kassongo, J., & Togo, C. A., (2010). The potential of whey in driving microbial fuel cells: A dual prospect of energy recovery and remediation. *African Journal of Biotechnology*, *9*(46), 7885–7890.

Khanafari, A., Sepahei, A., & Mogharab, M., (2006). Production and recovery of poly-hdroxybutyrate from whey degradation by *Azotobacter*. *Iran J. Environ. Health. Sci. Eng.*, *3*(3), 193–198.

Kim, B. S., (2000). Production of poly(3-hydroxybutyrate) from inexpensive substrates. *Enzyme and Microbial Technology*, *27*(10), 774–777.

Koller, M., Bona, R., Chiellini, E., Fernandes, E. G., Horvat, P., Kutschera, C., Hesse, P., & Braunegg, G., (2008). Polyhydroxyalkanoate production from whey by *Pseudomonas* hydrogenovora. *Bioresource Technology*, *99*(11), 4854–4863.

Koller, M., Hesse, P., Bona, R., Kutschera, C., Atlić, A., & Braunegg, G., (2007). Biosynthesis of high quality polyhydroxyalkanoate Co- and terpolyesters for potential medical application by the archaeon *Haloferax mediterranei*. *Macromolecular Symposia*, *253*(1), 33–39.

Koller, M., Hesse, P., Salerno, A., Reiterer, A., & Braunegg, G., (2011). A viable antibiotic strategy against microbial contamination in biotechnological production of polyhydroxyalkanoates from surplus whey. *Biomass and Bioenergy*, *35*(1), 748–753.

Kosaric, N., & Asher, Y., (1985). The utilization of cheese whey and its components. In: *Agricultural Feedstock and Waste Treatment and Engineering* (pp. 25–60). Springer.

Kosikowski, F. V., & Wzorek, W., (1977). Whey Wine from Concentrates of Reconstituted Acid Whey Powder. *Journal of Dairy Science*, *60*(12), 1982–1986.

Kosikowski, F. V., (1979). Whey Utilization and Whey Products. *Journal of Dairy Science*, *62*(7), 1149–1160.

Kośmider, A., Drożdżyńska, A., Blaszka, K., Leja, K., & Czaczyk, K., (2010). Propionic acid production by *Propionibacterium freudenreichii* ssp. *shermanii* using industrial wastes: Crude glycerol and whey lactose. *Polish Journal of Environmental Studies*, *19*(6), 1249–1253.

Kumar, D., Jain, V. K., Shanker, G., & Srivastava, A., (2003). Utilization of fruits waste for citric acid production by solid state fermentation. *Process Biochemistry*, *38*(12), 1725–1729.

Kumar, N., & Vandna, (2015). *Fermented and Non-Fermented Whey Beverages*, *42*, 28–31.

Lee, P. C., Lee, S. Y., Hong, S. H., & Chang, H. N., (2003). Batch and continuous cultures of *Mannheimia succiniciproducens* MBEL55E for the production of succinic acid from whey and corn steep liquor. *Bioprocess and Biosystems Engineering*, *26*(1), 63–67.

Lee, P. C., Lee, W. G., Kwon, S., Lee, S. Y., & Chang, H. N., (2000). Batch and continuous cultivation of *Anaerobiospirillum succiniciproducens* for the production of succinic acid from whey. *Applied Microbiology and Biotechnology*, *54*(1), 23–27.

Lee, S. Y., Middelberg, A. P. J., & Lee, Y. K., (1997). Poly(3-hydroxybutyrate) production from whey using recombinant *Escherichia coli*. *Biotechnology Letters*, *19* (10), 1033–1035.

Liu, X., Chung, Y. K., Yang, S. T., & Yousef, A. E., (2005). Continuous nisin production in laboratory media and whey permeate by immobilized *Lactococcus lactis*. *Process Biochemistry*, *40*(1), 13–24.

Macfarlane, G., Steed, H., & Macfarlane, S., (2008). Bacterial metabolism and health-related effects of galacto-oligosaccharides and other prebiotics. *Journal of Applied Microbiology*, *104*(2), 305–344.

Mahoney, R. R., Nickerson, T. A., & Whitaker, J. R., (1975). Selection of strain, growth conditions, and extraction procedures for optimum production of lactase from *Kluyveromyces fragilis*. *Journal of Dairy Science*, *58*(11), 1620–1629.

Mansour, M. H., Ghaly, A. E., Benhassan, R. M., & Nassar, M. A., (1993). Modeling batch production of single cell protein from cheese whey. *Appl. Biochem. Biotechnol.*, *43*, 1–14.

Marchetti, J. M., (2012). A summary of the available technologies for biodiesel production based on a comparison of different feedstock's properties. *Process Safety and Environmental Protection*, *90*(3), 157–163.

Marcos, E., Castillo, F. A., Dimitrov, S. T., Gombossy, D. M. B. D., De Souza, R. P., (2013). Novel biotechnological applications of bacteriocins: A review. . *Food Control*, *32*, 134–142.

Marwaha, S. S., & Kennedy, J. F., (1998). Whey-pollution problem and potential utilization. *International Journal of Food Science and Technology*, *23*(4), 323–336.

Mayer, J., Conrad, J., Klaiber, I., Lutz-Wahl, S., Beifuss, U., & Fischer, L., (2004). Enzymatic production and complete nuclear magnetic resonance assignment of the sugar lactulose. *Journal of Agricultural and Food Chemistry*, *52*(23), 6983–6990.

McSweeney, P. L. H., Ottogalli, G., & Fox, P. F., (2004). Diversity of cheese varieties: An overview. In: Fox, P. F., McSweeney, P. L. H., Cogan, T. M., & Guinee, T. P., (eds.), *Cheese: Chemistry, Physics and Microbiology* (Vol. 2, pp. 1–23). Academic Press.

McSweeney, P. L., & Fox, P. F., (2009). *Advanced Dairy Chemistry: Volume 3: Lactose, Water, Salts and Minor Constituents* (Vol. 3). Springer.

Mockaitis, G., Ratusznei, S. M., Rodrigues, J. A. D., Zaiat, M., & Foresti, E., (2006). Anaerobic whey treatment by a stirred sequencing batch reactor (ASBR): Effects of organic loading and supplemented alkalinity. *Journal of Environmental Management*, *79*(2), 198–206.

Mollea, C., Marmo, L., & Bosco, F., (2013). Valorization of cheese whey, a by-product from the dairy industry. In: *Food Industry*, InTech. https://www.intechopen.com/books/food-industry/valorisation-of-cheese-whey-a-by-product-from-the-dairy-industry

Morin, P., Pouliot, Y., & Jiménez-Flores, R., (2006). A comparative study of the fractionation of regular buttermilk and whey buttermilk by microfiltration. *Journal of Food Engineering*, *77*(3), 521–528.

Moulin, G., & Galzy, P., (1984). Whey, a potential substrate for biotechnology. *Biotechnology and Genetic Engineering Reviews*, *1*(1), 347–374.

Mukhopadhyay, R., Chatterjee, S., Chatterjee, B. P., Banerjee, P. C., & Guha, A. K., (2005). Production of gluconic acid from whey by free and immobilized *Aspergillus niger*. *International Dairy Journal*, *15*(3), 299–303.

Muñoz-Páez, K., Poggi-Varaldo, H., García-Mena, J., Ponce-Noyola, M., Ramos-Valdivia, A., Barrera-Cortés, J., Robles-González et al., (2014). Cheese whey as substrate of batch hydrogen production: Effect of temperature and addition of buffer. *Waste Management and Research*, *32*(5), 434–440.

Nasirahmadi, S., & Safekordi, A. A., (2011). Whey as a substrate for generation of bioelectricity in microbial fuel cell using *E.coli*. *Int. J. Environ. Sci. Tech.*, *8*(4), 823–830.

Nath, A., Dixit, M., Bandiya, A., Chavda, S., & Desai, A. J., (2008). Enhanced PHB production and scale up studies using cheese whey in fed batch culture of *Methylobacterium* sp. ZP24. *Bioresource Technology*, *99*(13), 5749–5755.

Obruca, S., Marova, I., Melusova, S., & Mravcova, L., (2011). Production of polyhydroxyalkanoates from cheese whey employing *Bacillus megaterium* CCM 2037. *Annals of Microbiology*, *61*(4), 947–953.

O'Donnell, K., & Kearsley, M., (2012). *Sweeteners and Sugar Alternatives in Food Technology*. John Wiley & Sons.

Oliveira, R. P. D. S., Florence, A. C. R., Perego, P., De Oliveira, M. N., & Converti, A., (2011). Use of lactulose as prebiotic and its influence on the growth, acidification profile and viable counts of different probiotics in fermented skim milk. *International Journal of Food Microbiology*, *145*(1), 22–27.

P. J., (1992). Whey and lactose processing. In: J. G. Z., (ed.), *Whey Cheeses and Beverages* (pp. 157–193). Springer, Dordrecht: England.

Pais, J., Serafim, L. S., Freitas, F., & Reis, M. A. M., (2016). Conversion of cheese whey into poly(3-hydroxybutyrate-co-3-hydroxyvalerate) by *Haloferax mediterranei*. *New Biotechnology*, *33*(1), 224–230.

Palmer, G. M., &. M. R. F., (1978). Modern technology transforms whey into wine. *Food Prod. Development*, *12*, 31–34.

Panesar, P. S., Kennedy, J. F., Gandhi, D. N., & Bunko, K., (2007). Bioutilisation of whey for lactic acid production. *Food Chemistry*, *105*(1), 1–14.

Panesar Parmjit, S., Panesar, R., Singh, R. S., Kennedy, J. F., & Kumar, H., (2006). Microbial production, immobilization and applications of β-D-galactosidase. *Journal of Chemical Technology and Biotechnology*, *81*(4), 530–543.

Pantazaki, A. A., Papaneophytou, C. P., Pritsa, A. G., Liakopoulou-Kyriakides, M., & Kyriakidis, D. A., (2009). Production of polyhydroxyalkanoates from whey by *Thermus thermophilus* HB8. *Process Biochemistry*, *44*(8), 847–853.

Parasu, V. U., & Joshi, J. B., (1999). Measurement of gas hold-up profiles by gamma ray tomography: Effect of sparger design and height of dispersion in bubble columns. *Trans. Inst. Chem. Eng., 77*, 303–317.

Parrondo, J., Garcia, L. A., & Diaz, M., (2009). Whey vinegar. In: Solieri, L., & Giudici, P., (eds.), *Vinegars of the World* (pp. 273–288). Springer Milan: Milano.

Patel, S. R., & Murthy, Z. V. P., (2011). Waste valorization: Recovery of lactose from partially deproteinated whey by using acetone as anti-solvent. *Dairy Science and Technology, 91*(1), 53–63.

Pauli, T., & Fitzpatrick, J. J., (2002). Malt combing nuts as a nutrient supplement to whey permeate for producing lactic by fermentation with *Lactobacillus casei. Process Biochemistry, 38*(1), 1–6.

Pepper, D., & Orchard, A., (1982). Improvements in the concentration of whey and milk by reverse osmosis. *International Journal of Dairy Technology, 35*(2), 49–53.

Peppler, H. J., (1970). Food yeasts. In: Rose, A. H., & Harrison, J. S., (eds.), *The Yeasts* (Vol. 3, pp. 421–462). New York: Academic Press.

Pescuma, M., Hebert, E. M., Mozzi, F., & Font, D. V. G., (2008). *Whey Fermentation by Thermophilic Lactic Acid Bacteria: Evolution of Carbohydrates and Protein Content, 25*, 442–451.

Pescuma, M., Hébert, E. M., Mozzi, F., & Font, D. V. G., (2008). Whey fermentation by thermophilic lactic acid bacteria: Evolution of carbohydrates and protein content. *Food Microbiology, 25*(3), 442–451.

Pintado, M., Macedo, A. C., & Malcata, F. X., (2001). *Review: Technology, Chemistry and Microbiology of Whey Cheeses, 7*, 105–116.

Pirozzi, D., Ausiello, A., Zuccaro, G., Sannino, F., & Yousuf, A., (2013). Culture of oleaginous yeast in dairy industry wastewaters to obtain lipids suitable for the production of ii-generation biodiesel. *International Journal of Chemical and Molecular Engineering, 7*(4), 162–166.

Pizzillo, M., Salvatore, C., Cifuni, G., Fedele, V., & Rubino, R., (2005). *Effect of Goat Breed on the Sensory, Chemical and Nutritional Characteristics of Ricotta Cheese, 94*,33–40.

Pogoń, K., Sady, M., Jaworska, G., & Grega, T., (2014). *Characteristics of Orange-Whey Fermented Beverages, 1*, 1–8.

Powers, H. J., (2003). Riboflavin (vitamin B-2) and health. *The American Journal of Clinical Nutrition, 77*(6), 1352–1360.

Prazeres, A. R., Carvalho, F., & Rivas, J., (2012). Cheese whey management: A review. *Journal of Environmental Management, 110*, 48–68.

Quillaguamán, J., Guzmán, H., Van-Thuoc, D., & Hatti-Kaul, R., (2010). Synthesis and production of polyhydroxyalkanoates by halophiles: Current potential and future prospects. *Applied Microbiology and Biotechnology, 85*(6), 1687–1696.

Rao, M. V., & Dutta, S. M., (1977). Production of beta-galactosidase from *Streptococcus thermophiles* grown in whey. *Applied Environmental Microbiology, 34*(2), 185–188.

Rech, R., & Ayub, M. A. Z., (2007). Simplified feeding strategies for fed-batch cultivation of *Kluyveromyces marxianus* in cheese whey. *Process Biochemistry, 42*(5), 873–877.

Rizzolo, A., & Cortellino, G., (2017). *Ricotta Cheese Whey-Fruit-Based Beverages: Pasteurization Effects on Antioxidant Composition and Color, 3*, 15.

Romero, F. J., García, L. A., Salas, J. A., Díaz, M., & Quirós, L. M., (2001). Production, purification and partial characterization of two extracellular proteases from *Serratia marcescens* grown in whey. *Process Biochemistry, 36*(6), 507–515.

Roukas, T., & Kotzekidou, P., (1996). Continuous production of lactic acid from deproteinized whey by coimmobilized *Lactobacillus casei* and *Lactococcus lactis* cells in a packed-bed reactor. *Food Biotechnology, 10*(3), 231–242.

Roy, D., Goulet, J., Le Duy, A., (1987). Continuous production of lactic acid from whey permeate by free and calcium alginate entrapped *Lactobacillus helveticus*. *Journal of Dairy Science*, *70*(3), 506–513.

Ryan, K. N., Vardhanabhuti, B., Jaramillo, D. P., Van, Z. J. H., Coupland, J. N., & Foegeding, E. A., (2012). Stability and mechanism of whey protein soluble aggregates thermally treated with salts. *Food Hydrocolloids*, *27*(2), 411–420.

Ryder, D., (1980). Economic considerations of whey processing. *International Journal of Dairy Technology*, *33*(2), 73–77.

S. M. S., & F. K. J., (1988). Whey—pollution problem and potential utilization. *International Journal of Food Science and Technology*, *23*(4), 323–336.

Sakhale, B., Vn, P., & Ranveer, D. R., (2012). *Studies on the Development and Storage of Whey based RTS Beverage from Mango cv. Kesar, 3.*

Sandhu, D. K., &. Waraich, M. K., (1983). Conversion of cheese whey to single-cell protein. *Biotechnology and Bioengineering*, *XXV*, 797–808.

Schindler, D. W., (2006). Recent advances in the understanding and management of eutrophication. *Limnology and Oceanography*, *51*(1, part 2), 356–363.

Shah, N. P., (2007). *Functional Cultures and Health Benefits*, *17*, 1262–1277.

Shiby, V. K., & Mishra, H. N., (2013). Fermented milks and milk products as functional foods: A review. *Critical Reviews in Food Science and Nutrition*, *53*(5), 482–496.

Shukla, M., (2012). *Development of Probiotic Beverage from Whey and Pineapple Juice, 4.*

Shukla, P., & Kushwaha, A., (2017). *Development of Probiotic Beverage from Whey and Orange Juice, 7.*

Sienkiewicz, T., & Riedel, C. L., (1986). *Whey and Whey Utilization.* VEB Fachbuchverlag. Ref. 404.

Singh, R. S., Sooch, B. S., Kaur, K., & Kennedy, J. F., (2004). Optimization of parameters for citric acid production from cheddar cheese whey using *Metschnikowia pulcherrima* NCIM 3108. *Journal of Biological Sciences*, *4*(6), 700–705.

Siso, M. I. G., (1994). β-Galactosidase production by *Kluyveromyces lactis* on milk whey: Batch versus fed-batch cultures. *Process Biochemistry*, *29*(7), 565–568.

Siso, M. I. G., (1996). The biotechnological utilization of cheese whey: A review. *Bioresource Technology*, *57*(1), 1–11.

Skryplonek, K., & Jasinska, M., (2015). Fermented probiotic beverages based on acid whey. *Acta Scientiarum Polonorum, Technologia Alimentaria*, *14*(4), 397–405.

Smithers, G. W., (2008). Whey and whey proteins—from 'gutter-to-gold.' *International Dairy Journal, 18*(7), 695–704.

Sodini, I., Morin, P., Olabi, A., & Jiménez-Flores, R., (2006). Compositional and functional properties of buttermilk: A comparison between sweet, sour, and whey Buttermilk. *Journal of Dairy Science*, *89*(2), 525–536.

Spălăţelu (VICOL) & Constanța C., (2012). Biotechnological valorization of whey. *Innovative Romanian Food Biotechnology*, 10.

Torres, D. P., Gonçalves, M. D. P. F., Teixeira, J. A., & Rodrigues, L. R., (2010). Galacto-oligosaccharides: Production, properties, applications, and significance as prebiotics. *Comprehensive Reviews in Food Science and Food Safety*, *9*(5), 438–454.

Tunick, M. H., (2008). Whey protein production and utilization: A brief history. *Whey Processing, Functionality and Health Benefits*, 1–13.

Venetsaneas, N., Antonopoulou, G., Stamatelatou, K., Kornaros, M., & Lyberatos, G., (2009). Using cheese whey for hydrogen and methane generation in a two-stage continuous process with alternative pH controlling approaches. *Bioresource Technology*, *100*(15), 3713–3717.

Walsh, G., (2002). *Proteins: Biochemistry and Biotechnology*. John Wiley & Sons.

Wan, C., Li, Y., Shahbazi, A., & Xiu, S., (2008). In succinic acid production from cheese whey using *Actinobacillus succinogenes* 130 Z. In: Adney, W. S., McMillan, J. D., Mielenz, J., & Klasson, K. T., (eds.), *Biotechnology for Fuels and Chemicals* (pp. 111–119). Totowa, NJ.; Humana Press: Totowa, NJ.

Wei, C., Shuwen, Z., & Lansun, C., (2013). Impulsive state feedback control of cheese whey fermentation for single-cell protein production. *Journal of Applied Mathematics* (p. 10). Article ID 354095.

Willetts, A., & Unai, U., (1987). The production of single cell protein from whey. *Biotechnology Letters, 9*(11), 795–800.

Wong, H. H., & Lee, S. Y., (1998). Poly-(3-hydroxybutyrate) production from whey by high-density cultivation of recombinant *Escherichia coli. Applied Microbiology and Biotechnology*, *50*(1), 30–33.

Yadav, J. S. S., Bezawada, J., Elharche, S., Yan, S., Tyagi, R. D., & Surampalli, R. Y., (2014). Simultaneous single-cell protein production and COD removal with characterization of residual protein and intermediate metabolites during whey fermentation by *K. marxianus. Bioprocess Biosyst. Eng., 37*, 1017–1029.

Yadav, J. S. S., Yan, S., Pilli, S., Kumar, L., Tyagi, R. D., & Surampalli, R. Y., (2015). Cheese whey: A potential resource to transform into bioprotein, functional/nutritional proteins and bioactive peptides. *Biotechnology Advances*, *33*(6), 756–774.

Yalcin, A. S., (2006). Emerging therapeutic potential of whey proteins and peptides. *Current Pharmaceutical Design*, *12*(13), 1637–1643.

Yalcin, S. K., Bozdemir, M. T., & Ozbas, Z. Y., (2009). Utilization of whey and grape must for citric acid production by two *Yarrowia lipolytica* Strains. *Food Biotechnology*, *23*(3), 266–283.

Yellore, & Desai, (1998). Production of poly-3-hydroxybutyrate from lactose and whey by *Methylobacterium* sp. ZP24. *Letters in Applied Microbiology*, *26*(6), 391–394.

Young V. R., & Scrimshaw, N. S., (1975). In: Tannenbaum, S. R., & Wang, D. I. C., (eds.), *Single Cell Protein ZZ* (p. 564). MIT Press, Cambridge, MA.

Yves, V., (1979). Whey an admirable raw material for human and animal food manufacture. *Review Dairy French, 372*, 27–39.

Zadow, J. G., (1994). Utilization of milk components: Whey. In: Robinson, R. K., (ed.), *Robinson: Modern Dairy Technology: Volume 1: Advances in Milk Processing* (pp. 313–373). Springer US: Boston, MA.

Zafra-Stone, S., Yasmin, T., Bagchi, M., Chatterjee, A., Vinson, J., & Bagchi, D., (2007). *Berry Anthocyanins as Novel Antioxidants in Human Health and Disease Prevention, 51*, 675–683.

Zall, R. R., (1984). Trends in whey fractionation and utilization, a global perspective. *Journal of Dairy Science*, *67*(11), 2621–2629.

CHAPTER 9

Contributions of Biosurfactants in the Environment: A Green and Clean Approach

GEETA RAWAT and VIVEK KUMAR

Himalayan School of Biosciences, Swami Rama Himalayan University, Jolly Grant, Dehradun, India, E-mail: vivekbps@gmail.com

ABSTRACT

Environmental pollution by petrochemical contaminants is a main issue and challenge because unintentional discharge or spill have affected several ecosystems since they are less or poor biodegradable. Another major problem with petrochemical contaminants is that they are hydrophobic in nature, and hence they are less or poorly soluble in aqueous environments. Due to their higher interfacial tension, these pollutants are difficult to remove from the environment. The use of chemical or synthetic surfactants is a solution but it further leads to the production of another form of contaminants. Therefore, the application of surfactants having biological origin could be a better choice for removing these hydrophobic contaminants having petro-hydrocarbon origin. Surfactants from the biological origin are commonly known as 'biosurfactants' or 'biological surfactants,' which are produced by a wide variety of microorganisms. The benefit of using biosurfactants is that they perform a similar function to synthetic surfactants and causes no harm to the environment. In the future, the use of biosurfactants from efficient microorganisms involving cost-effective technology will certainly change the scenario of biodegradation of petro-hydrocarbons in environments. Moreover, these biosurfactants consist of unique special structure, therefore, may have a different type of properties to exploit them commercially. In this chapter, we will discuss the production of biosurfactants and their mechanisms as well as their application in the process of bioremediation.

9.1 INTRODUCTION

Petro hydrocarbons and their various products are being utilized all over the globe to suffice the energy requirement. We cannot imagine our life without a source of energy, and to move this world, petro-hydrocarbons are the main energy resources. During exploration, production, transportation, processing, and utilization of petro-hydrocarbon products, there are unintentional or intentional spills (Antoniou et al., 2015). These spills lead to hazardous conditions, threaten flora and fauna, even destroy a particular ecosystem, due to accidental spills (Karlapudi et al., 2018). There are several industries which release various highly hydrophobic pollutants like polycyclic aromatic hydrocarbons (PAHs) in the environment. Treating these hydrophobic contaminants using physical and chemical approaches further leads to the production of another form of contaminants. Biological treatment using efficient microbes is a better approach, but hydrophobicity of petrochemical products does not allow the microbes to come in contact. Therefore, the efficiency of microbes decreases, resulting in poor or no bioremediation of hydrophobic petro-hydrocarbons in the environment. This problem of hydrophobicity can be resolved by the application of surfactants (synthetic surfactants), these are the agents which help in solubilization of hydrophobic pollutants by reducing the surface and interfacial tension (Silva et al., 2014). The application of chemical surfactants results in the formation of new pollutants, which act as toxic materials since these are derived from chemicals. To overcome this problem, we prefer to adopt an eco-friendly approach that is, the use of biosurfactants. The biosurfactants are the most promising and effective molecules which are produced by the wide variety of microorganisms. They act like chemical surfactants but their source if biological, hence no production of secondary contaminants. The biosurfactants also decrease surface and interfacial tension and form the micro-emulsions (Vijaykumar and Saravanan, 2015). Owing to their distinctive properties such as less toxicity, amphipathic in nature, and easy to produce using microbes, therefore, they are biodegradable also. Due to these properties, the biosurfactants have attracted wide research interest all over the globe. Because of the above-mentioned properties, the biosurfactants are being employed in numerous industries including organic chemical, metallurgy, mining, petrochemical, food, cosmetics, beverages, fertilizers, pharmaceutical, and many others. Interestingly, different biosurfactants have demonstrated antimicrobial activities against various pathogenic bacterial, fungal, algal, and viral species (Vijaykumar and Saravanan, 2015).

Some bacterial, fungal species do produce extracellular ionic-surfactants, which emulsifies the CxHy substances in the microbial growth medium. The important example of this group of biosurfactant is the rhamnolipids, which are formed by numerous species of *Pseudomonas* (Potawary et al., 2017). Concerning studies on biosurfactants production, maximum work has been reported on marine microbes and bacterial species (Satpute et al., 2010; Bozo-Hurtado et al., 2012; Dhasayan et al., 2015). But there are fewer reports on yeast synthesizing biosurfactants. Accorsini et al., 2012 reported that yeasts *Pseudozima* sp., *Candida* sp., and *Yarrowia* sp. are able to produce more biosurfactants as compared to bacterial species.

The biosurfactants are organic components which are made up of critical groups of compounds, exhibiting distinguishing characteristics in terms of their chemical structure and specific composition (Ron and Rosenberg, 2002). The biosurfactant molecules perform by dipping interfacial energy (or interfacial tension) and the surface tension during the set-up of a well-organized molecular film at the hydrocarbon water interface (Zeppier et al., 2001). Biosurfactants can enhance the biological availability of petro-hydrocarbons towards the microbial cells by enhancing the aquatic hydrocarbon area of the interface. This leads to an increase in hydro-carbon dissolution rate and their consumption by microbes (Ebrahimi et al., 2012). The chief concern which affects directly the effectiveness of a biological remedy is the "bioavailability" of the pollutants. Probable surface assimilation of molecules into the soil matrix, the formation of the non-aqueous phase, leads to poor bioavailability of hydrocarbon molecules. Biosurfactants, which are the amphiphilic compounds, are able to dipping interfacial tension, disperse the oil particles, help the microbes to absorb or assimilate and finally biodegrade the hydrocarbon pollutants (Patel et al., 2019). The basic structure of a 'biosurfactant' molecule is shown in Figure 9.1.

Micelles formation plays an efficient role in the formation of microemulsion which is a very stable emulsion of oil and water separated by biosurfactants monolayer. The biosurfactants gather between the two immiscible fluids (oil and water) and also between soil and fluid. By reducing surface (air and liquid) and interfacial (liquid and liquid) tension they lessen the repulsive force between two dissimilar phases and allow these two phases to form a matrix. Figure 9.2 shows the micelle formation; the formation of a micelle is very important in the bioavailability of insoluble or unavailable hydrophobic substrates. The micelles capture the unavailable substrate and make it available to the microbial cell.

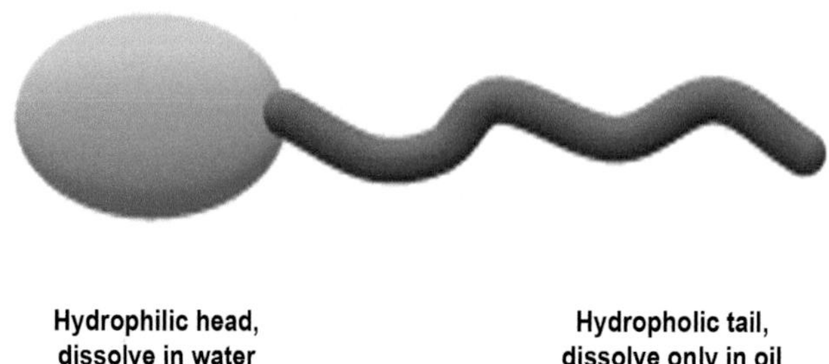

FIGURE 9.1 Biosurfactant molecule with hydrophobic and hydrophilic domains.

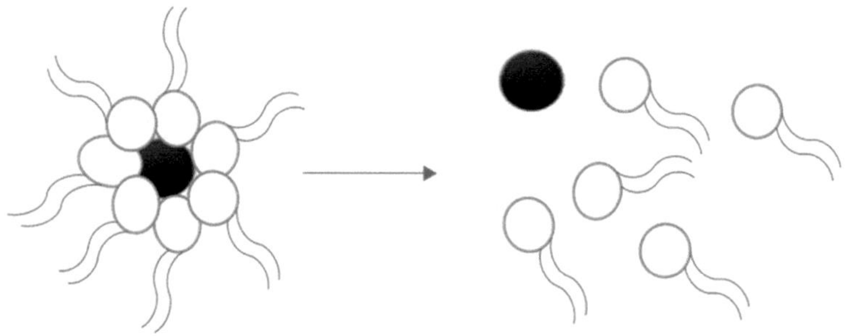

FIGURE 9.2 Micelle formation and capturing of substrate.

9.2 TYPES OF BIOSURFACTANTS

The biosurfactants are the tensioactive compounds which are broadly divided into two main categories on the basis of their molecular weight:

a. Low molecular weight like biosurfactants, for example, glycolipids and few short lipopeptides
b. High-molecular-weight compounds like bioemulsifiers, for example, lipoproteins or lipopolysaccharides (Varjani et al., 2017).

Based on the structure and chemical composition, biosurfactants can be divided into five types:

1) **Glycolipid:** They consist of sugar and lipids in which carbohydrates are linked to the long chain of aliphatic acids or hydroxyl-aliphatic acids

(fatty acid moiety) by an ester group. The well-known glycolipids are rhamnolipids, sophorolipids, and trehalolipids. The sources and properties of these different glycolipids are discussed below:

a. **Rhamnolipid:** These are encompassed of one or two units of rhamnose connected to one or two β-hydroxyl fatty acids (Figure 9.3). Glycolipids in which one or two molecules of rhamnose are linked to one or two hydroxy-decanoic acids. It is widely studied biosurfactants, which are the principal glycolipids produced by *P. aeroginosa.* The maximal production of rhamnolipids was 22.5 g/l, which was obtained in batch fermentation, at a dissolved oxygen concentration of 40% (Bazsefidpar et al., 2019). Rhamnolipid biosurfactants have huge possibilities and future in environmental applications including bioremediation of hydrocarbons, removal of heavy metals, anti-microbial activity, disruption of biofilm disruption, and synthesis of green nanoparticles (Kumar et al., 2010; Molnar et al., 2018).

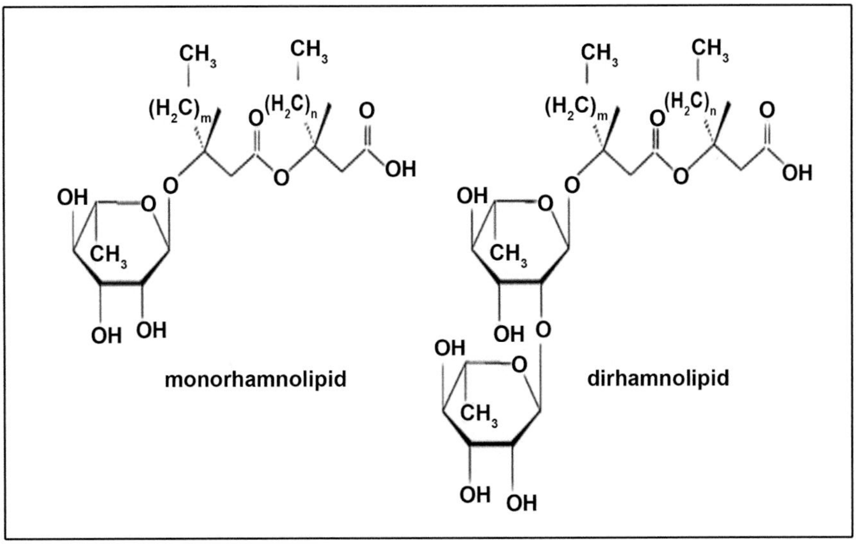

FIGURE 9.3 Structure of mono and di-rhamnolipids.

b. **Trehalolipid:** These consist of disaccharides trehalose linked at six position carbon to two β-hydroxybranched fatty acids. These are produced by many spp. of *Nocardia*, *Mycobacterium*, and *Corynebacterium* (Shao, 2011).

c. **Sophorolipids:** These are formed by dimeric carbohydrate sophoros molecule connected to a long chain hydroxy fatty acid by a glycosidic bond (Varjani et al., 2017) (Figure 9.4). Sophorolipds are the most extensively studied microbial biosurfactants produced by *Starmerella bombicola*. These are generally mixtures of at least 6–9 dissimilar hydrophobic sophrolipids. Lactonic and acidic form of a sophrolipids are generally preferred for many applications. It has shown good stability towards temperature and pH fluctuations. Its surface-active properties are consistent at pH values between 6–9 and at a temperature ranging from 20–90°C (Wang et al., 2019).

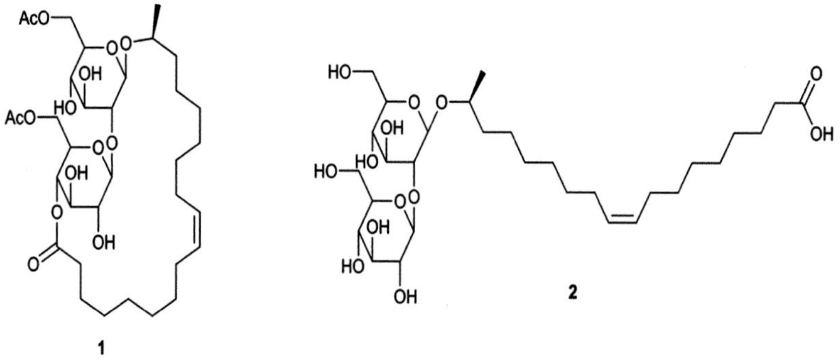

FIGURE 9.4 Structure of sophorolipids.

2) **Lipopeptide or lipoprotein**: These are mixtures of two different compound which are composed of lipid part (mostly hydroxyl fatty acid) and amino acid. The cyclic lipopeptide contains 8–17 amino acids and lipid part, composed of 8–9 methylene groups, as shown in Figure 9.5. They differ in the composition of amino acid and lactone ring position and lipid segment. Lipoprotein biosurfactants were produced from the lipid layers gathered from fish processing wastewater (Saranya et al., 2014).

The common examples of lipopeptides are lichenysin, viscosin, and surfactin (Moya et al., 2015).

a. **Surfactin:** This is one of the most potent biosurfactants, formed by the endospore-forming, Gram-positive, bacteria like *Bacillus subtilis*. The chemical assembly of surfactin molecule is comprised of seven amino acids which are connected to hydroxyl and carboxyl

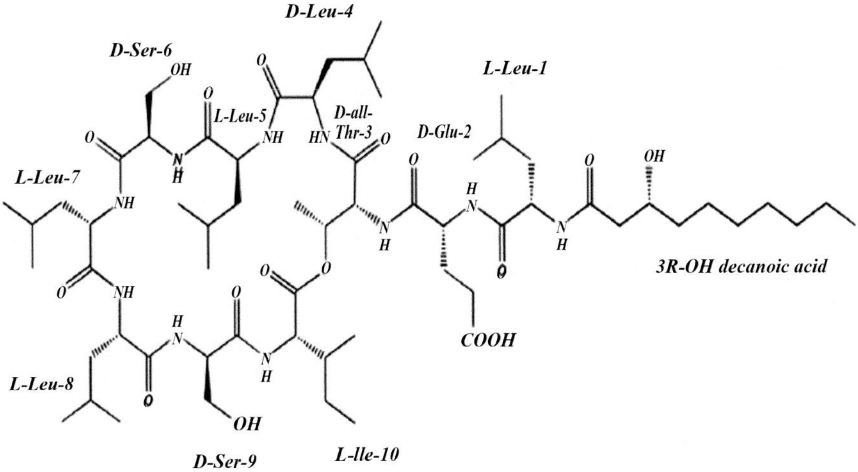

FIGURE 9.5 Structure of lipopeptides.

groups on a long-chain fatty acid (C_{13}–C_{15}) as shown in Figure 9.6. The prospective uses of these biosurfactants are curative and in environmental sustainability (Vijaykumar and Saravanan, 2015).

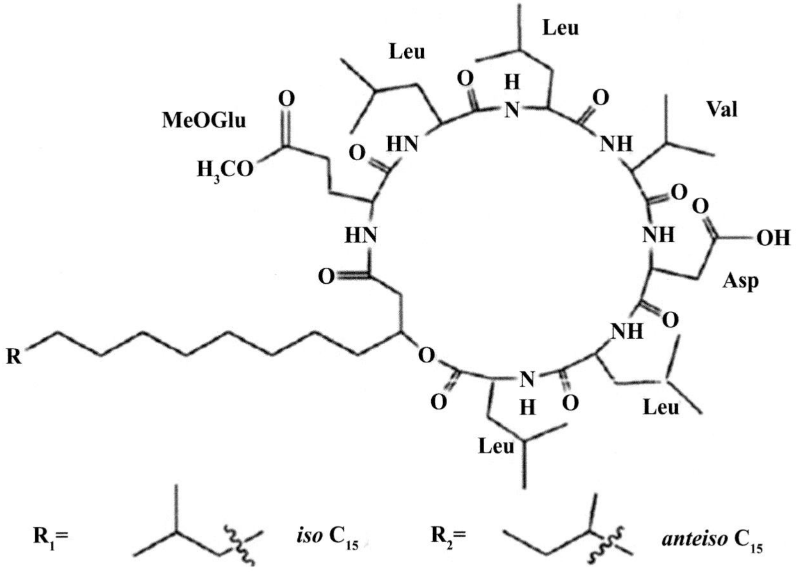

FIGURE 9.6 Structure of surfactin molecule.

b. **Lichenysin:** These biosurfactant molecules are manufactured by *Bacillus licheniformis,* which shows exceptional steadiness under high salt, pH, and temperature situations. Lichenysin biosurfactants are similar to the surfactin. These are anionic lipopeptide biosurfactants with cytotoxicity and antimicrobial activities which perform a very potential role in the chemical and biological application (Jonathan et al., 2017). A large no of cyclic lipopeptide including gramicidine, polymyxa, Ornithine, Taurine, Cerilipin, Viscosin, Arthrofactine have been reported (Sharma, 2009). Gramicidins (decapeptide antibiotic) produced by *B. brevis,* polymyxins (lipopeptide antibiotic) produced by *B. polymyxa*, ornithine, taurine, and cerilipin containing lipid from *Gluconobacter* sp. viscosin produced by *P. fluorecense* and arthrofactine is produced by *Arthrobacter spp.*

3) **Phospho-Lipids, Neutral-Lipids, and Certain Fatty Acids:** These are cell membrane constituents and possess surface activity. The phospholipid and fatty acids surfactants are created by numerous bacterial and yeast species in huge amount utilizing n-alkanes as a carbon source. These biological surfactants are capable to manufacture optically transparent microemulsion of alkane in aqua solution. These type of biosurfactants producing microorganisms are generally sulfur-reducing bacteria, such as *Thiobacillus thioxidans.* Such biosurfactants are consisted of complex C 12 to C 14 saturated fatty acid having alkyl and hydroxyl (Sharma, 2009)

4) **Polymeric Biosurfactants:** These consist of a backbone of polysaccharide to which are covalently linked are fatty acids side chain. The best investigated polymeric biosurfactants are liposan, emulsan and alasan. The emulsan is an effective emulsifying agent for petro-hydrocarbons in water. *Candida lipotyca* synthesize an extracellular water-soluble emulsifier and their significant role in food industries as an emulsifier. Different genera of bacterial species also produced exo-cellular polymeric biosurfactants which are composed of either lipopolysaccharides, proteins, polysaccharides or a multifaceted combination of these biological polymers (Rosenberg and Ron, 1999).

5) **Particulate Biosurfactants:** These biosurfactants forms a microemulsion, on the extracellular membrane of a microbial cell. These biosurfactants play a crucial part in taking up of alkane by cells of microbes. *Acinetobacter sp.* strains microemulsion vesicles have a diameter of 20–50 nm and thickness of 1.158 cg/cm^3. This vesicle is composed of proteins, phospholipid, and lipopolysaccharides (Tong and Thomas, 2004)

9.3 UNIQUE PROPERTIES OF BIOSURFACTANTS

- Temperature and pH tolerance
- Biodegradability
- Low toxicity
- Emulsion forming and emulsion breaking
- Surface and interface activity
- Antiadhesive property
- Biofilm formation
- Reduce the stress of PAH

– **High temperature and pH tolerance:** Extreme environmental microorganisms produce biosurfactants, due to this property, these extreme microbes have acquired consideration in the last many years because this property can be exploited commercially. Biosurfactants from extreme microbes do possess certain unique properties such as better surface activity, remain stable in adverse climatic factors such as pH, temperature, and pressure. Mcinerney et al., in 1990 reported that lichenys in a lipoprotein biosurfactants, produced by *Bacillus licheniformis* was able to remain stable at a temperature of 50°C, also a wide range of pH 4.5 to 9.0. Interestingly, the lichenysin was stable at higher concentration of NaCl and calcium up to 50 and 25 gL, respectively. Singh and Cameotra in 2004 reported that *Arthrobacter protophormiae* produced a biosurfactant, which was found to be stable at a wide range of pH (2 to 12) and wide range of temperature (30–100°C), therefore, such biosurfactants can be exploited commercially in adverse climatic conditions.

– **Biodegradability:** Microbial produced compounds are mainly self-degradable as compared to the chemically synthesized compounds, and therefore, are suitable for several environmental applications (Vijayakumar and Saravanan, 2015). Generally, the surfactants of biological origin are more efficient and effective since their Critical Micelle Concentration (CMC) is several times lesser than the synthetic surfactants, i.e., it helps in decreasing more surface tension, resulting in less application of biosurfactants as compared to synthetic one (Desai and Banat, 1997).

– **Low toxicity:** Biosurfactants, are normally thought to be nontoxic or having low toxicity. Such nontoxic biosurfactants are suitable for cosmetic, food and pharma uses (Poremba et al., 1991). On the other hand, the chemically synthesized surfactants are reported to have toxicity (e.g., Corexit) which exhibited a LC50 against *Photobacterium phosphoreum*.

The low toxic aspects of biological surfactants such as sophorolipids from *Candida bombicola* named them useful in food, pharma products as well as in skincare products (Vijayakumar and Saravanan, 2015).

– **Emulsion formation emulsion breaking:** The biosurfactants work as emulsifying or de-emulsifying agents (Nitschke and Costa, 2007). Emulsions are of usually two types: (a) oil in water or (b) water in oil (Ilia and Abdurahman, 2010). When emulsions are formed, they maintain a marginal firmness, this firmness or stability may be stabilized or alleviated by adding biosurfactants and therefore, can be kept as stable emulsions for several months to years. *Candida lipolytica* produces, liposan, which is a water-soluble biosurfactant. This is used to in emulsification of edible oils, consequently forms constant or firm emulsions. Biologically produced liposans are normally used in cosmetic and food products for preparing stable oil/water emulsions.

Emulsions can be categorized into macroemulsion (droplet size: 1.5–100 μm), microemulsion (droplet size: 3–50 nm) nanoemulsion (droplet size: 50–500 nm), the droplet size depends upon the dispersed particles (Piorkowski and McClements, 2013). In the case of microorganisms, the formation of a hydrophilic moiety of biosurfactants, by the water-soluble substrates like carbohydrate groups while the hydrophobic domain is made up of components like fatty acids (Silva et al., 2014)

9.4 SOURCES OF BIOSURFACTANTS

The biological surfactants are produced by several microorganisms, bacteria, fungi, and few yeasts. Table 9.1 shows the list of commonly used and produced biosurfactants by different microbes.

9.5 BIOSURFACTANTS V/S SYNTHETIC SURFACTANTS

Biosurfactants have been demonstrated to show several advantageous effects as compared to synthetic or chemical surfactants without showing any drawbacks. The biosurfactants are considered as non-toxic and biodegradable and several biosurfactants do not produce true micelles, therefore, they facilitate the direct transfer of surfactant associated hydrophobic poly-aromatic hydrocarbons to bacterial cells. The main differences between biological surfactants and synthetic surfactants are shown in Table 9.2.

TABLE 9.1 List of Biosurfactants Producing Microorganisms

Microorganisms	Biosurfactants/ emulsifiers	Group
Bacteria:		
Serratia marcescens	Serrawettin	Amino acid-containing lipoprotein, Lipopeptide
Rhodotorula glutinis, R. graminis	Polyol lipids	
Rhodococcus erythropolis,	Trehalose lipids	Glycolipids
Nocardia erythropolis, Corynebacterium spp., *Mycobacterium* spp.	Trehalose lipids	Glycolipids
Pseudomonas spp., *Thiobacillus thiooxidans, Ornithine* lipids *Agrobacterium spp.*	Ornithine lipids	Amino acid lipoprotein
Pseudomonas fluorescens	Viscosin	Amino acid Lipopeptide lipoprotein
Pseudomonas aeruginosa, Pseudomonas chlororaphis, Serratia rubidea	Rhamnolipids	Glycolipids
Lactobacillus fermentum	Di-glycerides	
Corynebacterium lepus	Fatty acid	Polar and neutral lipids
Acinetobacter spp.	Vesicles	Particulate BS.
Bacillus subtilis	Lipopeptides and lipoproteins	Surfactin
Fungai		
Torulopsis bombicola	Sophorolipid	Glycolipids
Candida bombicola	Sophorolipids	Glycolipids
Candida lipolytica	Protein lipopolysaccharides complex	Protein carbohydrate complex
Candida batistae	Sophorolipids	Glycolipids
Candida ishiwadae	Glycolipids	Glycolipids
Aspergillus ustus	Glycolipoprotein	
Trichosporon ashii	Sophorolipids	Glycolipids

TABLE 9.2 Main Difference Between Biosurfactants and Synthetic Surfactants

S.N.	Biosurfactants	Synthetic Surfactants
1.	Low toxic or non-toxic	Usually toxic
2.	Easy to degrade	Comparatively hard to degrade
3.	Produced biologically	Derived by the chemicals
4.	Mutagenic and toxicity effect are less	Mutagenic and toxicity effect are high
5.	More effective, having low critical micelle concentration (CMC)	Less effective, has higher critical micelle concentration (CMC)
6.	Not much cost-effective	Cost-effective

9.6 APPLICATIONS OF BIOSURFACTANTS

The biosurfactants have wide application, including food industries, cosmetic industries, removal of oil and petroleum contamination, bioremediation of toxic pollutants, pharmaceuticals and in the agriculture sector by removing the heavy metals. But in this chapter, we will focus on bioremediation aspects, especially contaminants caused by petrohydrocarbon products.

9.6.1 *BIOSURFACTANTS IN THE ENVIRONMENT*

Several beneficial properties of biosurfactants such as foaming, dispersion, coating, wetting, emulsification, and de-emulsification make them very advantageous in removing inorganic and organic contaminants, thus they play a significant role in physicochemical and bioremediation technologies. Biosurfactants enhance the biological availability of insoluble hydrocarbon molecules. In heavy-metal contaminated soils, biological surfactants form an intimate complex with the metals at the interface of soil (Almeida et al., 2016). After complex formation, there is desorption of the metal from the biosurfactant and removal from the soil surface leads to the upsurge of the metal amount and their biological availability in the soil solution. Table 9.3 shows the role of various biosurfactants, produced by different microbes in the removal of diverse pollutants. Here we have also mentioned the percentage removal of pollutants from the environment.

9.6.2 *BIODEGRADATION AND BIOREMEDIATION*

Bioremediation and biodegradation process of any pollutants and complex compounds can be done in two different ways:

- **Process *ex-situ***: It is carried out in a slurry reactor system or prepared bed.
- **Process *in-situ* processes**: This process is generally carried out by employing microbial culture and some nutrients into the soil, this technique allows the proliferation and growth of indigenous as well as inoculated microbial population.

The efficiency and effectiveness of the biodegradation process rely upon the number of petrohydrocarbon biodegrading microbes in an aquatic

TABLE 9.3 Microbes and Their Biosurfactants in Pollutants Remediation

S.N.	Biosurfactants	Microbes	Pollutants	Removal efficiency	References
1.	Rhamnolipid	*Shingomonas spp., Pseudomonas spp.*	Anthracene	52%, after 18 days	Cui et al., 2008
2.	Rhamnolipid	*Pseudomonas fluorescence*	Pyrene	98%, after 18 days	Husain et al., 2008
3.	Rhamnolipid	*Sphongomonas spp.*	Phenanthrene	99%, after 10 days	Pei et al., 2010
4.	Rhamnolipid	AM Fungi, microbial consortium & Alpha/Alpha	PAHs	61%, after 90 days	Zhang et al., 2010
5.	Rhamnolipid	Marine natural microflora	Crude hydrocarbon oil	25%, after 5 days	Mckew et al., 2007
6.	Rhamnolipid	*Candida tropicalis*	Phenol	99%, after 30 hrs	Liu et al., 2010
7.	Monorhamnolipid	*Candida tropicalis*	Hexadecane	93%, After 4 days	Zeng et al., 2011
8.	Rhamnolipid	*Pseudomonas putida*	Phenanthrene	91%, after 10 days	Gottefried et al., 2010
9.	Rhamnolipid	Ryegrass	Pyrene and phenanthrene	Addition of biosurfactants enhanced 5–6 times more uptake of phenanthrene and pyrene	Zhu and Zhang, 2008
10.	Rhamnolpid	*Sphingomonas*	Phenanthrene	47% after 40 days	Shin et al., 2006
11.	Rhamnolipid	*Bacillus spp.* And *Shingomonasspp*	Phenanthrene	23%, after 8 days	Shin et al., 2005
13.	Rhamnolipid (Di and mono)	*Bacillus algicola, Rhodoccocus Soli, Isoptericola Chiliensis* and *Pseudoalteromonas Agarivorans*	Crude oil	65%	Lee et al., 2018
14.	Non-specified	*Bacillus Subtilis*	Crude oil	Biosurfactants presence did not stimulate aromatic hydrocarbon biodegradation but accelerated biodegradation of aliphatic hydrocarbons	Cubitto et al., 2004
15.	Non-specified	*Bacillus theurengenesis* (ATS), *Actinomycetes* (AF-104), *P. earuginosa* (AF11-*GT*), *P. stutzeri* (AT3)	Crude oil	AF11-GT exhibit higher biodegradation 93.41% by the second day	Shanaby et al., 2015

system, soil, and sediments. On the other hand, the chemical composition and material state of petrohydrocarbon mixtures and availability of oxygen, and the thermal condition of water, temperature fluctuations, pH, and availability of inorganic nutrients. To make the process of bioremediation more effective, it can be achieved in two ways; by increasing the surface area of the hydrophobic substrate or by increasing the biological availability of the hydrophobic compound (Plociniczak et al., 2011).

9.6.2.1 BIOSURFACTANTS AND BIODEGRADATION STUDIES

The main challenge which directly influences the efficacy of bio-treatment of contaminants is its bioavailability. Probable absorption of molecules in the soil matrix, non-aqueous phase formation, organic matter interaction, biological transformation and finally the aging of contaminants, all these naturally occurring processes generally result in less, no or limited biological availability. All these things result in less efficiency or decreased efficacy of the bioremediation process (Azubuike, 2016). Therefore, the most common expected role of biological surfactants is to enhance the distribution of pollutants into the water phase and also increased their bioavailability. A way from interconnection with the contaminants the biosurfactants molecules may also rightly influence the efficacy of the adjoining biological remediation (plant or microbes), which is being used for the bioremediation process. Biosurfactants demonstrate strong bioactivity, particularly at the cell membrane level. These modifications might lead to increased hydrophobicity which is considered good for biodegradation aspects or modifies the permeability of cell membrane which could be beneficial during the process of bio extraction (Otzen DE 2017). Though it has been recognized that modification in cell properties might not be necessary with the capability to utilize certain hydrocarbon molecules, thus we cannot correlate this with biological remediation efficacy.

The ability of biosurfactants and biosurfactants manufacturing microbes to augment carbon pollutants, availability, and their bioremediation process has been reported by several workers (Calvo et al., 2009; Franzetti et al., 2010; Ron and Rosenberg, 2014). The biodegradation aspects of biosurfactants by using *Pseudomonas* strains to catabolize the hydrocarbons substances of diesel and crude oil was studied by (Obayori et al., 2009). The workers also reported that there was a degradation of 92.34% and 95.29% of crude oil and diesel oil, respectively. Reddy et al., 2010 reported the biodegradations aspects of *Brevibacterium* spp. strain PDM3, this strain produces a good amount of biosurfactants and it was observed that this strain was able to degrade phenanthrene up to 90–93%. Interestingly this bacterium was also capable of biodegrading fluorine and anthracene.

Sophorolipid a biosurfactant was used in biodegradation of aromatic, aliphatic hydrocarbons and Iranian crude and refined oil under laboratory conditions. The experiment was conducted creating artificial contamination conditions as mention above. It was observed that the addition of biosurfactants in the soil enhances the rate of biodegradations ranged for 85–94%. The results clearing demonstrated that sophorolipids have great possibilities in the enhancement of bioremediation of contaminated with petrohydrocarbon products. The presence of biosurfactants enhanced the solubility of insoluble hydrocarbons and therefore increase the bioavailability of insoluble organic contaminants for microbes. In another study, Kang et al., (2010), showed the beneficial aspects of biosurfactants manufacturing microbes in biological remediation of high polluted sites with petrohydrocarbnons. The workers used three biosurfactants synthesizing bacterial strains *P. aeruginosa* M., *Bacillus subtilis* DM-04, *Pseudomonas aeruginosa* NM to bioremediation the crude oil polluted soil samples. The soil samples were mixed with petrohydrocarbons and followed by inoculating the samples with biosurfactants solutions obtained from the above mentioned bacterial strains.

In another study, Patowary et al., (2017), used bacterial biosurfactants for separation of oil from petroleum sludge. In this study, the petroleum sludge was inoculated with *Bacillus* spp. and cell-free supernatant was also added. The sludge without bacterial inoculation was treated as control, after inoculation of cell-free supernatants to the sludge, separation of oil and total petroleum hydrocarbons (TPH) reduction was observed. Initially, the process of oil separation was very slow, but later the inoculation of bacteria enhanced the process of oil separation. Therefore, it was observed that bacterial biosurfactants production was enough to continue the process of oil separation from the petroleum sludge. Consequently, the application of biosurfactants exhibited the property to reduce interfacial as well as surface tension in both aquatic and hydrocarbons mixture and therefore has great potential for recovery of oil. Nievas et al., (2007) reported the biodegradation of "Bilge waste," this waste is produced during normal ship engine operation process, and this waste is very hazardous and consisted of a combination of marine water and residues of hydrocarbons. This residue is consisted of N-Alkanes, cyclic and branched aliphatic hydrocarbons along with some amount of aromatic hydrocarbons. These hydrocarbons residues generally exhibit great problems in bioremediation. The workers used a microbial consortium that produced emulsifier, the addition of this bacterial emulsifier resulted in the reduction of complex organic residue. It was observed that the microbial consortium producing emulsifier for bilge waste bioremediation, showed a reduction in the number of n-alkanes, resolvent hydrocarbons and an unsolvent mixture by 85%, 75%, and 58%, correspondingly.

9.6.3 CLEAN-UP TECHNOLOGY

The clean-up technology and combined technology was developed as a solution for efficient and inexpensive technology for clean-up of the contaminated soil by oil contaminants. This technology also involves phytoremediation along with bioremediation principles and also employed the physicochemical approach by washing the polluted soil (Kildisas et al., 2003; Baskeys et al., 2004).

This complex technique is constituted of two stages, in the first stage, with the help of biosurfactants the migrating part of contaminated were separated. In the second stage, the leftover non-migrating part was converted to harmless products using a biodegradation approach. In this technique, a phytoremediation approach is also used to enhance soil property. This technique was applied at pilot scale for cleaning of polluted soil in a space of 340 m^2. Initially, the oil-polluted concentration was between 180–270 gm/kg of soil, after washing this pollutant concentration reduced to 34–59 gm/kg. Interestingly, after the degradation process, the concentration of pollutant dropped down to 3.2–7.3 gm/kg soil (Baskys et al., 2004).

9.6.4 MICROBIAL ENHANCED OIL RECOVERY

The application of Biosurfactants in microbial enhanced oil recovery is the most promising method. Microbial enhanced oil recovery (MEOR) is a method by which, recovering the remaining oil from the reservoir after the primary (mechanical) and secondary (physical) procedures. The leftover or remaining oil is seldom located in those regions of the tank or reservoir that are difficult to access or recover. The leftover oil is imprisoned into the pores by capillary pressure and high interfacial tension. Application of biosurfactants decreases the interfacial tension between the oil and water or oil and rock. Due to the reduction in this interfacial tension the capillary forces are reduced which allow the movement of remaining oil through pores of rocks. The biosurfactants bind tightly to the oil and water surface and form an emulsion. Due to the formation of emulsion, the oil is desorbed from the rocks and solubilized in water, which is later removed and collected. One of the chief prospects of biosurfactants used could be in MEOR (Silva et al., 2014; Reis et al., 2013). In this technique, the microbial populations are energized to produce biosurfactants and other polymers to decrease the interfacial tension at the oil rock interface. This MEOR approach is usually applied in supply pipes, rocks, soil sediments, and tanks. Biosurfactants production by microbes at *in situ*, in the tanks, are usually provided with chief substrates, molasses, and inorganic nutrients (Banat et al., 2014). This is done to increase microbial growth and enhance biosurfactants production. Under *in situ* conditions, the applied microbes must be able to tolerate adverse climatic conditions such as low oxygen tension, high temperature, salinity, and pressure fluctuations (Sen, 2010). The effectiveness of MEOR using biosurfactants under field conditions was studied in hungry Poland, US former USSR, and former Soviet. It has been observed that in these countries significant oil recovery in the field was reported only in few cases (Banat et al., 1995a, 1995b). Figure 9.7 shows the effect of the addition of biosurfactant in rocks, the presence of biosurfactants mobilize the immobile crude oil.

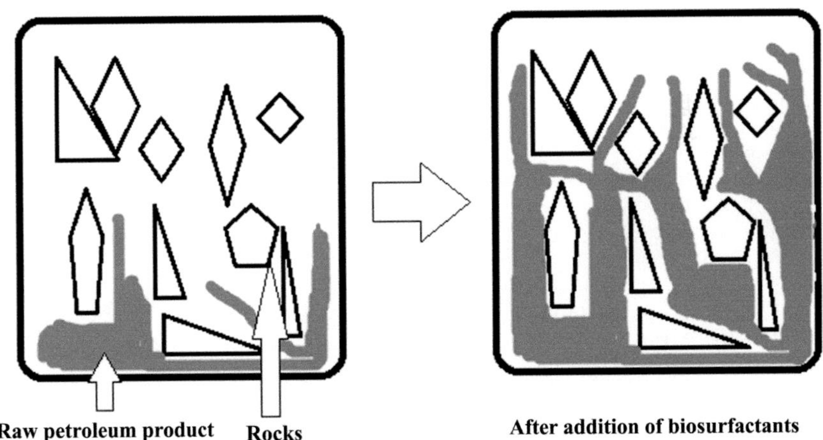

FIGURE 9.7 Role of biosurfactant in the separation of oil from rocks.

9.6.5 REMOVAL OF HEAVY METAL BY BIOSURFACTANTS

Biosurfactants have advantages in the removal of heavy metals from the contaminated soil, it is based on their ability to form complexes with the metals. Major toxic heavy metals, which are present in the environment are cadmium, arsenic, nickel, lead. The application of biosurfactants has a unique advantage because microbes producing these surface-active agents do not necessarily require to possess survival capability in heavy metal polluted soil. Though, application of biosurfactants only necessitates uninterrupted supply of these biomolecules, so that there is a constant decrease in surface and interfacial tension (Ayangbenro and Babalola, 2018).

The advantageousness of biological surfactants for remediation of heavy metal polluted soils is chiefly based on their capability to make metal complexes. The anionic biosurfactants form metal complexes in a non-ionic form by ionic bonds. These ionic bonds are stouter than metal's bond. The biosurfactant forms a complex of soil-metal-biosurfactant, which is later desorbed from the soil matrix into soil solution owing to the decrease of the surface and interfacial tension. The cationic biological surfactants can also substitute the same charged metal ions by the process of competition for some but not for all negatively charged surfaces. The metal ions can be detached from the surface of the soil by the biosurfactant micelles (Franzetti et al., 2014). The polar head group of micelles can attach with metal which mobilizes the metals in aquatic solution. Figure 9.8 shows the role of biosurfactants in desorption of metal from the soil matrix.

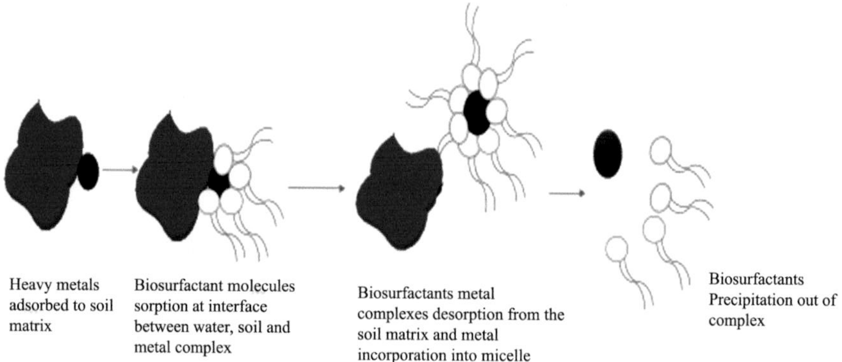

FIGURE 9.8 Process of heavy metal removal from soil matrix by biosurfactants.

9.7 CONCLUSION

Whatever has been discussed in above pages, from that it can be concluded that the conventional approaches used for bioremediation or biodegradation processes are not technically effective and also not cost-effective, thus resulting in the sometimes generation of secondary contaminants. Moreover, the hydrophobic nature of petrohydrocarbon contaminants creates another problem of non-bioavailability, due to their high interfacial tension. To solve this problem of high interfacial tension, some effective biomolecules can be used. These effective biomolecules are produced by most microbes and are known as 'biosurfactants.' These biosurfactants are effective molecules, which improves the solubility and availability of insoluble contaminants.

In this chapter, we have provided a reliable knowledge about the application of biosurfactants as a promising agent in the bioremediation technologies. Biosurfactants are molecules that are present in the microbial cell surface and maintains the cell surface hydrophobicity because it has both, hydrophobic and hydrophilic domains. Biosurfactants express various significant properties which could be helpful in several areas, bioremediation technologies are one of them. The ability of microorganisms to use petroleum hydrocarbons as their energy sources permits an alternative treatment approach based on biodegradation or bioremediation processes. Biosurfactants play a significant role in scavenging and availing of high degree hydrophobic pollutants from the environment due reduction of surface and interfacial tension. Owing to increased oil pollution and hydrophobic molecules in the environment, microbial metabolism is enhanced by the biosurfactants because of its tensioactive property. In this situation, biosurfactants act on oil-water interface and forms micelles, which increases the bioavailability of hydrocarbons to biological systems. This facilitates microbial metabolism and encourages the development of biomass, hence increasing the process of biodegradation. Generally, the biosurfactants are obtained from GRAS (Generally regards as safe) microorganisms.

9.8 FUTURE PROSPECTS

The regular discharge of pollutants inorganic, organic especially the hydrocarbon either though the industrial process or natural process resulting in accumulation and contamination of aquatic or soil ecosystem. People have tried many approaches including physical, chemical, and use of live microbes to restore and decontaminate polluted sites. It has been observed that the potency and efficacy of these processes are limited chiefly owing to the low aquatic solubility of pollutants and therefore less availability to physicochemical techniques and microbes acting as biodegrader. The application of biosurfactants to remediate the non-soluble contaminants offers an attractive choice due to its ecological safety, self-biodegradability, versatility and overall its environmental acceptance. The choice of using biological surfactants over synthetic surfactants is also important because surfactants having biological origin does not produce secondary contamination, moreover, they are low or non-toxic to environments. The major issue with biosurfactants is that, they are not cost-effective, also we are not able to produce sufficient quantity to full fill our requirement. Therefore, it becomes necessary for us to develop efficient microbial strain which can produce a huge amount of biosurfactants cheap raw material. Further maximum research work has been carried out under laboratory conditions, for concrete results. A positive report from the field is required to finally evaluate the effectiveness and efficiency of biosurfactants *in situ*.

KEYWORDS

- **bioremediation**
- **biosurfactants**
- **microbes**
- **pollutants**

REFERENCES

Accorsini, F. R., Mutton, M. J., Lemos, E. G., & Benincasa, M., (2012). Biosurfactants production by yeasts using soybean oil and glycerol as low-cost substrate. *Brazilian J. Microbiol., 43*(1), 116–125. doi: 10.1590/S1517–83822012000100013.

Almeida, D. G., De Cássia, R. D. C. F. S., Luna, J. M., Rufino, R. D., Santos, V. A., & Banat, I. M., (2016). Biosurfactants: Promising molecules for petroleum biotechnology advances. *Front Microbiol., 7*, 1718–1731.

Antoniou, E., Fodelianakis, S., Korkakaki, E., & Kalogerakis, N., (2015). Biosurfactant production from marine hydrocarbon-degrading consortia and pure bacterial strains using crude oil as carbon source. *Front. Microbiol., 6*, 274–283.

Ayangbenro, A. S., & Babalola, O. O., (2018). Metal (loid) bioremediation: Strategies employed by microbial polymers. *Sustainability, 10*, 3028; doi: 10.3390/su10093028.

Azubuike, C. C., Chikere, C. B., & Okpokwasili, G. C., (2016). Bioremediation techniques–classification based on site of application: Principles, advantages, limitations and prospects. *World J. Microbiol Biotechnol., 32*(11), 180. doi: 10.1007/s11274-016-2137.

Banat, I. M., (1995a). Characterization of biosurfactants and their use in pollution removal-state of the Art. (Review). *Engg Life Sci., 15*(3), 251–267.

Banat, I. M., (1995b). Biosurfactants production and possible uses in microbial enhanced oil recovery and oil pollution remediation: A review. *Bioresour Technol., 51*(1), 1–12.

Banat, I. M., Satpute, S. K., Cameotra, S. S., Satpute, S. K., Patil, R., & Nyayanit, N. V., (2014). Cost effective technologies and renewable substrates for biosurfactants' production. *Front Microbiol., 5*, 697–709.

Baskys, E., Grigiskis, S., Levisauskas, D., & Kildisas, V., (2004). A new complex technology of clean-up of soil contaminated by oil pollutants. *Environ. Res. Eng. Manage, 30*, 78–81.

Bazsefidpar, S., Mokhtarani, B., Panahi, R., & Hajfarajollah, H., (2019). Overproduction of rhamnolipid by fed-batch cultivation of *Pseudomonas aeruginosa* in a lab-scale fermenter under tight DO control. *Biodegradation, 30*(1), 59–69. doi: 10.1007/s10532–018–09866–3.

Bozo, H. L., Rocha, C. A., Malavé, R., & Suárez, P., (2012). Biosurfactant production by marine bacterial isolates from the Venezuelan Atlantic front. *Bull Environ. Contamin. Toxicol., 89*(5), 1068–1072.

Calvo, C., Manzanera, M., Silva-Castro, G. A., Uad, I., & González-López, J., (2009). Application of bioemulsifiers in soil oil bioremediation processes: Future prospects. *Sci. Total Environ., 407*, 3634–3640.

Cubitto, A. C., Marían, A., Marta, M., María, C. N., Mónica, C. D., & Siñeriz, B. F., (2004). Effects of Bacillus subtilis O9 biosurfactant on the bioremediation of crude oil-polluted soils. *Biodegradation, 15*(5), 281–287.

Cui, C. Z., Zeng, C., Wan, X., Chen, D., Zhang, J. Y., & Shen, P., (2008). Effect of rhamnolipids in degradation of anthracene by two newly isolated strains, *Sphingomonas* sp. 12A and *Pseudomonas* sp. 12B. *World J. Microbiol Biotechnol., 18*(1), 63–66.

Desai, J. D., & Banat, I. M., (1997). Microbial production of surfactants and their commercial potential. Microbiol Mol Biol Rev. 61, 47–64.

Dhasayan, A., Selvin, J., & Kiran, S., (2015). Biosurfactant production from marine bacteria associated with sponge *Callyspongia diffusa*. *Europe PMC, 5*(4), 443–454.

Ebrahimi, A., Tashi, N., & Lotfalian, S., (2012). Isolation of biosurfactant producing bacteria from oily skin areas of small animals. *Jundishapour J. Microbiol., 5*(2), 401–404.

Franzetti, A., Gandolfi, I., Bestetti, G., Smyth, T. J., & Banat, I. M., (2010). Production and applications of trehalose lipid biosurfactants. *European J. Lipid Sci. Technol., 112*, 617–627.

Franzetti, A., Gandolfi, I., Fracchia, L., Van, H. J., Gkorezis, P., Marchant, R., & Banat, I. M., (2014). Biosurfactant use in heavy metal removal from industrial effluents and contaminated sites. In: Kosaric, N., & Sukan, F. V., (eds.), *Biosurfactants: Production and Utilization-Processes, Technologies and Economics* (pp. 361–369). CRC Press.

Gottfried, A., Singhal, N., Elliot, R., & Swift, S., (2010). The role of salicylate and biosurfactant in inducing phenanthrene degradation in batch soil slurries. *Appl. Microbiol. Biotechnol., 86*, 1563–1571.

Husain, S., (2008). Effect of surfactants on pyrene degradation by *Pseudomonas fluorescens* 29 L. *World J. Microbiol. Biotechnol., 24*, 2411–2419.

Ilia, A. N., & Abdurahman, H. N., (2010). Effect of viscosity and droplet diameter on water-in-oil (w/o) emulsions: An experimental study. *Int. J. Chem. Mol. Eng., 4*, 02–25.

Jonathan, R., & Marques, A., (2017). Interaction of the lipopeptide biosurfactant lichenysin with phosphatidylcholine model membranes. *J. American Chem. Soc., 33*(38), 9997–10005. https://doi.org/10.1021/acs.langmuir.7b01827 (accessed on 24 July 2020).

Joseph, P. J., & Joseph, A., (2009). Microbial enhanced separation of oil from a petroleum refinery sludge. *J. Hazard Mater, 61*(1), 522–525. doi: 10.1016/j.jhazmat.2008.03.131.

Kang, S. W., Kim, Y. B., Shin, J. D., & Kim, E. K., (2010). Enhanced biodegradation of hydrocarbons in soil by microbial biosurfactant, sophorolipid. *Appl. Biochem. Biotechnol., 160*, 780–790.

Karlapudi, A. P., Venkateswarulu, T. C., Tammineedi, J., Kanumuri, L., Ravuru, K. B., Dirisala, V. R., & Kodali, V. P., (2018). Role of biosurfactants in bioremediation of oil pollution-A review. *Petroleum, 4*(3), 248–249.

Kildisas, V., Levisauskas, D., & Grigiskis, S., (2003). Development of clean-up complex technology of soil contaminated by oil pollutants based on cleaner production concepts. *Environ. Research Eng. Manage, 25*, 87–93.

Kumar, S., Devi, S. G., Karuna, M., & Balasubharmanian, S., (2010). Synthesis of biosurfactants-based silver nanoparticles with purified rhamnolipids isolated from *Pseudomonas aeruginosa*BS-161r. *J. Microbiol. Biotechnol., 20*(7), 1061–1068.

Lee, D. W., Lee, H., Kwon, B. O., Khim, J. S., Yim, U. H., Kim, B. S., & Kim, J. J., (2018). Biosurfactant-assisted bioremediation of crude oil by indigenous bacteria isolated from Taean beach sediment. *Environ. Pollut., 241*, 254–264. doi: 10.1016/j.envpol.2018.05.070.

Liu, Z. F., Zeng, G. M., Wang, J., Zhong, H., Ding, Y., & Yuan, X. Z., (2010). Effects of monorhamnolipid and tween 80 on the degradation of phenol by *Candida tropicalis*. *Process Biochem., 45*, 805–809.

McKew, B. A., Coulon, F., Yakimov, M. M., Denaro, R., Genovese, M., Smith, C. J., Osborn, A. M. et al., (2007). Efficacy of intervention strategies for bioremediation of crude oil in marine systems and effects on indigenous hydrocarbonoclastic bacteria. *Environ. Microbiol., 9*, 1562–1571.

Molnár, Z., Bódai, S. G. V., Erdélyi, B., Fogarassy, Z., Sáfrán, G., Varga, T., Kónya, Z. et al., (2018). Green synthesis of gold nanoparticles by thermophilic filamentous fungi. *Sci. Rep., 8*, 3943. doi: 10.1038/s41598–018–22112–3.

Moya, R. I., Tsaousi, K., Rudden, M., Marchant, R., Alameda, E. J., Garcia, R. M., & Banat, I. M., (2015). Rhamnolipid and surfactin production from olive oil mill waste as sole carbon source. *Bioresour. Technol., 198*, 231–236.

Nievas, M. L., Commendatore, M. G., Esteves, J. L., & Bucalá, V., (2007). Biodegradation pattern of hydrocarbons from a fuel oil-type complex residue by an emulsifier-producing microbial consortium. *J. Hazard Mater., 154*(1–3), 96–104.

Nitschke, M., & Costa, O., (2007). Biosurfactants in food industry. *Trends Food Sci. Technol., 18*(5), 252–259. https://doi.org/10.1016/j.tifs.2007.01.002 (accessed on 24 July 2020).

Nurfarahin, A. H., Mohamed, M. S., & Phang, L. Y., (2018). Culture medium development for microbial derived surfactants production. *J. Molecules, 23*(5), 1049.

Obayori, O. S., Ilori, M. O., Adebusoye, S. A., Oyetibo, G. O., Omotayo, A. E., & Amund, O. O., (2009). Degradation of hydrocarbons and biosurfactant production by *Pseudomonas* sp. strain LP1. *World J. Microbiol. Biotechnol., 25*, 1615–1623.

Okoliegbe, I. N., & Agarry, O., (2012). Application of microbial surfactant (a review). *Scholarly J. Biotechnol., 1(*1), 15–23.

Otzen, D. E., (2017). Biosurfactants and surfactants interacting with membranes and proteins: Same but different. *Biochim. Biophys. Acta Biomembr., 1859*(4), 639–649. doi: 10.1016/j.bbamem.2016.09.024.

Patel, S., Homaei, A., Daverey, A., & Patil, S., (2019). Microbial biosurfactants for oil spill remediation pitfalls and potentials. *Appl. Microbiol. Biotechnol., 103*(1), 27–37.

Patowary, K., Patowary, R., Mohan, K. M. C., & Deka, S. K., (2017). Characterization of biosurfactant produced during degradation of hydrocarbons using crude oil as sole source of carbon. *Front Microbiol., 8*, 279. doi: 10.3389/fmicb.2017.00279.

Pei, X. H., Zhan, X. H., Wang, S. M., Lin, Y. S., & Zhou, L. X., (2010). Effects of a Biosurfactant and a synthetic surfactant on phenanthrene degradation by a *Sphingomonas* strain. *Pedosphere, 20*, 771–779.

Piorkowski, D. T., & McClements, D. J., (2013). Beverage emulsions: Recent developments in formulation, production, and applications. *Food Hydrocolloid*. doi: 10.1016/j.foodhyd.07.009.

Płociniczak, M. P., Płaza, G., Piotrowska, Z. S., & Cameotra, S. S., (2011). Environmental applications of biosurfactants: Recent advances. *Int. J. Mol. Sci., 12*(1), 633–654.

Poremba, K. W., Gunkel, S., & Wagner, F., (1991). Toxicity testing of synthetic and biogenic surfactants on marine microorganisms. *Environ. Toxicol. Water Qual., 6*, 157–163.

Reddy, M. S., Naresh, B., Leela, T., Prashanthi, M., Madhusudhan, N. C., Dhanasri, G., & Devi, P., (2010). Biodegradation of phenanthrene with biosurfactant production by a new strain of *Brevibacillus* spp. *Bioresour. Technol., 101*, 7980–7983.

Reis, R. S., Pacheco, G. J., & Pereira, A. G., (2013). Biosurfactants: Production and applications, biodegradation. Intech. doi: 10.5772/56144.

Ron, E. Z., & Rosenberg, E., (2002). Biosurfactants and oil bioremediation. *Curr. Opin. Biotechnol., 13*(3), 249–252.

Ron, E. Z., & Rosenberg, E., (2014). Enhanced bioremediation of oil spills in the sea. *Curr Opin Biotechnol., 27,* 191–194. doi: 10.1016/j.copbio.2014.02.004.

Rosenberg, E., & Ron, E. Z., (1999). High-and low-molecular-mass microbial surfactants. *Applied Microbiol. Biotechnol., 52*, 154–162.

Saranya, P., Swarnalathaa, S., & Sekaran, G., (2014). Lipoprotein biosurfactant production from an extreme acidophile using fish oil and its immobilization in nanoporous activated carbon for the removal of Ca^{2+} and Cr^{3+} in aqueous solution. *RSC Adv., 4*, 34144–34155. doi: 10.1039/C4RA03101F.

Sen, R., (2010). *Biosurfactants* (p. 331). Springer New York.

Shahaby, A. F., Alharthi, A. A., El Tarras, A. E., (2015). Bioremediation of petroleum oil by potential biosurfactant producing bacteria using gravimetric assay. *Int. J. Curr. Microbiol. Appl. Sci., 4*(5), 390–403.

Shao, Z., (2011). Trehalolipids. In: Soberón-Chávez, G., (eds.), *Biosurfactants-From Genes to Applications* (pp. 121–143). Springer, Berlin; Heidelberg.

Sharma, D., (2009). *Production of Microbial Biosurfactants Using Cost Effective Resources*. Doctoral thesis, Indian Institute of Technology, Roorkee, India.

Shin, K. H., Ahn, Y., & Kim, K. W., (2005). Toxic effect of biosurfactants addition on the biodegradation of phenanthrene. *Environ. Toxicol. Chem., 24*, 2768–2774(A).

Shin, K. H., Kim, K. W., & Ahn, Y., (2006). Use of biosurfactant to remediate phenanthrene-contaminated soil by the combined solubilization biodegradation process. *J. Hazard. Mater. 137*, 1831–1837(B)

Silva, R. D. C. F. S., Almeida, D. G., & Rufino, R. D., (2014). Applications of biosurfactants in the petroleum industry and the remediation of oil spills. *Intl. J. Mol. Sci., 15*, 12523–12542.

Singh, P., & Cameotra, S. S., (2004). Potential applications of microbial surfactants in biomedical sciences. *Trends Biotechnol., 22*, 142–146.

Tong, J., & Thomas, J. M., (2004). Structure of supported bilayers composed of lipopolysaccharides and bacterial phospholipids: Raft formation and implications for bacterial resistance. *Biophys J., 86*(6), 3759–3771. doi: 10.1529/biophysj.103.037507.

Varjani, S. J., & Upasani, V. N., (2017). Review on biosurfactants analysis, purification and characterization using rhamnolipid as a model biosurfactants. *Bioresour. Technol., 047*(2), 389–397.

Vijayakumar, S., & Saravanan, V., (2015). Biosurfactants-types, sources and applications. *Res. J Microbiol., 10*(5), 181–192.

Wang, H., Roelants, S., Patria, R. D., Kaur, G., Lau, N. S., Lau, C. Y., Lin, C. S., & Bogaert, V., (2019). *Starmerella bombicola*: Recent advances on sophorolipids production and prospects of waste stream utilization. *J. Chem. Technol. Biotechnol., 94*(4), 999–1007 https://doi.org/10.1002/jctb.5847 (accessed on 24 July 2020).

Youssef, N. H., Dunacn, K. E., Nagle, D. P., Savage, K. N., Knapp, R. M., & McInerney, M. J., (2004). Comparison of methods to detect biosurfactant production by diverse microorganism. *Microbiol. Meth., 56*, 339–347.

Zeng, G., Liu, Z., Zhong, H., Li, J., Yuan, X., Fu, H., Ding, Y., Wang, J., & Zhou, M., (2011). Effect of monorhamnolipid on the degradation of N-hexadecane by *Candida tropicalis* and the association with cell surface properties. *Appl. Microbiol. Biotechnol., 90*, 1155–1161.

Zeppier, S., Rodrigaug, J., & Ramos, D., (2001). Interfacial tension of alkane and water system. *J. Chem. Eng., 46*(5), 1086–1088 https://doi.org/10.1021/je000245 (accessed on 24 July 2020).

Zhang, J., Yin, R., Lin, X., Liu, W., Chen, R., & Li, X., (2010). Interactive effect of biosurfactant and microorganism to enhance phytoremediation for removal of aged polycyclic aromatic hydrocarbons from contaminated soils. *J. Health Sci., 56*, 257–266.

Zhu, L., & Zhang, M., (2008). Effect of rhamnolipids on the uptake of PAHs by ryegrass. *Environ. Pollut., 156*, 46–52.

CHAPTER 10

Fungal Solid-State Bioprocessing of Grapefruit Waste

RAMÓN LARIOS-CRUZ,[1] ROSA M. RODRÍGUEZ-JASSO,[1] JUAN BUENROSTRO-FIGUEROA,[2] ARELY PRADO-BARRAGÁN,[3] HÉCTOR RUIZ,[1] and CRISTÓBAL N. AGUILAR[1]

[1]*Research Group of Bioprocesses and Bioproducts, Food Research Department, School of Chemistry, Universidad Autónoma de Coahuila, Saltillo, Coahuila, 25280, México, E-mail: cristobal.aguilar@uadec.edu.mx*

[2]*Research Center for Food and Development, CIAD, A.C. Ciudad Delicias, Chihuahua, México*

[3]*Departamento de Biotecnología, Universidad Autónoma Metropolitana Unidad Iztapalapa, Delegación Iztapalapa, Distrito Federal, 09340, México*

ABSTRACT

Solid-state bioprocessing is a microbial process on a support material in the absence of free water. Among its principal application are found the production of metabolites as enzymes, colorants, bioactive compounds. In the process, it can use a great variety of supports, emphasizing the use of agroindustrial wastes. In the present study, the fungal growth was evaluated by respirometry and the percentage of solids reduction was determined. Also, the recovery of fermented extracts with antimicrobial activity was analyzed. The fermentation was carried out with grapefruit wastes and *Aspergillus niger* GH1, and the measure of CO_2 production was made by gas chromatography. Results showed a reduction of 40 and 56% of solids for systems of 50 and 70% of humidity, respectively, at the end of the fermentation; besides, ethanolic extracts showed antimicrobial activity.

10.1 INTRODUCTION

Solid-state bioprocessing or fermentation (SSF) is a microbiological process commonly occurring on the surface of solids that have the property of absorbing water, with and without nutrients (Viniegra-González, 1997). Filamentous fungi are the most widely used in this system because of their ability to penetrate the solid substrate for nutrient absorption. The main use of these microorganisms in FES is the use of agro-industrial waste (Raimbault, 1998). Within the fermentation process, oxygen consumption and release of carbon dioxide are metabolic activities that can be used to estimate biomass synthesis (Saucedo-Castañeda et al., 1994). On the other hand, there are variables of the process that can influence not only the production of metabolites but the performance of the bioreactor. One of these is the water content, which influences the exchange of nutrients and the biosynthesis of intracellular metabolites (osmotic pressure). In filamentous fungi can change the orientation and radius of growth of the hyphae, also related to the sporulation and germination (Gervais and Molin, 2003).

Several agro-industrial wastes can be applied in SSF to increase its value. One of them is the citrus residue. In the case of grapefruit, approximately 55% of the weight of the fruit corresponds to the residues generated after the extraction of the juice (shell and Bagasse). With the various industrial processes of harvesting the fruit, it is important to look for waste-use procedures. An alternative is the SSF, which allows the reduction of waste at the end of the fermentation process, besides it is possible to recover compounds of high added value as enzymes and others with antioxidant activity to name a few. For this reason, in the present work, the objective was to evaluate the microbial growth via respiration and to determine the percentage of reduction of solids as well as recovery of fermented extracts with antimicrobial activity.

10.2 MATERIALS AND METHODS

The fungal strain was *Aspergillus niger* GH1 which belongs to the fungal collection of Research Group of Bioprocesses and Bioproducts (DIA-UAdeC). Grapefruit residues were donated by a juice producer from the local market at Saltillo, Mexico. After harvesting they were dried in a stove at 60 °C for 48 h, ground, and sieved to a particle size of 2.00-1.84 mm.

Raimbault glass columns were used as bioreactors for the growth of the *A. niger* GH1 on grapefruit residues to evaluate the growth and production of metabolites with antimicrobial activity. The influence of initial moisture on the bioprocess was evaluated at two levels (50% and 70%). Grapefruit residues were not sterilized. The inoculation was with $2x10^7$ spores/gdm. The fermentation was carried out by 120 h, with a sampling every 24 h. The fungal growth was monitored by gas chromatography to monitor in real-time the production of CO_2 (Saucedo-Castañeda et al., 1994).

For extraction two solvents were used, water and 70% ethanol. Fermented residues were divided into three fractions previous to the extraction process. The pH analysis and antimicrobial activity assay were evaluated in the extracts. For antimicrobial activity, *Escherichia coli* strain PPF4-UAMI was selected using the assay reported by Mejia *et al* (1996). However, for that ethanolic extracts, the solvent was firstly removed by evaporation and the solids were resuspended in distilled water.

10.3 RESULTS AND DISCUSSION

During fermentation, the simpler degrading substrate was firstly consumed by the fungus, and then a phase of re-adaptation (diauxic phase) to the medium was presented and then another type of substrate was developed. This same trend can be seen in Figure 10.1. In which CO_2 production continues to accumulate at the end of fermentation, this suggests that the fungus continues to degrade substrate.

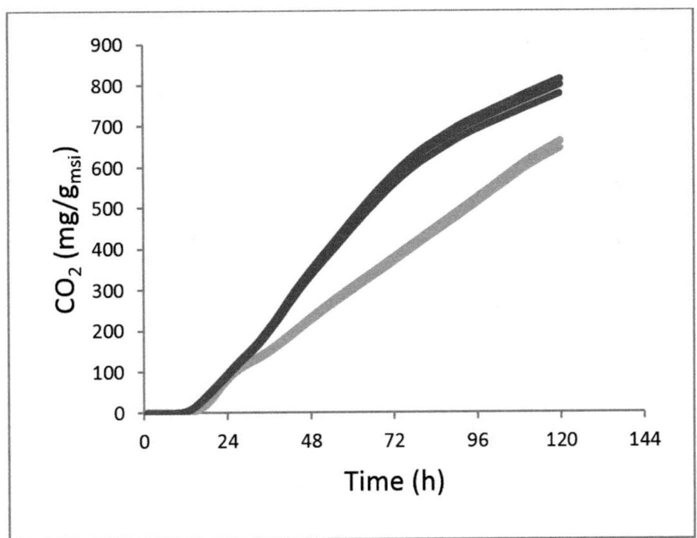

FIGURE 10.1 Total CO_2 production. The results of the three repetitions are displayed. CO_2 is expressed to standard pressure and temperature conditions. (a) 50% humidity. (b) 70% humidity.

For the 50% humidity system, a CO2 conversion of 657.56 mg gmsi-1 and 70% humidity 797.81 mg gmsi-1 (Figure 10.1) was performed. Both correspond to the milligrams of carbon sources that have grapefruit residues. This result can be related to weight loss or residue degradation after the fermentation process. It is possible to appreciate that moisture influences the growth of the fungus; therefore, it is to be expected that there will be greater degradation of solids at greater growth of the micro-organism (Figure 10.2).

The delay phase was cut short 12.34 h to 0.31 and 10.72 h to 0.15 for 50% and 70% humidity, respectively, which could be due both to the nutrient nutrients and trace elements necessary for the growth of the fungus and to the high concentration of inoculum (Nagel et al., 1999; Saucedo-Castañeda et al., 1994). During the first phase of growth, simple sugars contained in grapefruit residues such as sucrose, glucose, and fructose, and others that are easy to degrade such as pectin may be consumed.

For the second stage of growth, the consumption of polysaccharides such as cellulose may be a possibility, even if its content is low in grapefruit residues (Liu et al., 2012).

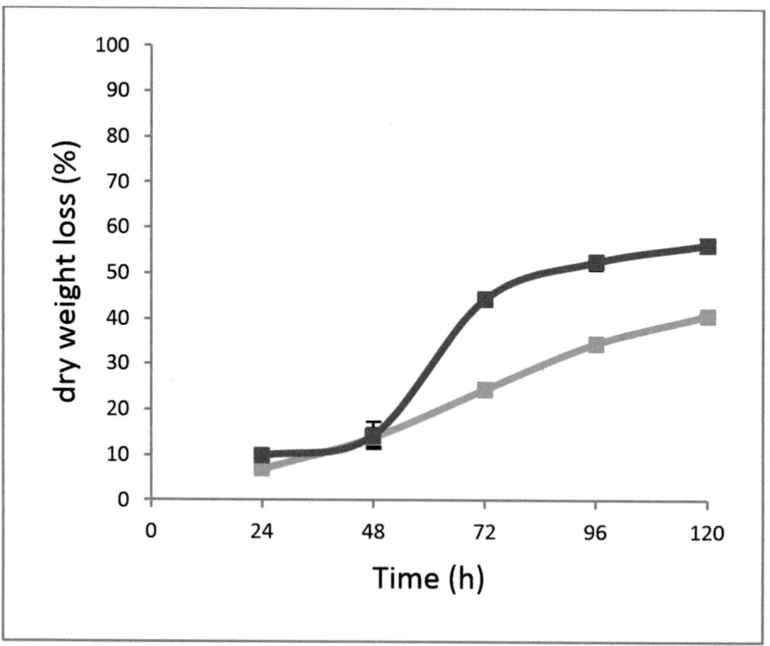

FIGURE 10.2 Dry weight loss. The value corresponds to the degradation of solids by SSF. 50% humidity. 70% humidity.

For aqueous extracts, the pH was measured, and their behavior is expressed in Figure 10.3. It is possible to appreciate a decrease in the first few hours of fermentation corresponding to the maximum growth of *A. niger* GH1. For a subsequent increase in pH after 72 h fermentation.

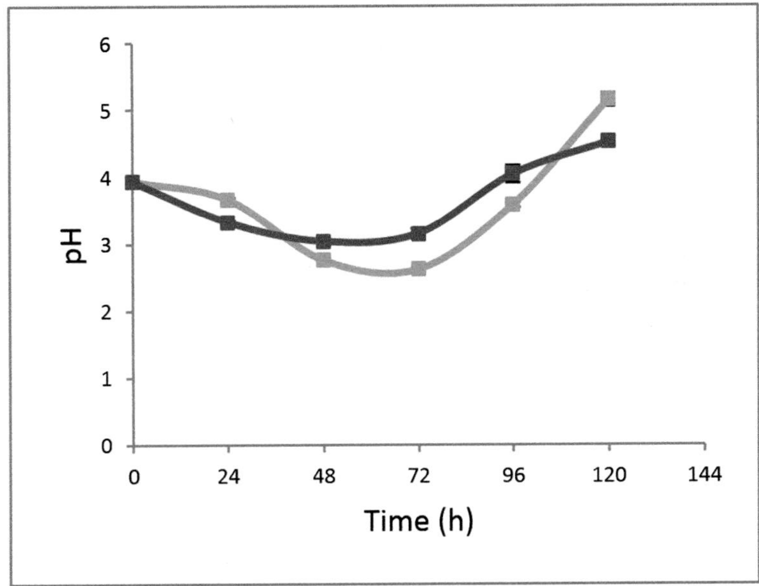

FIGURE 10.3 Effect of moisture on pH. (a) 50% humidity. (b) 70% humidity.

The pH may influence metabolite recovery. However, it is difficult to control during the SSF process and its online measurement is limited due to its dependence on the water content (exchange of protons with the electrode). It is usually regulated at the beginning of fermentation but its changes are not known or controlled during fermentation (Chen, 2013).

The acidification of the medium is mainly due to the growth of mycelium (Raimbault and Alazard, 1980). On the other hand, its increase to final fermentation times may be due to lysis of the microorganism which increases the value of pH by the products generated (Gooday et al., 1986).

For the antimicrobial activity of ethanol extracts against *E. coli*, the results showed an increase in inhibition halos at longer fermentation times (Figure 10.4). This is clear evidence of the effect of bioactive compound release during fungal degradation of cell walls of grapefruit residues.

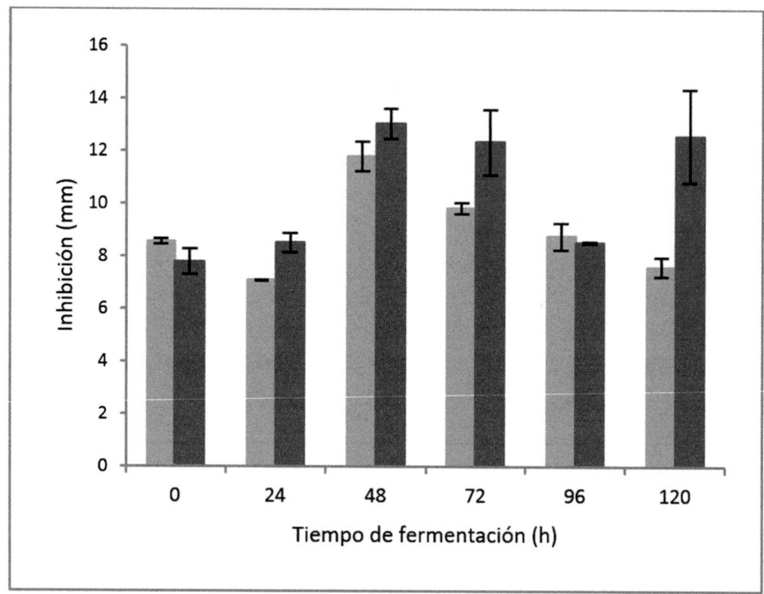

FIGURE 10.4 Halos inhibiting extracts on *E. coli* at 50% humidity and 70% humidity.

The type of cell wall of microorganisms (Gram-positive or Gram-negative) influences the mechanism of action of flavonoids to inhibit their growth. Some phenolic compounds can damage the cell wall (Gram-negatives) and others pass directly into the cells to cause DNA damage (Puupponen-Pimia et al., 2001).

10.4 CONCLUSION

The respirometry profile shown in total CO_2 production demonstrates the affinity of *A. niger* GH1 in grapefruit residue degradation. The process not only allowed the release and recovery of metabolites of antimicrobial interest; but degraded the grapefruit residue. A 50% reduction in solids suggests the fermenting process as an alternative in the treatment of grapefruit residues by SSF.

ACKNOWLEDGMENTS

The author thanks the National Council of Science and Technology (CONACYT-Mexico) for the financial support.

KEYWORDS

- **solid-state fermentation**
- ***Aspergillus niger***
- **metabolic respiration**

REFERENCES

Chen, H., (2013). *Modern Solid State Fermentation-Theory and Practice.* Springer, Londres.

Gervais, P., & Molin, P., (2003). The role of water in solid-state fermentation. *Biochem. Eng. J., 13*, 85–101. http://dx.doi.org/10.1016/S1369-703X(02)00122-5 (accessed on 24 July 2020).

Gooday, G. W., Humphreys, A. M., & McIntosh, W. H., (1986). In: Muzzarelli, R., Jeuniaux, C., & Gooday, G. W., (eds), *Roles of Chitinases in Fungal Growth* (pp. 83–91). Chitin in Nature and Technology. Springer US.

Liu, Y., Heying, E., & Tanumihardjo, S. A., (2012). History, global distribution, and nutritional importance of citrus fruits. *Compr. Rev. Food Sci. Food Saf., 11*, 530–545. doi: 10.1111/j.1541-4337.2012.00201.x.

Mejía, A. G., Ramírez-Gama, R. M., & Velázquez, O., (1996). Influence of environmental factors on the growth and death of microorganisms. In: Ramírez-Gama, R. M., Luna, B., Mejía, A. G., Velázquez, O., Tsuzuki, G., Vierna, L., Hernández, L., & Müggenburg, I., (eds.), *Manual De Prácticas De Microbiología General* (pp. 222–259). Laboratorio de microbiología experimental. UNAM, México.

Nagel, F. J., Oostra, J., Tramper, J., & Rinzema, A., (1999). Improved model system for solid-substrate fermentation: Effects of pH, nutrients and buffer on fungal growth rate. *Process Biochem., 35,* 69–75. http://dx.doi.org/10.1016/S0032-9592(99)00034-5 (accessed on 24 July 2020).

Puupponen-Pimiä, R., Nohynek, L., Meier, C., Kähkönen, M., Heinonen, M., Hopia, A., & Oksman-Caldentey, K. M., (2001). Antimicrobial properties of phenolic compounds from berries. *Journal of Applied Microbiology, 90*, 494–507. doi: 10.1046/j.1365-2672.2001.01271.x.

Raimbault, M., (1998). General and microbiological aspects of solid substrate fermentation. *Electron. J. Biotechnol., 1*, 26–27.

Raimbault, M., & Alazard, D., (1980). Culture method to study fungal growth in solid fermentation. *European Journal of Applied Microbiology and Biotechnology, 9*, 199–209. 10.1007/bf00504486.

Saucedo-Castañeda, G., Trejo-Hernández, M. R., Lonsane, B. K., Navarro, J. M., Roussos, S., Dufour, D., & Raimbault, M., (1994). On-line automated monitoring and control systems for CO_2 and O_2 in aerobic and anaerobic solid-state fermentations. *Process Biochem., 29*, 13–24. http://dx.doi.org/10.1016/0032-9592(94)80054-5 (accessed on 24 July 2020).

Viniegra-González, G., (1997). Solid state fermentation: Definition, characteristics, limitations, and monitoring. In: Roussos, S., Lonsane, B. K., Raimbault, M., & Viniegra-González, G., (eds.), *Advances in Solid State Fermentation* (pp. 5–22). Springer Netherlands.

CHAPTER 11

Valorization of Ataulfo Mango Seed Byproduct Based on Its Nutritional and Functional Properties

CRISTIAN TORRES-LEÓN,[1,3] JUAN A. ASCACIO-VALDÉS,[1]
MÓNICA L. CHÁVEZ-GONZÁLEZ,[1] LILIANA SERNA-COCK,[2]
NATHIELY RAMIREZ-GUZMAN,[1,3] ALCIDES CINTRA,[3]
CLAUDIA LÓPEZ-BADILLO,[4] ROMEO ROJAS,[5]
RUTH BELMARES-CERDA,[1] and CRISTÓBAL N. AGUILAR[1]

[1]*Food Research Department, School of Chemistry, Universidad Autónoma de Coahuila, Saltillo, México, E-mail: cristobal.aguilar@uadec.edu.mx*

[2]*School of Engineering and Administration, Universidad Nacional de Colombia, Palmira, Colombia*

[3]*Departamento de Bioquímica, Centro de Ciencias Biológicas, Universidad Federal de Pernambuco, Recife, Brazil*

[4]*School of Chemistry, Universidad Autónoma de Coahuila, Saltillo, México*

[5]*School of Agronomy, Universidad Autonoma de Nuevo Leon, General Escobedo, Mexico*

ABSTRACT

The mango seed is usually discarded as waste. However, this byproduct is an excellent source of bioactive compounds and nutrients with a high potential for valorization as novel food ingredients. In this chapter, Ataulfo mango seed was investigated for its nutritional and functional properties. The yield, protein, fatty acids (GC-MS), minerals, dietary fiber as well as total polyphenols, antioxidant activity, and HPLC-MS analysis were determined. The results indicated that mango processing generated 9.5% of seed as a byproduct; this has a high dry matter content, protein, ash, and lipids. Mango seed is a good source of essential minerals such as potassium, calcium, and phosphorus and lipids as palmitic acid, stearic acid, and oleic acid (free of trans fatty acids). Mango seed has a high content of polyphenols compounds (612.8 mg/g) and a high antioxidant activity ($ABTS^{+}$ = 1899.15 ± 124.38 and FRAP= 986.61 ± 226.72 mg Trolox/g). These results were higher than previously reported in mango seeds and other natural extracts with high power bioactive. The IC_{50} (r^2 = 0.99) was higher (0.07 mg/mL) previously registered in the commercial anti-oxidants. HPLC-MS analysis reveals the presence of families of compounds with high antioxidant power. This allows concluding that the mango seed the variety Ataulfo has a high potential of valorization. This work unveils several investigative perspectives to promote new technological applications of this byproduct, based on its functional and nutritional properties.

11.1 INTRODUCTION

The agro-food industries generate huge quantities of biodegradable solid wastes and consist of organic residues of the processed raw materials (Nayak and Bhushan, 2019). Food waste and byproducts from different fruit processing industries have attracted special interest in recent decades (Torres-León et al., 2018). These biomaterials are a source of important molecules such as fibers (Nawirska and Kwaśniewska, 2005), minerals (Xu et al., 2008), and antioxidants (da Silva and Jorge, 2014). Although the use of food waste and byproducts has been proposed to meet the growing demand for products rich in bioactive compounds (Mirabella et al., 2014), currently the greater part of the waste is disposed, generating a serious economic and environmental problems like toxicity to aquatic life, pollution of surface and ground waters, altered soil quality, phytotoxicity, and odor (Nayak and Bhushan, 2019).

Mexico is the first mango producer in America and the fifth in the world (FAOSTAT, 2017). Mango processing waste constitutes approximately 600,000 t. per year; a medium industry that can process 200 t. per day can obtain 84 t. of byproducts (García et al., 2013). These byproducts are mainly peels (13-16%) (Serna, Torres, and Ayala, 2015) and seeds (10-25%) (Solís and Durán, 2011). Although, several studies have reported the use of Mango byproducts (*Mangifera Indica* L.) as a good source of fiber (Ajila, Aalami, Leelavathi, and Rao, 2010; Ajila, Rao, and Rao, 2010; Ajila and Rao, 2013), pectins (Koubala et al., 2008; Rojas et al., 2018, 2015), antioxidants (Ajila, Naidu, Bhat, and Rao, 2007; Ajila, Rao, et al., 2010; Dorta, Lobo, and González, 2012; Kim et al., 2010), minerals (Sonia Ribeiro and Schieber,

2010) and flours with applications in the formulation of animal feeds (Fontes et al., 2008; Odunsi, 2005) and biodegradable packaging (Torres-león et al., 2018); today most of these byproducts are discarded as waste after industrial processing. Poor knowledge of the properties and composition of byproducts reduces the possibility of generating added value. The chemical and functional characterization of the mango byproducts is necessary to identify business opportunities. The economics of processing tropical fruits could be improved by developing higher-value use for their byproducts (Morais et al., 2014).

Our bioprocess research group (bio-uadec.com) recently published a complete review article on the nutritional and functional properties of mango seeds (*Mangifera indica* L.) of the main varieties grown in the world (Torres-León et al., 2016). However, Ataulfo mango seed (*Mangifera caesia,* endemic to Mexico), has not been characterized. The aim of this study was to identify the nutritional and functional composition of the Ataulfo mango seed, specifically minerals, a fatty acid composition by GC-MS, protein, dietary fiber, total phenols, antioxidant activity, IC_{50} and major phenolic compounds by HPLC-MS.

11.2 MATERIALS AND METHODS

11.2.1 RAW MATERIAL

Thirty-five mangoes the Ataulfo variety were obtained from the local market in Saltillo, Coahuila, México. Mangoes were selected and classified by size (16 gauge), maturity a color uniform (NOM-188-SCFI, 2012). After that, were disinfected (200 ppm sodium hypochlorite) and the seed was manually separated from the fruit. The seed kernel was cut into small pieces.

11.2.1.1 YIELD MEASUREMENT

The yields of the fruit components were determined by Eq. (1):

$$\% \text{ Yield} \frac{\text{Component weight}}{\textit{Fruit weight}} * 100 \tag{1}$$

11.2.1.2 EVALUATION OF THE NUTRITIONAL COMPOSITION OF MANGO SEED

The seed was dried in an oven (Napco Model 322) at 60°C and ground in an attritor mill discs (Pulvex 100 MINI-México). The dry matter was determined by the method AOAC 934.01 (1990), the content of crude protein (CP) by titration (Kjeldahl, 1883), the ash content (A) according to the method AOAC 942.05, (1990), lipids content (L) according to the method and AOAC 930.09, (2005), contents cellulose, hemicellulose and lignin were calculated by determining neutral detergent fiber and acid detergent fiber's according to the method described by Van Soest, Robertson, and Lewis, (1991), and total carbohydrates (Tc) were obtained by Eq. (2).

$$Tc = 100 - \%CP + \%A + \%L \tag{2}$$

11.2.2.2.1 Minerals in Mango Seed

The minerals in the mango seed were determined by X-ray fluorescence (Epsilon 1, UK).

11.2.2.2.2 Fatty Acid Composition by Gas Chromatography (GC-MS)

Five grams of mango seed was extracted with 50 mL of Hexane, in an ultrasonic bath sonicator for 60 minutes; then the samples were centrifuged (1000 rpm, 10 min). One μL of the extract was injected into a gas chromatograph equipped with a polar capillary column (30 m × 0.32 mm × 0.25 μm) Agilent HP-Innowax. Injector and detector (FID) temperatures were 250°C. The oven temperature was maintained at 50°C for 2 min. Subsequently, it reached a temperature of 220°C with a ramp of 30°C/min, keeping at that temperature for 25 min. Finally, the temperature was increased to 255°C, maintaining this temperature for 7 min. The different esters were identified by comparing their retention times with those of the Supelco 37 FAME Mix standard.

11.2.1.3 EVALUATION FUNCTIONAL OF MANGO SEED

The seed kernel was lyophilized (Labconco Freezone 4.5, USA) for two days. The dried mango seeds were ground in a mill (Pulvex 100MINI, México) and were stored until the respective analysis. Extraction was carried out using (MARS 6-USA, 600W), with ethanol (90%). The extraction was carried out in closed vessels, with a built-in safety valve according to the method of Torres-León et al. (2017). The extract was measured: scavenging activity of free radicals $ABTS^+$, ferric ion reducing antioxidant power (FRAP), total polyphenol compounds, the inhibitory concentration of 50% (IC_{50}) and RP-HPLC-ESI-MS analysis.

The activity of free radical scavenging ABTS·+ was determined according to the methodology developed by Re et al. (1999). FRAP analysis was performed according to the method described by Benzie and Strain (1996), the two results were expressed as Trolox equivalent antioxidant capacity (TEAC) with a calibration curve with Trolox (6-hydroxy-2,5,7,8-tetramethylchroman-2-carboxylic acid) in different dilutions. The total polyphenolic compounds were determined as the sum between the hydrolyzable polyphenols and condensed polyphenols, hydrolyzable polyphenols were determined according to the methodology described by Wong-Paz et al. (2015), condensed polyphenols content was evaluated by the method HCl-butanol described by Swain and Hillis (1959). The inhibitory concentration of 50% was determined with a linear regression of the lipid oxidation inhibition (LOI) as a function to concentration (0.0031 to 1 $mg.mL^{-1}$). This assay was done according to the method reported by Martinez et al. (2011).

11.2.1.3.1 *Compound Identification by HPLC-MS*

The analyses by HPLC-MS was developed as described by Torres-León et al. (2017).Varian HPLC system includes a Denali C18 column (150 mm × 2.1 mm, 3μm, Grace, USA), an autosampler (VarianProStar 410, USA), a ternary pump (Varian ProStar 230I, USA) and a PDA detector (Varian-ProStar 330, USA). A liquid chromatography ion trap mass spectrometer (Varian 500-MS IT Mass Spectrometer, USA) equipped with an electrospray ion source also was used.

11.2.1.3.2 *Statistical Analysis*

Analyses are presented as mean ± SD. The results were analyzed through the Statistica 7.0 software (StatSoft, Tulsa, OK, USA). All assays were performed in triplicate.

11.3 RESULTS AND DISCUSSION

11.3.1 *YIELD MEASUREMENT*

Ataulfo mango seed is composed of a woody outer shell encasing a kernel (seed proper) (Masibo and He, 2009). The seed represents the 9.5 ± 1.3% and kernel 4.9 ± 0.4% of the total weight of the fruit (51.5% of total seed weight). Similar results were reported in other mango varieties (Arogba, 1997; Solís and Durán, 2011).

11.3.2 EVALUATION OF THE NUTRITIONAL COMPOSITION OF MANGO SEED

Table 11.1 shows the composition of macronutrients of mango seed kernel; fresh kernels have moisture content similar to those reported by Nzikou et al. (2010) in mango seed kernels from Nigeria. The moisture content can promote biochemical and microbial spoilage; the drying technique is recommended to stabilize the material. The ash content was higher than reported in mango seed kernels (1.4-3.2%) (Odunsi, 2005; Zein et al., 2005), this suggests that the mango seed of the variety Ataulfo could be a good source of minerals. The lipid content (Table 11.1) was lower than previously reported in seeds of other varieties (8.15 to 13.16%) (Abdalla et al., 2007; Ashoush and Gadallah, 2011; Nzikou et al., 2010), this phenomenon may be attributed to weather conditions, cultivation, and variety. Solís and Durán (2011) reported that lipids mango seed kernels are considered as a substituent of saturated fats harmful to human health.

TABLE 11.1 The Chemical Composition of Mango Seed

Component	Mango seed kernel	Mango seed endocarp and testa
Moisture	40.0 ± 0.01	66.06 ± 6.672
Ash	5.7 ± 0.44	1.7 ± 0.32
Protein	7.3 ± 0.19	4.0 ± 0.12
Lipids	5.4 ± 0.07	0.8 ± 0.01
Carbohydrates	81.9 ± 0.08	93.6 ± 0.08
Crude fiber	22.1 ± 0.58	52.6 ± 0.35
Hemicellulose	14.1 ± 0.02	9.6 ± 0.03
Cellulose	2.9 ± 0.39	22.4 ± 0.02
Lignin	2.0 ± 0.47	9.77 ± 0.50

Results reported in% on a dry basis. Data are mean ± SD.

Source: Modified from: Torres-León et al. (2019).

As shown in Table 11.1, the seed has a high protein content. This is higher than reported in mango seeds of other varieties (6–6.7%) (Elegbede, Achoba, and Richard, 1995; Odunsi, 2005). In this regard, Abdalla et al. (2007) say that even though the protein content in mango seed is low compared to other plant sources, mango seed has all the essential amino acids at higher levels than those referenced by FAO. The starch of mango seed is similar to tapioca starch (Saadany et al., 1980).

The analysis of the dietary fiber present in the mango seed is shown in Table 11.1. The kernel of the mango seed presented a fiber content of 22%, and the endocarp and testa (that covers the kernel) presented a fiber content of 52%. In the scientific literature, works on the valuation of mango seeds refer to the use of kernel only (Torres-León et al., 2016), so the endocarp and testa are not taken into account. Table 11.1 demonstrates that the endocarp of the mango seed is a good source of dietary fiber, especially cellulose (22%) which can enhance its use in the food and biotechnology industry. The content of dietary fiber in the kernel was higher than previously reported in the almond of the command seed (4.6%) (Odunsi, 2005). The high fiber content is attributed to the presence of tegument joined kernel.

11.3.2.1 MINERALS IN MANGO SEED

Mineral nutrients are essential for the normal development of human metabolism and the maintenance of good health. Table 11.2 shows the main minerals present in the mango seed. K was the main mineral with a percentage of 70%, Ca and P presented a concentration of 15 and 5.5%, respectively. S, Cl, Si, and Fe were also present in the mango seed, although with smaller proportions (< 3%). Previously it was reported that potassium, phosphorus, and magnesium are the major minerals in the mango seeds (Ashoush and Gadallah, 2011; Nzikou et al., 2010). Potassium and phosphorus are essential nutrients in the diet; the necessary consumption is 10 g/day (Gupta and Gupta, 2014). This study showed that Ataulfo mango seed is also a good source of Ca. Calcium plays a pivotal role in sustaining prime bone and teeth health and other normal body functions (Waheed et al., 2019). Calcium deficiency is responsible for diseases such as hypertension and osteoporosis.

TABLE 11.2 Minerals in Mango Seed

Mineral	Concentration (%)
K	70.3 ± 0.15
Ca	15.0 ± 0.47
P	5.5 ± 0.04
S	3.0 ± 0.01
Cl	2.4 ± 0.05
Si	0.9 ± 0.04
Fe	0.7 ± 0.04

Data are mean ± standard deviation.
Percentage in relation to the mineral fraction.

11.3.2.2 FATTY ACID COMPOSITION BY GAS CHROMATOGRAPHY (GC-MS)

Table 11.3 shows the major fatty acids present in Ataulfo mango seed, GC-MS revealed the presence of palmitic acid, stearic acid, oleic acid, linoleic acid, linolelaidic acid, and stearidonic acid. These results are consistent with what was previously reported in mango seed from Egypt (Abdalla et al., 2007; Abdel et al., 2012), Thailand (Sonwai, Kaphueakngam, and Flood, 2012), Congo (Nzikou et al., 2010), Taiwan (Wu et al., 2015) and Malasia (Jahurul et al., 2014). Mango seed lipids did not contain "trans" fatty acids (molecules that have adverse health effects), so these have a great potential for valorization. Ataulfo mango seed lipids have characteristics closely resemble those of Cocoa butter. According to the European Union Directive (2000/36/EC), by its composition, the mango seed is one of the only six fats which are allowed to be used in chocolate fat production (Jin et al., 2019). We recommend continuing with new studies that investigate the industrial application of Mexican mango seed lipids.

TABLE 11.3 Fatty Acid Composition of Ataulfo Mango Seed Kernel by GC-MS

Retention time (min)	Molar mass	Molecular formula	Common Name	ω-n
7.298	126.3	Unidentified	–	
7.784	135	Unidentified	–	
8.772	263.8	Unidentified	–	
13.356	256.4	$C_{16}H_{32}O_2$	16:0 Palmitic acid	ω–7
18.256	284.48	$C_{18}H_{36}O_2$	18:0 stearic acid	
20.202	263.7	Unidentified	–	
20.202	282.5	$C_{18}H_{34}O_2$	18:1 oleic acid	ω–9
20.900	280.4	$C_{18}H_{32}O_2$	18:2 linoleic acid	ω–6
20.900	280.45	$C_{18}H_{32}O_2$	18:2 Linolelaidic acid	ω–6
24.392	104.8	Unidentified	–	
25.869	276.4	$C_{18}H_{28}O_2$	18:4 Stearidonic acid	ω–3

11.3.3 FUNCTIONAL EVALUATION OF MANGO SEED

The antioxidant capacity analysis was performed by methods that assess the ability to deactivate free radicals by electron transfer ($ABTS^{\cdot +}$ y FRAP) and transfer hydrogens (LOI); these are widely used and have great relevance to biologically by reliability, repeatability and low cost. As shown in Table 11.4, the extracts have high antioxidant activity expressed in equivalent Trolox, this is higher than previously reported in mango seeds and other parts of the plant mango (Chong et al., 2013; Dorta et al., 2012; Ma et al., 2011; Sogi et al., 2013). Also, the results are higher than those reported in plant extracts considered good source of antioxidants, as nails (*Syzygium aromaticum*), true cinnamon tree (*C. cassia*) and black mustard (*B. nigra*) (Radha et al., 2014) and peel of pomegranate (*Punica granatum*) (Amyrgialaki et al., 2014).

TABLE 11.4 Antioxidant Activity and Polyphenols in Mango Seed Kernel

ABTS$^{\cdot+}$ (TEAC/g)	FRAP (TEAC/g)	Total polyphenols (mg/g)	Condensed tannins (%)	Hydrolyzable tannins (%)
1899.1 ± 124.38	986.6 ± 226.72	612.8 ± 13.20	2.35	97.65

[a] *Percentage of participation in the total polyphenol content. Values are expressed on a dry basis.*

The total polyphenol content (Table 11.4) is higher than reported in other mango seeds (21.9–399 mg.g^{-1}) (Abdel et al., 2012; Ashoush and Gadallah, 2011; Dorta, Lobo, and González, 2013; Khammuang and Sarnthima, 2011; Ribeiro et al., 2008; Sogi et al., 2013). The extraction method shows a low content of condensed tannins (2.35%); this promotes their use in the food industry by antinutritional characteristics associated with high levels of condensed tannins. The high content of total polyphenolic compounds is associated with a significant increase in antioxidant activity (Table 11.4). Dorta et al. (2012) demonstrated that the polyphenol content is responsible for antioxidant activity in mango seeds. Natural antioxidants are compounds that remove free radicals and protect the body from oxidative reactions (Yedhu and Rajan, 2016), have anti-inflammatory and anti-carcinogenic activity, provides cardiovascular protection and prevent aging (Hoensch and Oertel, 2015; Warpe et al., 2015). As shown in Table 11.4, the mango seed has high antioxidant activity and a high content of polyphenols; therefore is a good source of natural antioxidants.

LOI assay was used to evaluate the antioxidant activity of the phenolic compounds from Ataulfo mango seed. The influence of antioxidants on percentage inhibition is related to the ability of their hydrogen donation. Figure 11.1 shows that the percentage inhibition increases with the increases in the concentrations of dry kernel extract. This has the highest values in concentrations above 0.4 mg/mL. The percentage inhibition of 50% was determined with a linear regression of the values below to 0.1 mg/mL (R^2=0.99). Mango seed extract has an IC_{50} de 0.07 mg/mL. It is required a lower concentration than with Trolox® (IC_{50}=0.14 mg/mL). These results indicate that the phenolic compounds from the mango seed kernel have greater antioxidant activity than Trolox®. Trolox® is the standard antioxidant used to express antioxidant activity in several spectrophotometric techniques. The antioxidant activity of kernel also was very higher than reported in Butylated hydroxytoluene BHT (0.34 mg/mL) (Cerqueira, Souza, Martins, Teixeira, and Vicente, 2010), this is one of the most synthetic antioxidants incorporations in feed (Carocho and Ferreira, 2013) which has reported negative health effects (Lorenzo et al., 2013; Sarafian et al., 2002).

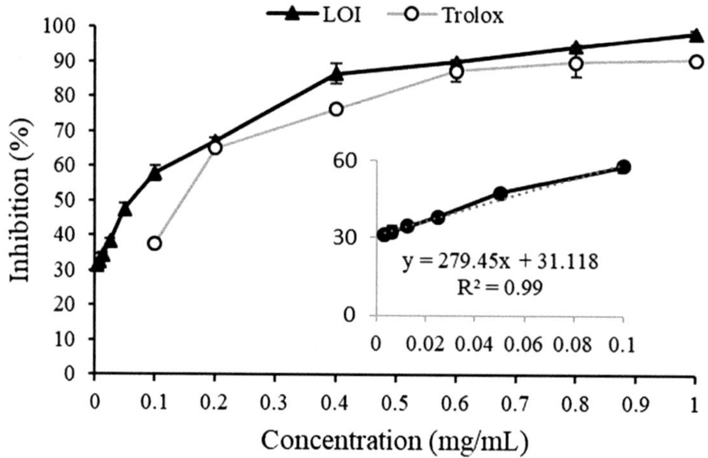

FIGURE 11.1 IC_{50} of active extracts of mango seed kernel Ataulfo variety in the function of the concentration.

11.3.4 RP-HPLC-ESI-MS ANALYSIS OF EXTRACTS

The identification of phenolic compounds was performed by comparing the retention time of HPLC analysis with MS; the molecular weight was compared with literature and database food research department (DIA-UAC). Ten phenolic compounds have been separated and identified (Table 11.5). The phenolic group was composed of flavonoids, gallates and gallotannins and ellagitannins.

Flavonoids are plant pigment that is synthesized from phenylalanine (Havsteen, 2002), and are produced in the cytosol and the vacuole of vegetal cells (Samanata, Das, and Das, 2011), these are antioxidants and free radical scavengers protecting against the oxidative reactions taking place in the body (Yedhu and Rajan, 2016). These compounds have anti-cancer and

TABLE 11.5 Phenolic Compounds in Ataulfo Mango Seed by HPLC-MS

No.	Retention time (min)	Name of compound	Molecular formula	Family	Molecular weight (MW)-
1.	2.6	Caffeic acid 4-O-glucoside	$C_{15}H_{18}O_9$	Hydroxycinnamic acids	341.1
2.	5.8	Galloyl glucose	$C_{13}H_{16}O_{10}$	Hydroxycinnamic acids	331.1
3.	29.8	Gallic acid	$C_7H_6O_5$	Hydroxycinnamic acids	169.01
3.	29.8	Ethyl gallate	$C_9H_{10}O_5$	Hydroxybezoic acids	197
5.	32.2	Tetra-O-galloyl-glucoside	$C_{34}H_{28}O_{22}$	Hydroxybezoic acids	787.1
6.	33.5	Penta-O-galloyl-glucoside	$C_{41}H_{32}O_{26}$	Hydroxybezoic acids	939.2
7.	33.5	Valoneic acid dilactone	$C_{24}H_{22}O_{10}$	Hydroxybezoic acids	469.1
8.	34.8	Rhamnetin-3-[6″-2-butenoil-hexoside] dimerized	N.N.	Flavone	1091.1
9.	34.8	Rhamnetin-3-[6″-2-butenoil-hexoside]	$C_{26}H_{16}O_{13}$	Flavone	545.1
10.	36.1	Apigenin 7-O-glucuronide	$C_{27}H_{26}O_{17}$	Flavone	621.1
11.	38.1	Ethyl 2,4-dihydroxy-3-(3,4,5-rihydroxybenzoyl) oxybenzoate	$C_{16}H_{14}O_9$	Hydroxybezoic acids	349

anti-inflammatory effects, so they have a wide application in the pharmaceutical industry (Hoensch and Oertel, 2015). Was identified as flavonoids Rhamnetin-3-[6″-2-butenoil-hexoside] (8) with a [M − H]− ion at *m/z* 545.1; at the same time retaining (34.8 min), a compound with [M − H]− ion at *m/z* 1091.1 (7) was separated, this compound was tentatively identified as Rhamnetin-3-[6″-2-butenoil-hexoside] dimerized, since this compound is the sum of the two molecular weights. Peak **9** showed a [M − H]− ion at *m/z* 621.1 that corresponded with the molecular formula $C_{27}H_{26}O_{17}$ this compound was tentatively identified as Apigenin 7-O-diglucuronide, this molecule is attributed to anticancer effect (recycling molecules) and an anti-inflammatory efficacy (Hoensch and Oertel, 2015), and had not been identified in mango seed.

Gallotannins, a class of hydrolysable tannins, are esters of 3,4,5-trihydroxy benzoic acid (gallic acid) (Barbehenn and Constabel, 2011), by high biological activity, interest in gallotannins has grown over the past decade (Luo et al., 2014), Preclinical studies have shown that GA possesses effects antioxidant, anticancer, antimicrobial and anti-inflammatory (Nikbakht et al., 2015; Sohi, Mittal, Hundal, and Khanduja, 2003). Galloyl glucose was identified as peak 2 showed a [M − H]− ion at m/z 331.1, Also, peak 27 gave a M − H] − ion at m/z 197 was identified as Ethyl gallate and the main fragment at m/z 169.01, typical fragment that determines the presence of gallic acid. Peaks 4 and 5, were identified as Tetra-O-galloyl-glucoside and Penta-O-galloyl-glucoside showing [M − H]− ion at m/z of 787.1 and 939.2, Penta-O-galloyl-glucoside is the major compound identified in the mango seed, a result similar to reported by Luo et al. (2014), mango seeds from China, the authors say antioxidant activity of the mango byproducts may be partially caused by Penta-O-galloyl-glucoside or its analogs. Penta-galloyl glucose (PGG) is a valuable phenolic compound with functional properties as anticarcinogenic, antimicrobial, anti-inflammatory, antioxidant and anti-diabetic (Torres-León et al., 2017).

Ellagitannins, are also derived biosynthetically from PGG by reactions between the gallic acid units (Mueller, 2001), has been reported these compounds have chemo-preventive, anti-apoptotic, anti-inflammatory, cardioprotective, gastroprotective, ulcer healing, antifibrotic, antidiabetic, hypolipidemic and anti-atherosclerotic (Priyadarsini et al., 2002; Warpe et al., 2015). Compound of peak **6**, with [M − H]− ion at m/z 469.1 of molecular formula $C_{24}H_{22}O_{10}$, was identified as Valoneic acid dilactone, this compound is formed by the union of ellagic acid m/z 301 and gallic acid m/z 169, was recently reported by Dorta et al., (2014) waste of various varieties of mangoes from Spain.

The results of HPLC-MS analysis were similar to those reported previously in seeds of other varieties of mango (Dorta et al., 2014). However, mango seed kernel is variations in the concentrations of tannins and the presence of some compounds (phenolic acids and flavonoids). This may be due to genotype, stage of development of tissues, and environmental conditions (Barbehenn and Constabel, 2011).

11.4 CONCLUSION

The Ataulfo mango seed has a great potential of valorization as a source of value-added ingredients, it is rich in essential minerals, free of trans fatty acids, protein, and has a high biological activity so it can be used as a source of natural antioxidants. HPLC-MS analysis reveals that the mango seed kernel is a source of valuable phytochemicals. This work unveils several investigative perspectives to promote new technological applications of Ataulfo mango seed, based on its functional and nutritional properties. With the valorization of this byproduct can be obtained economic, nutritional, and environmental benefits.

ACKNOWLEDGMENTS

The author Cristian Torres-León thanks the Mexican Council for Science and Technology (CONACYT) for his post-graduate scholarship.

KEYWORDS

- **antioxidants**
- **byproducts**
- **Food Security and Nutrition**
- **food waste**
- ***Mangifera indica* L.**
- **mango seed**
- **phenolic compounds**
- **sustainability**

REFERENCES

Abdalla, A., Darwish, S., Ayad, E., & El-Hamahmy, R., (2007). Egyptian mango by-product 1. Compositional quality of mango seed kernel. *Food Chemistry, 103*(4), 1134–1140. https://doi.org/10.1016/j.foodchem.2006.10.017 (accessed on 24 July 2020).

Abdel, M., Ashoush, I., & Nessrien, M., (2012). Characteristics of Mango Seed Kernel Butter and its Effects on Quality Attributes of Muf fi ns. *Food Science and Technology, 9*(2), 1–9.

Ajila, C., Aalami, M., Leelavathi, K., & Rao, P., (2010). Mango peel powder: A potential source of antioxidants and dietary fiber in macaroni preparations. *Innovative Food Science and Emerging Technologies, 11*(1), 219–224. https://doi.org/10.1016/j.ifset.2009.10.004 (accessed on 24 July 2020).

Ajila, C., Naidu, K., Bhat, S., & Rao, U., (2007). Bioactive compounds and antioxidant potential of mango peel extract. *Food Chemistry, 105*(3), 982–988. https://doi.org/10.1016/j.foodchem.2007.04.052 (accessed on 24 July 2020).

Ajila, C., Rao, J., & Rao, P., (2010). Characterization of bioactive compounds from raw and ripe *Mangifera indica* L. peel extracts. *Food and Chemical Toxicology : An International Journal Published for the British Industrial Biological Research Association, 48*(12), 3406–3411. https://doi.org/10.1016/j.fct.2010.09.012 (accessed on 24 July 2020).

Ajila, C., & Rao, P., (2013). Mango peel dietary fibre: Composition and associated bound phenolics. *Journal of Functional Foods, 5*(1), 444–450. https://doi.org/10.1016/j.jff.2012.11.017 (accessed on 24 July 2020).

Amyrgialaki, E., Makris, D. P., Mauromoustakos, A., & Kefalas, P., (2014). Optimization of the extraction of pomegranate (Punica granatum) husk phenolics using water/ethanol solvent systems and response surface methodology. *Industrial Crops and Products, 59*, 216–222. https://doi.org/10.1016/j.indcrop.2014.05.011 (accessed on 24 July 2020).

AOAC 930.09, (2005). *Official Methods of Analysis of the Association of Official Analytical Chemists International* (18th edn.). Maryland, USA.

AOAC 934.01, (1990). *Official Methods of Analysis of the Association of Official Analytical Chemists* (15th edn.). Method 934.01. Virginia, USA.

AOAC 942.05, (1990). *Official Methods of Analysis of the Association of Official Analytical Chemists.* (15th edn.). Arlington, Virginia, USA. METODO número 942.05.

Arogba, S., (1997). Physical, chemical and functional properties of Nigerian mango (Mangifera indica) kernel and its processed flour. *Journal of Agriculture Food Chemistry, 73*, 321–328.

Ashoush, I., & Gadallah, M., (2011). Utilization of mango peels and seed kernels powders as sources of phytochemicals in biscuit. *World Journal of Dairy and Food Sciences, 6*(1), 35–42. http://idosi.org/wjdfs/wjdfs6%281%29/6.pdf (accessed on 24 July 2020).

Barbehenn, R., & Constabel, C., (2011). Tannins in plant–herbivore interactions. *Phytochemistry, 72*(13), 1551–1565. https://doi.org/10.1016/j.phytochem.2011.01.040 (accessed on 24 July 2020).

Benzie, I., & Strain, J., (1996). The ferric reducing ability of plasma (FRAP) as a measure of "antioxidant power": The FRAP assay. *Analytical Biochemistry, 239*, 70–76.

Carocho, M., & Ferreira, I., (2013). A review on antioxidants, prooxidants and related controversy: Natural and synthetic compounds, screening and analysis methodologies and future perspectives. *Food and Chemical Toxicology, 51*, 15–25. https://doi.org/10.1016/j.fct.2012.09.021 (accessed on 24 July 2020).

Cerqueira, M. A., Souza, B., Martins, J. T., Teixeira, J. A., & Vicente, A. A., (2010). Seed extracts of *Gleditsia triacanthos*: Functional properties evaluation and incorporation into galactomannan films. *Food Research International, 43*(8), 2031–2038. https://doi.org/10.1016/j.foodres.2010.06.002 (accessed on 24 July 2020).

Chong, C., Law, C., Figiel, A., Wojdyło, A., & Oziembłowski, M., (2013). Colour, phenolic content and antioxidant capacity of some fruits dehydrated by a combination of different methods. *Food Chemistry, 141*(4), 3889–3896. https://doi.org/10.1016/j.foodchem.2013.06.042 (accessed on 24 July 2020).

Da Silva, A., & Jorge, N., (2014). Bioactive compounds of the lipid fractions of agro-industrial waste. *Food Research International, 66*, 493–500. https://doi.org/10.1016/j.foodres.2014.10.025 (accessed on 24 July 2020).

Dorta, E., González, M., Lobo, M. G., Sánchez-Moreno, C., & De Ancos, B., (2014). Screening of phenolic compounds in by-product extracts from mangoes (Mangifera indica L.) by HPLC-ESI-QTOF-MS and multivariate analysis for use as a food ingredient. *Food Research International, 57*, 51–60. https://doi.org/10.1016/j.foodres.2014.01.012 (accessed on 24 July 2020).

Dorta, E., Lobo, G., & González, M., (2012). Using drying treatments to stabilise mango peel and seed : Effect on antioxidant activity. *Food Science and Technology, 45*(2), 261–268. https://doi.org/10.1016/j.lwt.2011.08.016 (accessed on 24 July 2020).

Dorta, E., Lobo, G., & González, M., (2013). Optimization of factors affecting extraction of antioxidants from mango seed. *Food and Bioprocess Technology, 6*(4), 1067–1081. https://doi.org/10.1007/s11947-011-0750-0 (accessed on 24 July 2020).

Elegbede, J., Achoba, I., & Richard, H., (1995). Nutrient composition of mango (mangnifera indica) seed kernel from Nigeria. *Journal of Food Biochemistry, 19*(5), 391–398. https://doi.org/10.1111/j.1745-4514.1995.tb00543.x (accessed on 24 July 2020).

FAOSTAT, (2017). *Food and Agriculture Organization of the United Nations*. Retrieved from Statistical Database—Agriculture website: http://www.fao.org/faostat/es/#data/QC/visualize (accessed on 24 July 2020).

Fontes, P., Queiroz, J., Teixeira, L., Kling, G., Barbosa, A., Sialino, E., & Dos, S. M., (2008). Revista brasileira de zootecnia Efeitos da inclusão de farelo do resíduo de manga no desempenho de frangos de corte de 1 a 42 dias 1 Effects of inclusion of mango residues on performance of broilers chickens from 1 to 42 days material e métodos O experime. *Revista Brasileira De Zootecnia, 37*(12), 2173–2178.

García, M., García, H., Bello, L., Sáyago, S., & Oca, M., (2013). Functional properties and dietary fiber characterization of mango processing by-products (Mangifera indica L, cv Ataulfo and Tommy Atkins). *Plant Foods for Human Nutrition, 68*(3), 254–258. https://doi.org/10.1007/s11130-013-0364-y (accessed on 24 July 2020).

Havsteen, B., (2002). The biochemistry and medical significance of the flavonoids. In: *Pharmacology and Therapeutics* (Vol. 96). https://doi.org/10.1016/S0163-7258(02)00298-X (accessed on 24 July 2020).

Hoensch, H., & Oertel, R., (2015). The value of flavonoids for the human nutrition: Short review and perspectives. *Clinical Nutrition Experimental, 3*, 8–14. https://doi.org/10.1016/j.yclnex.2015.09.001 (accessed on 24 July 2020).

Jahurul, M., Zaidul, I., Nik, N., Sahena, F., Kamaruzzaman, B., Ghafoor, K., & Omar, A., (2014). Cocoa butter replacers from blends of mango seed fat extracted by supercritical carbon dioxide and palm stearin. *Food Research International, 65*, 401–406. https://doi.org/10.1016/j.foodres.2014.06.039 (accessed on 24 July 2020).

Jin, J., Jin, Q., Akoh, C. C., & Wang, X., (2019). Mango kernel fat fractions as potential healthy food ingredients: A review. *Critical Reviews in Food Science and Nutrition, 59*(11), 1794–1801. https://doi.org/10.1080/10408398.2018.1428527 (accessed on 24 July 2020).

Khammuang, S., & Sarnthima, R., (2011). Antioxidant and antibacterial activities of selected varieties of Thai mango seed extract. *Pakistan Journal of Pharmaceutical Sciences, 24*(1), 37–42. https://doi.org/10.1016/j.ifset.2009.10.004 (accessed on 24 July 2020).

Kim, H., Moon, J., Kim, H., Lee, D., Cho, M., Choi, H., & Cho, S., (2010). Antioxidant and antiproliferative activities of mango (Mangifera indica L.) flesh and peel. *Food Chemistry, 121*(2), 429–436. https://doi.org/10.1016/j.foodchem.2009.12.060 (accessed on 24 July 2020).

Kjeldahl, J., (1883). Neue methode zúr bestimmung der stickstoffs in organischen körpern. *Z Anal. Chem., 22*, 366–382.

Koubala, B. B., Kansci, G., Mbome, L. I., Crépeau, M. J., Thibault, J. F., & Ralet, M. C., (2008). Effect of extraction conditions on some physicochemical characteristics of pectins from "améliorée" and "mango" mango peels. *Food Hydrocolloids, 22*(7), 1345–1351. https://doi.org/10.1016/j.foodhyd.2007.07.005 (accessed on 24 July 2020).

Lorenzo, J., González, R., Sánchez, M., Amado, I., & Franco, D., (2013). Effects of natural (grape seed and chestnut extract) and synthetic antioxidants (buthylatedhydroxytoluene, BHT) on the physical, chemical, microbiological and sensory characteristics of dry cured sausage "chorizo." *Food Research International, 54*(1), 611–620. https://doi.org/10.1016/j.foodres.2013.07.064 (accessed on 24 July 2020).

Luo, F., Fu, Y., Xiang, Y., Yan, S., Hu, G., Huang, X., & Chen, K., (2014). Identification and quantification of gallotannins in mango (Mangifera indica L.) kernel and peel and their antiproliferative activities. *Journal of Functional Foods, 8*(1), 282–291. https://doi.org/10.1016/j.jff.2014.03.030 (accessed on 24 July 2020).

Ma, X., Wu, H., Liu, L., Yao, Q., Wang, S., Zhan, R., & Zhou, Y., (2011). Polyphenolic compounds and antioxidant properties in mango fruits. *Scientia Horticulturae, 129*(1), 102–107. https://doi.org/10.1016/j.scienta.2011.03.015 (accessed on 24 July 2020).

Martinez, C., Aguilera, A., Rodriguez, R., & Aguilar, C., (2011). Fungal enhancement of the antioxidant properties of grape waste. *Ann. Microbiol, 62*, 923–930.

Masibo, M., & He, Q., (2009). Mango Bioactive Compounds and Related Nutraceutical Properties—A Review. *Food Reviews International, 25*(4), 346–370. https://doi.org/10.1080/87559120903153524 (accessed on 24 July 2020).

Mirabella, N., Sala, S., Mirabella, N., Castellani, V., & Sala, S., (2014). Current options for the valorization of food manufacturing waste: A review. *Journal of Cleaner Production, 65*, 28–41. https://doi.org/10.1016/j.jclepro.2013.10.051 (accessed on 24 July 2020).

Morais, L., Altina, E., Figueiredo, T. D., Maria, N., Silva, P., Gusmao, I., & Montenegro, I., (2014). Quantification of bioactive compounds in pulps and by-products of tropical fruits from Brazil. *Food Chemistry, 143*, 398–404.

Mueller, I., (2001). Analysis of hydrolysable tannins. *Animal Feed Science and Technology, 91*(1, 2), 3–20. https://doi.org/10.1016/S0377-8401(01)00227-9 (accessed on 24 July 2020).

Nawirska, A., & Kwaśniewska, M., (2005). Dietary fibre fractions from fruit and vegetable processing waste. *Food Chemistry, 91*(2), 221–225. https://doi.org/10.1016/j.foodchem.2003.10.005 (accessed on 24 July 2020).

Nayak, A., & Bhushan, B., (2019). An overview of the recent trends on the waste valorization techniques for food wastes. *Journal of Environmental Management, 233*, 352–370. https://doi.org/10.1016/j.jenvman.2018.12.041 (accessed on 24 July 2020).

Nikbakht, J., Hemmati, A. A., Arzi, A., Mansouri, M. T., Rezaie, A., & Ghafourian, M., (2015). Protective effect of Gallic acid against bleomycin-induced pulmonary fibrosis in rats. *Pharmacological Reports*. In press. https://doi.org/10.1016/j.pharep.2015.03.012 (accessed on 24 July 2020).

NOM-188-SCFI, (2012). *Mango Ataulfo del Soconusco, Chiapas (Mangifera caesia Jack ex Wall)Especificaciones y métodos de prueba.*

Nzikou, J., Kimbonguila, A., Matos, L., Loumouamou, B., Pambou-Tobi, N., Ndangui, C., & Desobry, S., (2010). Extraction and characteristics of seed kernel oil from mango (Mangifera indica). *Research Journal of Environmental and Earth Sciences, 2*(1), 31–35.

Odunsi, A., (2005). Response of laying hens and growing broilers to the dietary inclusion of mango (Mangifera indica L.) seed kernel meal. *Tropical Animal Health and Production, 37*(2), 139–150. https://doi.org/10.1023/B:TROP.0000048455.96694.85 (accessed on 24 July 2020).

Priyadarsini, K., Khopde, S., Kumar, S., & Mohan, H., (2002). Free radical studies of ellagic acid, a natural phenolic antioxidants. *J. Agric. Food Chem., 50*, 2200–2206.

Radha, K., Babuskin, S., Azhagu, P., Sasikala, M., Sabina, K., Archana, G., & Sukumar, M., (2014). Antimicrobial and antioxidant effects of spice extracts on the shelf life extension of raw chicken meat. *International Journal of Food Microbiology, 171*, 32–40. https://doi.org/10.1016/j.ijfoodmicro.2013.11.011 (accessed on 24 July 2020).

Re, R., Pellegrini, N., Proteggente, A., Pannala, A., Yang, M., & Rice, C., (1999). Antioxidant activity applying an improved ABTS radical cation decolorization assay. *Free Radical Biology and Medicine, 26*, 1231–1237.

Ribeiro, S., Barbosa, L., Queiroz, J., Knödler, M., & Schieber, A., (2008). Phenolic compounds and antioxidant capacity of Brazilian mango (Mangifera indica L.) varieties. *Food Chemistry, 110*(3), 620–626. https://doi.org/10.1016/j.foodchem.2008.02.067 (accessed on 24 July 2020).

Ribeiro, S., & Schieber, A., (2010). Bioactive compounds in mango (Mangifera indica L.). *Bioactive Foods in Promoting Health, 34*, 507–523. https://doi.org/10.1016/B978-0-12-374628-3.00034-7 (accessed on 24 July 2020).

Rojas, R., Alvarez-Pérez, O. B., Contreras-Esquivel, J. C., Vicente, A., Flores, A., Sandoval, J., & Aguilar, C. N., (2018). Valorization of mango peels: Extraction of pectin and antioxidant and antifungal polyphenols. *Waste and Biomass Valorization*, 1–10. https://doi.org/10.1007/s12649-018-0433-4 (accessed on 24 July 2020).

Rojas, R., Contreras, J., Orozco, M., Muñoz, C., Aguirre, J., & Aguilar, C., (2015). Mango peel as source of antioxidants and pectin: Microwave assisted extraction. *Waste and Biomass Valorization, 6*(6), 1095–1102. https://doi.org/10.1007/s12649-015-9401-4 (accessed on 24 July 2020).

Saadany, R., Foda, Y., & Saadany, F., (1980). Studies on starch extracted from mango seeds (Mangifira indica) as a new source of starch. *Starch, 32*(4), 113–116.

Samanata, A., Das, G., & Das, S., (2011). Roles of flavonoids in plants. *Int. J. Pharm. Sci. Technol.*, (6), 12–35.

Sarafian, T., Kouyoumjian, S., Tashkin, D., & Roth, M., (2002). Synergistic cytotoxicity of delta(9)-tetrahydrocannabinol and butylated hydroxyanisole. *Toxicology Letters, 133*(2, 3), 171–179.

Serna, L., Torres, C., & Ayala, A., (2015). Evaluation of food powders obtained from peels of mango (Mangifera indica) as sources of functional ingredients. *Información Tecnologica.,*

26(2), 41–50. https://doi.org/10.4067/S0718-07642015000200006 (accessed on 24 July 2020).

Sogi, D., Siddiq, M., Greiby, I., & Dolan, K., (2013). Total phenolics, antioxidant activity, and functional properties of "Tommy Atkins" mango peel and kernel as affected by drying methods. *Food Chemistry, 141*(3), 2649–2655. https://doi.org/10.1016/j.foodchem.2013.05.053 (accessed on 24 July 2020).

Sohi, K., Mittal, N., Hundal, M., & Khanduja, K., (2003). Gallic acid, an antioxidant, exhibits antiapoptotic potential in normal human lymphocytes: A Bcl-2 independent mechanism. *Journal of Nutritional Science and Vitaminology*, *49*, 221–227.

Solís, J., & Durán, M., (2011). Chapter 88-mango (Mangifera indica L.) seed and its fats. In: *Nuts and Seeds in Health and Disease Prevention* (pp. 741–748).

Sonwai, S., Kaphueakngam, P., & Flood, A., (2012). Blending of mango kernel fat and palm oil mid-fraction to obtain cocoa butter equivalent. *Journal of Food Science and Technology, 51*(10), 1–13. https://doi.org/10.1007/s13197-012-0808-7 (accessed on 24 July 2020).

Swain, T., & Hillis, E., (1959). The phenolic constituents of *Prunus domestica*. The quantitative analysis of phenolic constituents. *J. Sci. Food Agric.*, *10*, 63–68.

Torres-León, C., Ramirez-Guzman, N., Ascacio-Valdes, J., Serna-Cock, L., Dos, S. C. M. T., Contreras-Esquivel, J. C., & Aguilar, C. N., (2019). Solid-state fermentation with *Aspergillus niger* to enhance the phenolic contents and antioxidative activity of Mexican mango seed: A promising source of natural antioxidants. *LWT-Food Science and Technology*, *112*, 108236. https://doi.org/10.1016/j.lwt.2019.06.003 (accessed on 24 July 2020).

Torres-León, C., Ramírez-Guzman, N., Londoño-Hernandez, L., Martinez-Medina, G. A., Díaz-Herrera, R., Navarro-Macias, V., & Aguilar, C. N., (2018). Food waste and byproducts: An opportunity to minimize malnutrition and hunger in developing countries. *Frontiers in Sustainable Food Systems*, *2*. https://doi.org/10.3389/fsufs.2018.00052 (accessed on 24 July 2020).

Torres-León, C., Rojas, R., Contreras-Esquivel, J. C., Serna-Cock, L., Belmares-Cerda, R. E., & Aguilar, C. N., (2016). Mango seed: Functional and nutritional properties. *Trends in Food Science and Technology, 55*, 109–117. https://doi.org/10.1016/j.tifs.2016.06.009 (accessed on 24 July 2020).

Torres-León, C., Rojas, R., Serna-Cock, L., Belmares-Cerda, R., & Aguilar, C. N., (2017). Extraction of antioxidants from mango seed kernel: Optimization assisted by microwave. *Food and Bioproducts Processing*, *105*, 188–196. https://doi.org/10.1016/j.fbp.2017.07.005 (accessed on 24 July 2020).

Torres-león, C., Ventura-Sobrevilla, J., Serna-Cock, L., Ascacio-Valdés, J. A., Contreras-Esquivel, J., Aguilar, C. N., & Aguilar, C. N., (2017). Pentagalloylglucose (PGG): A valuable phenolic compound with functional properties. *Journal of Functional Foods*, *37*, 176–189. https://doi.org/10.1016/j.jff.2017.07.045 (accessed on 24 July 2020).

Torres-león, C., Vicente, A. A., Flores-lópez, M. L., Rojas, R., Serna-cock, L., Alvarez-pérez, O. B., & Aguilar, C. N., (2018). Edible films and coatings based on mango (var. Ataulfo) by-products to improve gas transfer rate of peach. *LWT-Food Science and Technology*, *97*, 624–631. https://doi.org/10.1016/j.lwt.2018.07.057 (accessed on 24 July 2020).

Van, S. P., Robertson, J., & Lewis, B., (1991). Methods for dietary fiber, neutral detergent fiber and non-starch polysaccharides in relation to animal nutrition. *J. Dairy Sci.*, *74*, 3583–3597.

Waheed, M., Butt, M. S., Shehzad, A., Adzahan, N. M., Shabbir, M. A., Rasul, S. H. A., & Aadil, R. M., (2019). Eggshell calcium: A cheap alternative to expensive supplements. *Trends in Food Science and Technology, 91*, 219–230. https://doi.org/10.1016/j.tifs.2019.07.021 (accessed on 24 July 2020).

Warpe, V., Malik, V., S, A., Bodhankar, S., & Mahadik, K., (2015). Cardioprotective effect of ellagic acid on doxorubicin-induced cardiotoxicity in Wistar rats. *Journal of Acute Medicine, 5*(1), 1–8. https://doi.org/10.1016/j.jacme.2015.02.003 (accessed on 24 July 2020).

Wong-Paz, J. E., Contreras-Esquivel, J. C., Rodríguez-Herrera, R., Carrillo-Inungaray, M. L., López, L. I., Nevárez-Moorillón, G. V., & Aguilar, C. N., (2015). Total phenolic content, in vitro antioxidant activity and chemical composition of plant extracts from semiarid Mexican region. *Asian Pacific Journal of Tropical Medicine, 8*(2), 104–111. https://doi.org/10.1016/S1995-7645(14)60299-6 (accessed on 24 July 2020).

Wu, S., Tokuda, M., Kashiwagi, A., Henmi, A., Okada, Y., Tachibana, S., & Nomura, M., (2015). Evaluation of the fatty acid composition of the seeds of *Mangifera indica* L. and their application. *Journal of Oleo Science, 484*(5), 479–484.

Xu, G., Chen, J., Liu, D., Zhang, Y., Jiang, P., & Ye, X., (2008). Minerals, phenolics compounds, and antioxidant capacity of citrus peel extract by hot water. *Journal of Food Science, 1*(73), 11–18.

Yedhu, R., & Rajan, K., (2016). Microwave-assisted extraction of flavonoids from *Terminalia bellerica*: Study of kinetics and thermodynamics. *Separation and Purification Technology, 157*, 169–178. https://doi.org/10.1016/j.seppur.2015.11.035 (accessed on 24 July 2020).

Zein, R., El-Bagoury, A., & Kassab, H., (2005). Chemical and nutritional studies on mango see kernels. *Journal of Agricultural Science, 30*(6), 3285–3299.

CHAPTER 12

Kinetic Parameters of the Carotenoids Production by *Rhodotorula glutinis* under Different Concentration of Carbon Source

AYERIM HERNÁNDEZ-ALMANZA,[1] VÍCTOR NAVARRO-MACÍAS,[1] OSCAR AGUILAR,[2] JUAN C. CONTRERAS-ESQUIVEL,[1] JULIO C. MONTAÑEZ,[1] GUILLERMO MARTÍNEZ AVILA,[3] and CRISTÓBAL N. AGUILAR[1]

[1]*Bioprocesses and Bioproducts Research Group (BBG-DIA), Food Research Department, School of Chemistry, Universidad Autónoma de Coahuila, Saltillo, 25280, Coahuila, México, Tel: +52 01 (844) 416-12-38; E-mail: cristobal.aguilar@uadec.edu.mx*

[2]*Center of Biotechnology-FEMSA, Tecnológico de Monterrey, Campus Monterrey, 64849, NL, Mexico*

[3]*Laboratory of Chemistry and Biochemistry, School of Agronomy, Universidad Autónoma de Nuevo León, 66050, Nuevo León, Mexico*

ABSTRACT

Carotenoids are lipophilic isoprenoid pigments, its antioxidant, anti-inflammatory, and anticancer properties have been reported. The industrial production of carotenoids using biotechnological ways is of great interest due to the possibility of obtaining high productivity and to the variety of applications as food and cosmetic coloring agents or such as supplements and vitamins. In the present study, the influence of carbon source concentration in the carotenoids production by *Rhodotorula glutinis* P4M422 under submerged culture and the kinetic parameters of this bioprocess were evaluated. The highest carotenoid concentration was obtained with the addition of dextrose at concentrations of 40 and 80 g/L (1.85 and 1.88 mg/g, respectively). However, the best productivity was achieved when dextrose at 40 g/L was added. Therefore, in order to improve the biotechnological production of carotenoids by *R. glutinis*, the dextrose 40 g/L is the optimal concentration, however, the evaluation of other culture conditions and the influence of C/N ratio is necessary.

12.1 INTRODUCTION

Carotenoids are secondary metabolites synthesized by plants, filamentous fungi, microalgae, and bacteria (Kumar Saini and Young-Soo, 2017). Carotenoids are lipophilic isoprenoid molecules containing double bonds responsible to form a chromophore (Alves et al., 2017). Its range in color from yellow to red and protect the cells against damage by light, oxidation, and free radicals (Thanapimmetha et al., 2017). Also, some carotenoids have provitamin A activity, and the antioxidant, anti-inflammatory, and anticancer properties have been attributed.

On the other hand, is it expected that the global market of carotenoids increases to US$1.4 billion by 2018 (Cardenas-Toro et al., 2015). Nowadays, there are some microbial carotenoids produced on an industrial scale; for example, ankaflavin (*Monascus* sp.), anthraquinone (*Penicillium oxalicum*), riboflavin (*Ashbya gossypi*) and β-carotene (*Blakeslea trispora*) and other more are under research (Alves et al., 2017) such as *Rhodotorula glutinis* to produce carotenoids. *R. glutinis* is a red and oleaginous yeast producer of lipids and carotenoids such as β-carotene, torulene, torularhodin, currently, the production of lycopene has been reported (Cutzu et al., 2013; Hernández-Almanza et al., 2014). However, the industrial obtaining of these molecules is not developed yet. In order to improve this bioprocess, the detailed study of culture conditions is necessary and special attention should be focused on carbon sources.

The industrial production of carotenoids, under biotechnological ways, is of great interest due to the possibility of obtaining high productivity and to the variety of applications as food and cosmetic coloring agents or such as supplements and vitamins (Cardoso et al., 2016; Colet et al., 2014). Parameters estimation have an important role in developing appropriate mathematical models used in the simulation, control, optimization and improvement of biochemical processes (Rivera et al., 2013; Zhao et al., 2013).

This study aims to evaluate various concentrations of carbon source in the carotenoids production by *R. glutinis* and estimate the kinetic parameters of the bioprocess.

12.2 MATERIALS AND METHODS

12.2.1 REAGENTS AND MICROORGANISM

All reagents used were of analytical grade, obtained from Sigma Aldrich® Mexico. *Rhodotorula glutinis* P4M422 was obtained from DIA-U.A.de C. Collection, which was preserved lyophilized until use. The inoculum was prepared in YM broth (g/L: dextrose 20, peptone 5, yeast extract 4) incubated at 30°C, 150 rpm during 48 h.

12.2.2 FERMENTATION CONDITIONS

Fermentation for carotenoids production was performed into Erlenmeyer flask of 500 mL containing 50 mL of YM broth and inoculated with 1×10^8 cell/mL of *R. glutinis* P4M422. In this study different concentration of carbon source were evaluated (0, 10, 20, 40, 80, 160, and 320 g/L). The flasks were incubated at 30°C in a rotatory incubator at 150 rpm for 120 h. Each 24 h one sample was taken and biomass production, substrate consumption, and total carotenoids production were measured.

12.2.3 ANALYTICAL PROCEDURES

Once the fermentation was carried, the biomass was recovered by centrifugation at 13,000 rpm during 5 min, then it was washed with distilled water and was dried at 37°C. After, the production of biomass was determined by dry weight method. Then, the carotenoids extraction was performed; the DMSO at 60°C was added to the biomass and incubated during 1 h with occasional vortex, finally, acetone was added and centrifuged for the quantification at 475 nm. The substrate consumption was determined according to the anthrone method for total sugar quantification reported by Dreywood (1946) with some modifications. Briefly, 0.5 mL of sample was transferred into assay tubes, the tubes were put in an ice bath and 1 mL of anthrone solution was added. The mixture was collocated at 80°C for 15 min, after this time, the tubes were cooled in an ice-water bath for 5 min. The total sugar concentration was measured in a spectrophotometer at 530 nm.

12.2.4 STATISTICAL ANALYSIS

In the present study, an experimental design with mono-factorial array was employed and the data were analyzed by comparison of means and calculation of kinetic parameters. All the experiments were realized by triplicate.

12.2.5 KINETIC PRODUCTION OF CAROTENOIDS

For the calculation of the kinetic parameters of the microbial production of total carotenoids, the next equations were employed:

The biomass was evaluated using the Logistics model:

$$X = Xmax / 1 + C)^{e-\mu t} \tag{1}$$

Substrate consumption was evaluated with the Piret equation:

$$S = S_0 - \left\{ \left(\frac{X - X_0}{Yx / s} \right) + \left[\left[\frac{m * Xmax}{\mu} \right] * LN \left[\frac{(Xmax - X_0)}{Xmax - X} \right] \right] \right\} \tag{2}$$

The total carotenoids production was measured using the Luedeking-Piret model:

$$P = P_0 + a(X - X_0) + \frac{\beta * Xmax}{\mu} LN \left[\frac{Xmax - X_0}{Xmax - X} \right] \tag{3}$$

where P_0 is total carotenoids produced at an initial time of fermentation.

$$\mu_{max} = \frac{1}{x} \frac{dX}{dt} \tag{4}$$

where μ_{max} is the specific growth rate obtained during the exponential growth phase (h^{-1}), *dX/dt* is the growth rate and X is the biomass concentration (mg/mL).

$$Y_{X/S} = \frac{\Delta X}{\Delta S} \tag{5}$$

where Yx/s is degrowth yield based on the substrate consumption (g biomass/g substrate).

$$Y_{P/S} = \frac{\Delta P}{\Delta S} \tag{6}$$

where $Y_{P/S}$ is the total carotenoids yield based on substrate consumption (mg total carotenoids/g substrate).

$$q_S = \frac{1}{X} \frac{dS}{dt} \tag{7}$$

where q_S is the specific substrate consumption rate (g substrate/g de biomass) and dS/dt is the substrate consumption rate (g/L/h).

$$q_P = \frac{1}{X}\frac{dP}{dt} \tag{8}$$

where q_p is the specific total carotenoids production (mg total carotenoids/g de biomass) and dP/dt is the carotenoids formation rate (mg/g/h).

12.3 RESULTS AND DISCUSSION

Biotechnological production of carotenoids is an alternative to replace the chemical synthesis of pigments since the use of toxic compounds can be reduced and the growth rate of microorganisms is relatively short. The biomass production, substrate consumption, and total carotenoids were modeled in order to define the theoretical kinetic parameters. Figure 12.1 shows the theoretical and experimental values about the biomass production by *R. glutinis* P4M422 employing different concentration of carbon source. The highest biomass accumulation (0.016 and 0.019 mg/mL) was obtained with dextrose at 160 and 320 g/L, respectively. However, the major carotenoids production was obtained when dextrose at 40 and 80 g/L was used (Figure 12.2).

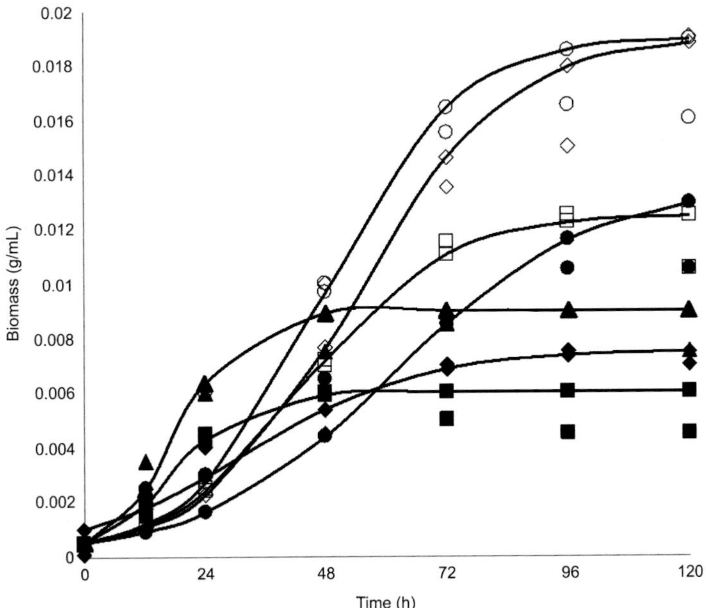

FIGURE 12.1 Correlation between modeled values of biomass production and values obtained experimentally when different concentrations of carbon source were employed: 0 g/L (■), 10 g/L (◆), 20 g/L (▲), 40 g/L (□), 80 g/L (●), 160 g/L (○) and 320 g/L (◇). The continuous line indicates the theoretical data.

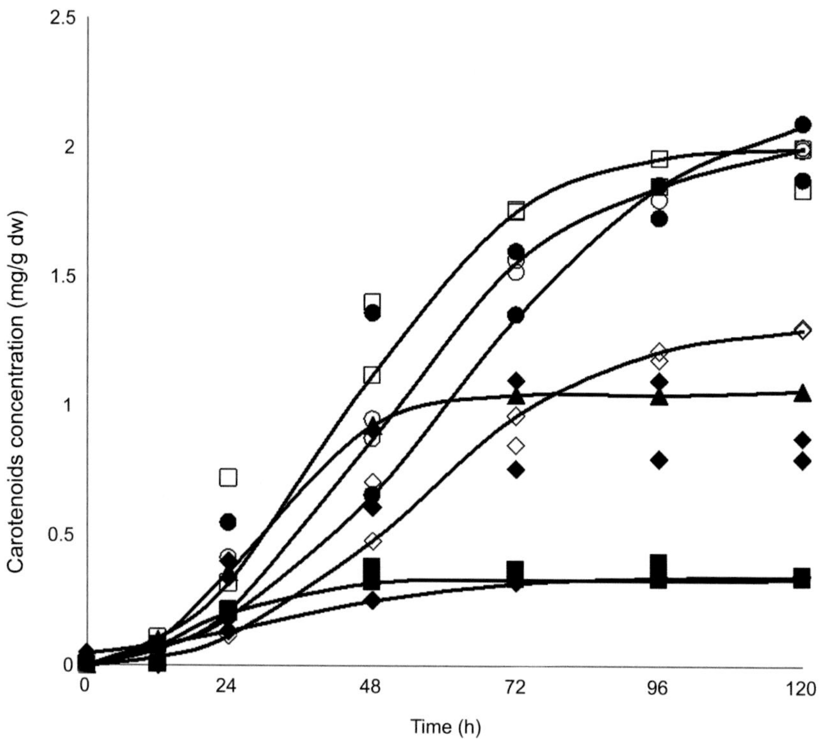

FIGURE 12.2 Correlation between modeled values of total carotenoids production and values obtained experimentally when different concentrations of carbon source were employed: 0 g/L (■), 10 g/L (◆), 20 g/L (▲), 40 g/L (□), 80 g/L (●), 160 g/L (○) and 320 g/L (◇). The continuous line indicates the theoretical data.

On the other hand, Figure 12.3 shows the percentage of substrate consumption and the correlation between theoretical data. The specific growth rate based in substrate concentration (Figure 12.4) showed that at major substrate concentration the microorganism grows at the maximum rate, while at lowest concentration the growth rate decrease but the substrate is not consumed totally.

Several reports indicate that the carotenoids production is important in protection against photo-oxidative damage and some microorganism rely on carotenoids for light and air protection (Marova et al., 2012; Zhang et al., 2014). *R. glutinis* is a yeast producer of lipids and carotenoids, the carbon source has an important role in the biosynthesis of secondary metabolites, determining the kind and the yields of production.

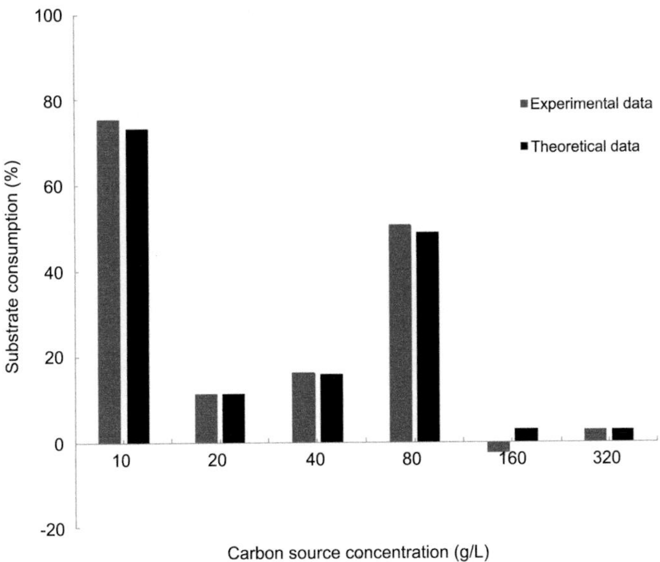

FIGURE 12.3 Substrate consumption (%) by *R. glutinis* P4M422.

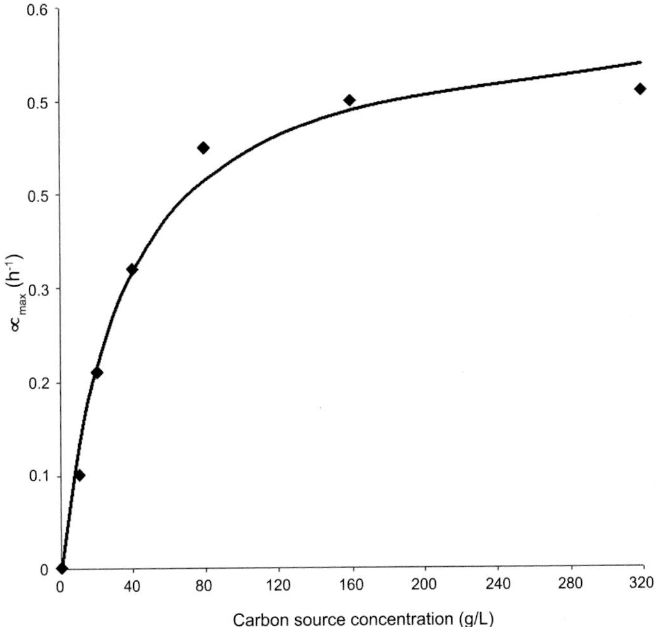

FIGURE 12.4 Specific growth rate during cultivation of *R. glutinis* P4M422 under different concentration of carbon source.

Furthermore, the productivity and specific production (Figure 12.5) of carotenoids indicate that dextrose at 40 g/L allows the major yield; therefore, Table 12.1 shows the kinetic parameters of growth and productivity of carotenoids by *R. glutinis* P4M422 using dextrose 40 g/L like carbon source. The $Y_{p/x}$ value decrease when the carbon source is increased, this fact could be due to the excess of a substrate and few cells to convert it into carotenoids (Colet et al., 2014).

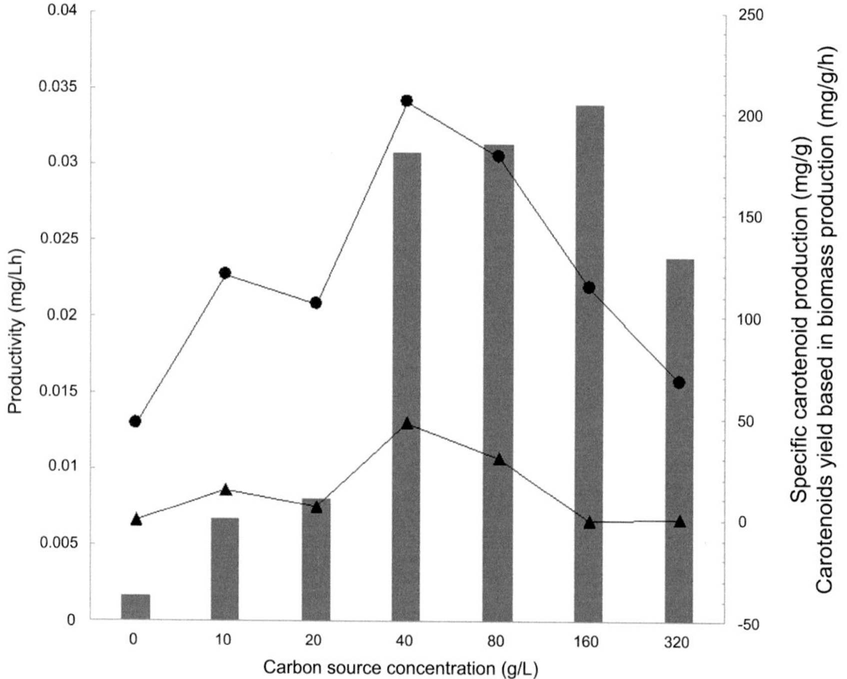

FIGURE 12.5 Productivity (g/L h), specific total carotenoids production (▲), and the total carotenoids yield based on biomass production (●) during cultivation of *R. glutinis* P4M422 under different concentration of carbon source.

TABLE 12.1 Kinetic Parameters of Growth and Substrate Consumption by *R. glutinis* P4M422

	X_{max}	X_0	□	$Y_{x/s}$(g/g)	q_s	$Y_{p/s}$
Carotenoids production	0.01250	0.00050	0.23130	0.00166	0.00038	0.34257

12.4 CONCLUSION

Carbon source concentration is an important factor in microbial production of carotenoids. The highest concentrations of carbon source can inhibit the carotenoid production due to the osmotic pressure present in the culture broth. In the present study, the effect of dextrose concentration has not a proportional relation between the biomass concentration and carotenoid production. We found that the optimal carbon concentration by this experiment was at 40 g/L, however, to industrial-scale this bioprocess the evaluation of other factors is necessary.

ACKNOWLEDGMENTS

The author Hernández-Almanza thanks to CONACYT (National Council of Science and Technology) for the financial support given to her postgraduate studies at DIA-FCQ (UAdeC).

KEYWORDS

- **glucose**
- **growth rate**
- ***Rhodotorula glutinis***

REFERENCES

Alves, L., Cardoso, D. C., Yuri, K., Kanno, F., & Karp, S. G., (2017). *African Journal of Biotechnology Microbial Production of Carotenoids: A Review, 16*, 139–146.

Cardenas-Toro, F. P., Alcázar-Alay, S. C., Coutinho, J. P., Godoy, H. T., Forster-Carneiro, T., & Meireles, M. A. A., (2015). Pressurized liquid extraction and low-pressure solvent extraction of carotenoids from pressed palm fiber: Experimental and economical evaluation. *Food Bioprod. Process., 94*, 90–100.

Cardoso, L. A. C., Jäckel, S., Karp, S. G., Framboisier, X., Chevalot, I., & Marc, I., (2016). Improvement of *Sporobolomyces* ruberrimus carotenoids production by the use of raw glycerol. *Bioresour. Technol., 200*, 374–379.

Colet, R., Di Luccio, M., & Valduga, E., (2014). Fed-batch production of carotenoids by *Sporidiobolus salmonicolor* (CBS 2636): Kinetic and stoichiometric parameters. *Eur. Food Res. Technol., 240*, 173–182.

Cutzu, R., Clemente, A., Reis, A., Nobre, B., Mannazzu, I., Roseiro, J., & Lopes, D. S. T., (2013). Assessment of β-carotene content, cell physiology and morphology of the yellow yeast *Rhodotorula glutinis* mutant 400A15 using flow cytometry. *J. Ind. Microbiol. Biotechnol., 40*, 865–875.

Dreywood, R., (1946). Qualitative test for carbohydrate material. *Ind. Eng. Chem. Analytical Ed., 18*, 499–505.

Hernández-Almanza, A., Montañez-Sáenz, J., Martínez-Ávila, C., Rodríguez-Herrera, R., & Aguilar, C. N., (2014). Carotenoid production by *Rhodotorula glutinis* YB-252 in solid-state fermentation. *Food Biosci., 7*, 31–36.

Kumar, S. R., & Young-Soo, K. B., (2017). Progress in microbial carotenoids production. *Indian J. Microbiol., 57*, 129–130.

Marova, I., Carnecka, M., Halienova, A., Certik, M., Dvorakova, T., & Haronikova, A., (2012). Use of several waste substrates for carotenoid-rich yeast biomass production. *J. Environ. Manage., 95*, S338–S342.

Rivera, E. C., Yamakawa, C. K., Garcia, M. H., Geraldo, V. C., Rossell, C. E. V., Filho, R. M., & Bonomi, A., (2013). A procedure for estimation of fermentation kinetic parameters in fed-batch bioethanol production process with cell recycle. *Chem. Eng. Trans., 32*, 1369–1374.

Thanapimmetha, A., Suwaleerat, T., Saisriyoot, M., Chisti, Y., & Srinophakun, P., (2017). Production of carotenoids and lipids by Rhodococcusopacus PD630 in batch and fed-batch culture. *Bioprocess Biosyst. Eng., 40*, 133–143.

Zhang, Z., Zhang, X., & Tan, T., (2014). Lipid and carotenoid production by Rhodotorula glutinis under irradiation/high-temperature and dark/low-temperature cultivation. *Bioresour. Technol., 157*, 149–153.

Zhao, C., Xu, Q., Lin, S., & Li, X., (2013). Hybrid differential evolution for estimation of kinetic parameters for biochemical systems. *Chinese J. Chem. Eng., 21*, 155–162.

Index

α

α-amylases, 88, 144
α-carotene, 72
α-galactosidase, 129, 142

β

β-1,4 glycosidic bonds, 4, 9
β-carotene, 254
β-cryptoxantin, 72
β-galactosidase, 166
β-glucosidase, 4
β-hydroxybranched fatty acids, 205
β-mannanase, 142, 150
β-thioglucosides, 130
β-xylosidades, 5

γ

γ-cyclodextrin, 21
γ-terpinene, 76

2

2,2-diphenyl-1-picrylhydrazyl free radical scavenging method (DPPH), 75

3

3, 5-diglucoside, 119
3D modeling, 149
3-hydroxy-2-metilvalerate, 180
3-hydroxybutyrate (3HB), 179
3-hydroxyvalerate (3HV), 179

4

4-O-(b-galactopyranosyl)-D-sorbitol, 165
4-O-bD-galactopyranosyl-D-fructose, 165
4-θ-β-D-galactopyranosyl-D-glucose, 164

5

5,7,3'-trihydroxyflvanone, 90
5,7,3-trihydroxy-3-isobutyroylflavanonol, 90
5,7-dihydroxy-3'-methoxyflavanone, 90

A

Acetic aldehyde, 69
Acetobacter pasteurans, 186
Acetogenins, 86
Acetone, 51, 70, 76, 120, 255
Acidic phenols, 110
Acinetobacter, 208, 211
Acoustic cavitation, 69
Acquired immunodeficiency diseases, 50
Actinobacillus succinogenes, 187
Aerobic
  anaerobic fermentation, 180
  digestion, 180, 189
Aglycones, 89, 137
Agri-food
  activities, 14
  residues, 17
Agrobacterium spp., 211
Agro-food
  cell structures, 3
  residues, 2–4, 6, 8–11, 13, 14, 22, 145–148, 150, 151
  waste, 141, 145, 146, 149
Agro-industrial, 35, 36, 51, 52, 69, 82, 107, 117, 226, 125
  byproducts, 35, 36, 51, 54
  residues, 10, 13
  waste, 107, 225
Akkermansia, 92
Alanine, 131
Albedo, 62
Albumins, 128
Alcoholic drinks (whey beer, whey wine, and whey champagne), 173
Aldehyde dehydrogenase, 89
Alkaline isomerization, 165
Alkaloids, 16, 86, 89, 109–111, 116, 117, 120, 145
Alzheimer's disease, 42, 50
Amentoflavone, 84
Amino acids, 34, 88, 108, 128, 146, 185, 187, 206, 238

Amphiphilic compounds, 203
Amphoteric properties, 169
Amylase, 14, 142, 143, 183, 184
Anacardiaceae, 34
Anaerobic
  culture, 176
  digester, 178
Anaerobiospirillum succiniciproducens, 187
Analytical procedures, 255
Androgendependent and independent prostate, 134
Animal feed, 84, 116, 141, 142, 144, 145, 147–149, 151, 159, 163, 180
  enzyme, 151
  industry, 141
  supplementation, 142
Animal nutrition, 147, 151
Ankaflavin (*Monascus* sp.), 254
Annona muricata, 86
Anthocyanidins, 10, 91, 110
Anthocyanins, 10, 62, 51, 65, 72, 83, 108, 110, 117, 119, 120, 130, 132, 133
Anthraquinone, 254
Anthrone method, 255
Anti-apoptotic, 44, 244
Anti-atherosclerotic, 244
Antibacterial
  agents, 45
  effects, 71
Anti-biotic, 45
Anticancer, 34, 35, 40, 42, 43, 53, 54, 66, 73, 83, 85, 88, 90, 112, 117, 121, 242, 244, 253, 254
  activity, 43
  capacities, 42
  effects, 42
Anticarcinogenic, 40, 42, 61, 109, 117, 120, 133, 241, 244
Antidiabetic, 34, 40, 45, 49, 53, 75, 83, 88, 89, 244
Antidiarrheal, 49, 53, 83, 89, 116, 117, 120
  activities, 34
Antidisenteric, 116
Antifibrotic, 244
Antigenic epitopes, 129
Antihemolytic, 34, 45, 53
Antihepatotoxic, 42
Anti-inflammatory, 40, 42, 61, 63, 65, 71–74, 83, 85, 88, 90, 120, 121, 126, 134, 135, 137, 241, 244, 253, 254,
Anti-litogenetic, 88
Antimicrobial, 9, 34, 35, 40, 43, 45, 47, 50, 53, 54, 65, 75, 76, 83, 88, 89, 109, 112, 116, 117, 120, 126, 202, 205, 208, 225–227, 229, 230, 244
  activity, 45, 75, 225, 227
  agents, 35, 43
  resistance, 43
Anti-nutrient compounds, 142
Antinutrients, 141
Antinutritional characteristics, 241
Antioxidant, 2, 9, 10, 34, 35, 40, 42, 43, 49–54, 61, 63, 64, 70, 75, 82, 83, 86, 88–91, 107, 108, 110, 112, 117, 119–121, 126, 130, 135, 137, 138, 185, 226, 233–237, 240–242, 244, 253, 254
  activity, 43, 51, 53, 75, 135, 137, 240–242, 244
  capacity, 70, 75, 108, 120, 121
  effects, 91
  properties, 40
Antiparasitic, 83, 84
Antipyretic, 84, 89
Antiseptic, 83
Antiviral, 9, 40, 42, 45, 73
Aqueous two-phase systems (ATPS), 14–18, 20–22
Arabinose, 5, 129, 131
Arabinoxylans, 148, 150
  degradation, 5
Arabinoxylanases, 5
Arachidonic, 120
Aroma compounds, 171
Aromatic
  capacities, 66
  plants, 65
  ring, 64, 110, 131
Artemisia absinthium, 111
Arthrobacter protophormiae, 209
Arthrofactine, 208
Ascaris lumbricoides, 118
Ascorbic acid, 35, 135
Ashbya gossypi, 254
Aspergillus, 17–19, 52, 71, 147, 149, 180, 185, 186, 225, 226

niger, 145, 147, 183, 184, 186, 231
oryzae, 145, 149, 150, 180
terreus, 184
ustus, 211
Ataulfo, 34, 38, 233–235, 237–242, 245
mango seed, 235, 240, 245
variety, 235, 242
Aurontioideae, 62
Azohydro monaslata, 181
Azotobacter chrococcum, 181

B

Bacillus, 18, 21, 46, 76, 145, 180, 181, 213, 215
cereus, 21, 46, 76
licheniformis, 208, 209
subtilis, 76, 206, 211
Bacteria inhibition, 93
Bacterial
deterioration, 170
emulsifier, 215
resistance, 43
species, 137, 203, 208
Bacteriocin, 159, 160, 182, 183, 191
Bakery products, 164
Benzene rings, 73
Benzophenones, 39
Beverages, 91, 116, 160, 169–173, 190, 202
B-glucosidase, 10, 13
Bifidobacterium, 92, 93, 172
animalis, 172
Bioactive, 2–4, 9–11, 13, 14, 22, 34–36, 45, 49–51, 53, 61, 63, 67, 69, 70, 72, 74, 76, 81–87, 90, 107–109, 116, 119, 121, 126, 134–137, 171, 225, 229, 233, 234
characteristics, 50
compound, 2, 3, 10, 11, 14, 22, 35, 36, 45, 49–51, 53, 61, 63, 67, 69, 70, 72, 74, 81–87, 90, 107–109, 116, 119, 121, 126, 134–136, 225, 229, 233, 234
hydrophilic compounds, 70
molecules, 14, 69, 135, 137, 171
phenolics, 9
source, 11
substances, 63, 109
Bioavailability, 43, 53, 85, 89, 136, 138, 203, 214, 215, 219
Biocatalyst, 21
Biochemical
oxygen demand (BOD), 162, 163, 175, 191
processes, 254
Bioconversion, 20–22, 138, 159, 191
Biodegradability, 209
Biodegradable
packaging, 235
substrate, 162
Biodegradation, 1, 3, 4, 11, 62, 201, 212–216, 218, 219
bioremediation, 212
Biodiesel, 176
Biodiversity, 162
Bioelectricity, 159, 189–191
Bioethanol, 63, 66, 71
production, 63, 71
Biofuels, 16, 21, 159, 160, 175, 179, 190, 191
industrial purposes, 21
industries, 16
Bioinformatics analysis, 149
Biological
activities, 9, 34–36, 53, 125
methods, 8
remediation, 214, 215
transformation, 214
waste, 36
Biomass, 1–3, 14, 17, 22, 149, 177, 180, 185, 188, 219, 226, 255–257, 260, 261
based feedstock, 146
concentration, 256
production, 1, 188, 255, 257, 260
Biomolecules, 15–17, 21, 182, 218, 219
Biopolymers, 159, 160, 179, 180, 182, 191
Bioprocess, 1–3, 8, 10, 12–14, 17, 22, 75, 180, 182, 183, 191, 225, 226, 235, 253, 254, 261
Bioprocessed agro-food residues, 3
Bioproducts, 2, 3, 15–18, 21, 22, 34, 146, 159
Bioremediation, 201, 202, 205, 212, 214–216, 218–220
Biosurfactants, 201–220
biodegradation studies, 214
environment, 212
producing microorganisms, 208, 211
v/s synthetic surfactants, 210
Biosynthesis, 226, 258

Biotechnological, 2, 14, 17, 22, 63, 71, 142, 145–147, 151, 159, 160, 162, 163, 175, 180, 253, 254
applications, 71
process, 17, 22, 175, 147, 160, 162, 163, 180
valorization, 14
value, 145
Biotechnology, 2
Biotransformation, 16, 182
Blakeslea trispora, 254
Blood cholesterol levels, 120
Brassica
genus plants, 127
juncea, 148
oleracea var
botrytis, 130
capitata, 130
gemmifera, 130
italica, 130
phytochemicals against cancer, 133
rapa subsp rapa, 130
species, 132, 135
vegetables, 126, 129, 133, 135
Brassicaceae, 130, 134
Breast cancer cells, 44
Brettanomyces anomalus, 188
Brevibacterium, 214
Broccoli, 125–127, 129, 130, 132, 137
Bronchial conditions, 116
Bulgaricus, 172, 185
Butanol fractions, 75
Butyl-hydroxy-toluene (BHT), 53, 108, 242
Byproducts, 35, 52, 90, 234, 235, 245

C

Cabbage, 125–127, 130, 132, 136
Caffeine, 16, 111, 148
Calamondin, 75, 76
peel, 75
pulp, 75
Calpain quantities, 75
Cancer cells, 43, 44, 117, 133–135
Candida, 175, 176, 185, 203, 208, 213
batistae, 211
bombicola, 210, 211
Candida
ishiwadae, 211
lipolytica, 186, 210, 211
Cantaloupe, 84
Capparaceae, 130
Capsicum annuum, 88
Carbohydrases production, 149
Carbohydrate
groups, 210
monomers, 85
Carbohydrates, 15, 21, 34, 42, 62, 66, 71, 83, 84, 88, 94, 129, 145, 204, 236
Carbohydrotases, 148
Carbon source, 4, 17, 159, 179, 180, 182, 208, 253–255, 257–261
Carboxylic acid, 6, 237
Carboxymethyl cellulose, 172
Carcinogenesis, 43, 75, 134
Cardiac glycosides, 110
Cardioprotective, 244
Cardiovascular
diseases, 117, 120, 126
disorders, 50
protection, 40, 45, 241
system, 63, 121
Carotenoids, 62, 65, 72, 75, 88, 126, 130, 171, 253–258, 260, 261
production, 253, 255, 257
Carpellary membranes, 117
Casein coagulation, 160
Cassia fistula, 89
Catalases, 64
Catalytic hydrogenation, 165
Cauliflower, 125–127, 129, 130, 137
Cell cycle arrest, 133, 134
Cell's molecules, 64
Cellobiohydrolase, 4
Cellobiose, 4
Cellulase, 4, 8, 10, 13, 17, 21, 142, 148, 149
Cellulolytic, 71
Cellulose, 4, 5, 7, 8, 14, 21, 37, 66, 68, 129, 141, 143, 145, 149, 169, 228, 236, 238, 239
Cerastes cerastes, 45
Cerilipin, 208
Chalcones, 64
Chelate, 65
Chemical
oxygen demand (COD), 159, 162, 163, 175, 179, 185, 188–191
processes, 8, 108

Chemokines, 135
Chemo-preventive, 134, 244
  effects, 134
  properties, 125, 138
Chili (*Capsicum annuum*), 88
Chitin, 4, 7–9, 14
Chitinases, 8
Chitin-deacetylase, 9
Chitooligomers, 9
Chitooligosaccharides, 8, 9
Chitosan, 4, 8, 9
Chitosanase, 9
Chlorogenic acid, 90, 93, 110
Chondroitin, 9
Chromatography, 37, 88, 90, 225, 227, 237, 240
Chymotrypsin, 144
Citric acid, 62, 116, 161, 171, 185, 186
Citrus
  aurantifolia, 62, 111
  limon, 62, 76
  paradisi, 62
  peels waste, 75
  reticulata, 62
  sinensis, 62
  waste, 63, 70, 71, 74–76
    human health, 74
Clean-up technology, 216
Cloacibacillus, 92
Cnidoscolus aconitifolius, 111
Coffee, 90
Cohobation, 15
Colchicine, 111
Colorectal cancer, 126
Concentrated whole whey (CW), 163, 164
Condensed tannins, 12, 40, 83, 85, 110, 145, 241
Conventional extraction techniques, 51, 52
Coprinopsis cinérea, 149
Corynebacterium, 205, 211
Cosmetic coloring agents, 253, 254
Cosmetics, 10, 42, 61, 63, 66, 109, 160, 182, 202
Coumarins, 62, 72, 75
Creosote bush (*Larreta tridentata*), 81, 89
Critical
  humidity point (CHP), 118
  micelle concentration (CMC), 209, 211
Cruciferae, 125, 132
Cruciferous, 125–127, 129, 130, 132–136, 138
Crude
  oil polluted soil samples, 215
  protein (CP), 236
Cryptococcus
  curvatus, 177
  laurentii, 177
Crystallization, 164, 170, 174
Cultivation, 70, 112, 114, 127, 238, 259, 260
Curcuminoids, 91
Cyanidin, 90, 119, 133
  3-glucoside, 119
Cyanogenic glycosides, 110, 111
Cyclic lipopeptide, 206, 208
Cyclooxygenase-2 expression, 134
Cysteine, 128
Cytokines, 135
Cytosol, 242
Cytotoxic
  activity, 120
  effects, 43, 49
Cytotoxicity mechanisms, 135

D

Dairy
  cheese industries, 175
  products, 160, 174, 182, 185
Deacetylated glucosamine, 9
Deacetylation, 9
Decarboxylation, 137
Degradation products, 1, 4
Delocalization, 64
Delphinidin 3-glucoside and 3,5-diglucoside, 119
Demethoxylated pectin, 6
Demineralization, 164
Dephytinization, 148
Deproteinization, 173
Diafiltration (DF), 164, 169
Diarrhea, 49, 109, 116, 118
Dietary fiber, 66, 72–75, 86, 129, 233, 235, 239
Dihydrogen phosphate, 132
Diosmetin, 62
Diphenylpyranes, 65
Disaccharidases, 3
Drosinos, 136

E

Eco-friendly approach, 202
Ecological protocols, 70
Effleurage, 68
Eicus coloratus, 45
Electrodialysis (ED), 164
Electrospray, 37, 237
Electrostatic, 16
Ellagic acid, 3, 10, 11, 13, 14, 20, 34, 37, 39, 40, 88, 117, 118, 121, 244
Ellagitannase, 13, 14
Ellagitannins, 10–12, 40, 83, 92, 117, 119, 145, 242, 244
Emulsifier, 164, 208, 215, 216
Emulsion formation emulsion breaking, 210
Endogenic myrosinase, 137
Endoglucanase, 4, 14, 150
Endo-inulinase, 7
Endoxylanases, 5
Energy
  requirements, 15, 146
  sources, 163, 219
Enfleurage, 15
Enterobacter cloacae, 190
Enterococcus faecalis, 46, 76, 183
Environmental
  impact, 159, 160, 163, 175, 191
  pollution, 201
  problem, 163, 189
  protection agency, 52
Environment-friendly, 146
Enzymatic
  activity, 7, 145, 150
  antioxidants, 64
  depolymerization, 12
  modification, 163
  non-enzymatic mechanisms, 64
Enzymes, 2–5, 7, 10–14, 16–18, 22, 49, 52, 62, 71, 88, 92, 129, 130, 133–137, 141, 142, 144–151, 159, 160, 180, 182, 183, 191, 225, 226
  activity, 142, 148, 150
  animal nutrition, 142
  assisted extraction (EAE), 52
  degradation, 2
  metabolite production, 3
  produced by whey, 184
  producers, 183
  production, 2, 3, 10, 13, 147–149
  selection, 149
  trypsin, 144
Epicatechin, 88, 90, 145
Epigallocatechin gallate, 145
Eriocitrin, 65
Escherichia coli, 47, 48, 76, 119, 180, 181, 190, 227, 229, 230
Essential oils, 65
Essential or non-essential compound, 109
Ethanol, 21, 35, 51, 70, 71, 175, 176, 186, 227, 229, 236
Ethanolic extracts, 76, 225, 227
Ethnobotanical studies, 35
Ethyl acetate fraction, 75
Eukaryotes, 94
Eutrophication, 162
Evaluation functional of mango seed, 236
Exo-cellular polymeric biosurfactants, 208
Exoglucanase, 4
Exo-inulinase, 7
Extracellular ionic-surfactants, 203
Extractive bioconversion method, 22

F

Fatty acids, 34, 65, 74, 81, 108, 117, 119, 120, 142, 177, 179, 205, 208, 210, 233–235, 240, 245
Fatty tissues, 120
Fermentation, 2, 8, 17, 52, 71, 92, 94, 125, 126, 129, 136–138, 141, 145–151, 163, 172, 173, 176–180, 183, 185, 186, 188, 191, 205, 225–229, 255, 256
  conditions, 255
  deproteinization, 8
  procedure, 149
  process, 137, 146, 188, 226
  products, 71
Fermented whey beverages, 172
Fertility, 163
Fertilizers, 202
Filamentous fungi, 146, 149, 150, 226, 254
Final comments, 22
Fisetin, 110
Fish diet, 162
Flavanols, 64, 65
Flavanone, 65, 72, 110, 133
Flavans, 10

Flavedo, 62
Flavones, 3, 10, 64, 65, 72, 73, 85, 91, 110, 133
Flavonoid, 3, 10, 37, 40–42, 53, 62, 64, 65, 71–73, 75, 76, 83–91, 107, 110, 126, 130, 132, 133, 135, 230, 242, 244, 245
  aglycone, 90
  derivatives, 85
  glucosides, 137
  type, 87
Flavonols, 10, 72, 91, 132
Flavonone, 110
Flourensia rethinophylla, 90
Folin-Ciocaleteu
  assay, 72
  method, 88
Food
  applications, 161, 164, 191
  consumption, 91
  feed ingredients, 49
  industries, 4, 16, 17, 66, 82, 145, 208, 212, 234
  industry, 1, 2, 6, 35, 43, 81, 84, 111, 120, 163, 164, 183, 185, 241
  ingredients, 49, 125, 233
  pharma, 209
  security and nutrition, 245
  waste, 36, 63, 234, 245
    streams, 75
Forecast enzymes production, 151
Fortunella crassifolia, 76
Fructooligosaccharides, 7
Fructose, 6, 7, 62, 71, 129, 166, 186, 228
Fruit
  beverages, 170
  fermentations, 126
  industry, 114
  juices mixed with whey, 170
Functional ingredients, 70
Fungal
  degradation, 229
  species, 203
Fusarium moniliforme, 186

G

Galactooligosaccharides (GOS), 166, 167
Galactose, 5, 129, 165–167, 183
Galactosidase, 19, 165, 173, 183, 188
Galacturonic acid, 6, 66, 129
Gallate, 39, 145, 244
Gallic, 3, 11, 20, 34, 40, 41, 44, 53, 84, 86, 88, 90, 108, 117, 118, 145, 244
  acid (GA), 40, 41, 45, 46, 244
Gallotannins, 10, 11, 37, 39, 40, 53, 83, 242, 244
  ellagitannins, 11
Gas chromatography (GC), 37, 233, 235, 236, 240
  mass spectrometry (GC-MS), 37
Gastric cancer, 134
Gastrointestinal tract, 85, 93, 129, 171
Gastroprotective, 244
Gelation effect, 6
Gene fusion strategy, 149
General chemical composition, 161
Genetic engineering, 150
Genotypic, environmental factors, 37
Geraniales, 62
Glioblastoma, 134
Global production, 162
Globose shape, 112
Globulins, 127, 128
Glomerular structure, 128
Glucanase, 142, 143, 150
Glucoiberin decomposition, 134
Glucoraphanin, 131, 132
Glucosamine, 8, 9
Glucose, 4, 5, 7, 10, 37, 39–41, 66, 71, 75, 84, 86, 88, 129, 131, 145, 147, 165–167, 178, 183, 184, 186, 187, 189, 228, 243, 244, 261
  galactose, 166
  isomerase, 166
  metabolism, 84
Glucosinolates (GLS), 110, 125, 126, 130–132, 134–138
  found in *brassica*, 131
Glucosyl moiety termination, 6
Glucuronic acid, 5
Glucuronoxylans, 5
Glutathione, 64, 134
  peroxidases, 64
Glycolipid, 204
Glycosidases, 3, 13
Glycoside, 73, 88, 109, 110, 130
  hydrolases, 10

Glycosidic, 4, 5, 8–10, 66, 110, 206
  bonds, 4, 5, 8, 66
  core, 10
Glycosyl hydrolases, 137
Grains, 81, 84, 85, 91, 92, 95, 109, 112, 116
Gramicidine, 208
Gram-negative, 45, 230
  bacteria, 45
Grampositive, 45, 206, 230
Granato pommel, 112
Grano grain melo, 112
Grape pomace, 90, 91
Grapefruit residues, 228
Growth rate, 182, 256–259, 261

H

Haloferax mediterranei, 181
Helical conformation, 129
Helicobacter pylori, 134
Hemicellulose, 4, 5, 21, 37, 66, 129, 141, 145, 236, 238
Hemicellulosic structures, 5
Hemolysis, 45, 49
Hepatic encephalopathy, 165, 166
Hepatoprotective agents, 50
Herbalism, 108, 109, 118
Herbicides, 108
Hesperidin, 62, 65, 72, 73
Heterocycle, 110, 111
Heterogeneous, 62, 66, 110
Heteropolysaccharide, 5
Hexoses, 66
High performance liquid chromatography (HPLC), 35–37, 39, 88, 90, 233–237, 242, 243, 245
Homogalacturonan, 66
Homogeneous, 70, 188
Homopolysaccharide, 4
Hondrodimou, 136
Horticultural production, 126
Human
  consumption, 163
  diseases, 91
  health, 35, 40, 43, 74, 167, 238
  immunodeficiency virus, 40
  platelet aggregation, 73
Husks, 13, 61, 83, 95, 119, 146
Hydraulic retention, 178, 179
Hydro diffusion, 70
Hydrocarbon
  dissolution rate, 203
  molecules, 203, 212, 214
Hydrodistillation, 51, 65, 69
Hydrogen, 16, 43, 129, 177–179, 241
  atom transfer (HAT), 43
Hydrogenogenic reactor, 178
Hydrolytic enzymes, 151
Hydrolyzable tannins, 83, 110
Hydrolyzation, 125
Hydrophobic
  comportment, 16
  contaminants, 201, 202
  domain, 210
  pollutants, 202, 219
  substrates, 203
Hydrophobicity, 128, 202, 214, 219
Hydroquinone type, 87
Hydroxybenzoic, 40
  acid, 40, 65, 91
  hydroxycinnamic acids, 40
Hydroxycinnamic acid, 40, 65, 91, 132–134, 243
Hyperammonemia, 165
Hypocholesterolemic, 75
Hypoglycemic, 49, 72, 74
Hypolipidemic, 72, 74, 244
Hypotriglyceridemic effect, 75

I

Iberin, 134
Immune response, 66, 135
Indigenous bacteria, 190
Indole, 111, 131, 136
Industrial processing, 35, 235
Inflammatory mediators, 135
Infrared drying, 51
Inorganic nutrients, 214, 217
Inositol phosphate, 144
Insecticides, 108
Insulinotropic properties, 74
Intestinal
  microbiota, 93, 94
  parasites, 116, 118
Intracellular metabolites, 226
Inulin, 4, 6, 7
Inulinases, 7

Ion exchange, 164
  chromatography, 15
Isoflavones, 64, 65, 91, 133
Isoleucine, 131
Isomers, 117
Isopeletierine, 116
Isorhamnetin, 133
Isothiocyanates (ITCs), 132–135, 138

J

Juice extraction, 71

K

Kaempferol, 88, 133
Kernel cake, 149, 150
Kinetic
  parameters, 253–257, 260
  production of carotenoids, 256
Klebsiella, 47, 48, 92
  aerogenes, 47
  pneumoniae, 48
Kluyveromyces, 173, 175, 176, 185, 186, 188
  fragilis, 173, 184, 188
  lactis, 180, 184
  marxianus, 180, 184, 188
Kumquat, 75

L

Lactic acid, 8, 125, 136–138, 161, 172, 183, 185
  bacteria (LAB), 125, 136–138, 183
  fermentation, 126, 136, 137
Lactic fermentation, 8
Lactitol, 165, 166
Lactobacilli, 125, 137, 160
Lactobacillus, 8, 92, 93, 185
  acidophilus, 136, 172, 187
  brevis, 136, 137
  bulgaricus, 180, 187
  casei, 187
  delbrueckii, 172, 187
  fermentum, 137, 211
  helveticus, 186
  plantarum, 136, 137
  rhamnosus, 136, 172
  strains, 8
Lactochrome, 160
Lactococcus lactis, 180, 183, 187
Lactose, 161, 162, 164, 174, 177, 184, 186, 187
  hydrolysis, 173, 180
Lactulose, 165
Landfilling, 35
Larreta tridentata, 89
Leaves, 19, 65, 81, 83, 86–89, 95, 109, 112, 113, 116
Leucine, 131
Leuconostoc mesenteroides, 136
Leukemia cancer cells, 44
Leukotriene B4, 89
Lichenysin, 206, 208, 209
Lignans, 89, 91
Lignin, 37, 71, 110, 129, 145, 236
Lignocellulose, 145
Lignocellulosic
  feedstock, 71
  fibers, 3
Limonene, 65, 71, 75, 76
Limonic acid, 62
Limonin, 62
Limonoids, 62, 72, 75
Linoleic, 120, 240
Lipases, 3, 8, 142, 144
Lipid
  metabolism, 120
  oxidation inhibition (LOI), 237, 240, 241
Lipids, 42, 50, 62, 74, 84, 108, 142, 175, 177, 204, 211, 234, 236, 238, 240, 254, 258
Lipomyces starkey, 177
Lipopeptides, 204, 206, 207
Lipoprotein biosurfactants, 209
Liquid
  chromatography, 37
    ion trap mass spectrometer, 237
  cultures, 12, 13
  fermentation (LF), 17, 18, 22, 161
  liquid extraction methods, 16
Logistics model, 256
Low-density lipoprotein (LDL), 136
Luedeking-Piret model, 256
Lutein, 62, 72
Lycopene, 62, 72, 254
Lymphocyte, 91

M

Maceration, 15, 51, 52, 67, 68, 82
Macroemulsion, 210
Macular degeneration, 72
Malic acid, 62
Malonyl glycosides, 88
Mangifera, 34, 234, 235
  caesia, 34, 235
  foetida, 34
  indica, 34, 54, 245
Mango
  byproducts, 51, 234, 235, 244
  fruit, 35
  processing, 233
  seed, 33–37, 39–43, 45, 47, 49–54, 233–242, 244, 245
    kernels, 238
  waste, 35, 54
Mannanase production, 150
Mannheimia succiniciproducens, 187
Mannose, 5
Mass spectrometry (MS), 35–37, 39, 233–237, 240, 242, 243, 245
Mayonnaise, 190
Medical application, 50
Melogranato, 112
Mescaline or colchicine, 111
Mesophilic anaerobic bioreactor, 178
Metabolic respiration, 231
Metagenome sequencing, 94
Methanol, 47, 51, 70, 76
Methanolic extracts, 108, 118
Methionine, 127, 128, 131
Methoxyl groups, 6
Methyl
  gallate, 34, 41, 137, 145
  pelletierine, 116
Methylobacterium, 181
Metschnikowia pulcherrima, 186
Micelles capture, 203
Microbes, 1, 2, 6, 8, 12, 92, 190, 202, 203, 208–210, 212, 214, 215, 217–220
Microbial
  biomass, 2, 3, 14, 185, 188
  carotenoids, 254
  cell, 203, 208, 219
  consortium, 189, 213, 215, 216
  degradation process, 3
  depolymerization, 4
  development, 13
  enhanced oil recovery (MEOR), 216, 217
  enzyme
    inducers, 11
    production, 7
  fuel cell (MFC), 180, 189, 190
  growth medium, 203
  process, 1–4, 6, 10–12, 17, 225
  production, 8, 256, 261
  species, 136
  strains, 13
Microbiological biochemical stability, 36
Microbiota, 82, 85, 86, 91–94
Microfiltration (MF), 39, 164, 169
Microflora, 89, 213
Micronutrients, 85, 129
Microorganism
  rely, 258
  strain, 182
Microwave
  assisted extraction (MAE), 52, 53, 67, 70, 72, 82, 108
  gravity, 70
  vacuum drying, 51
Milk
  casein proteins, 160
  production, 162
  whey, 160
Minerals, 8, 34, 61, 86, 117, 126, 129, 132, 144, 145, 159, 160, 163, 164, 168, 170, 173, 188, 191, 233–236, 238, 239, 245
Minimum inhibitory concentration (MIC), 46–48, 76
Molecular distillation, 52
Monogastric animals, 142, 144, 145, 147
Monomeric sugars, 5
Monosaccharides, 3, 5, 21
Monoterpenes, 65, 76, 109
Moringa, 86–88
  oleifera, 86
Mouthfeel, 171–173
Mucor, 147
Mycobacterium, 205, 211
Myoinositol, 144
  1–6 hexaquis, 132
Myrosin, 132
Myrosinase, 132

N

N-acetyl glucosamine, 7
N-alkanes, 215
Nanofiltration (NF), 133, 135, 164
Nanowires, 189
Narginine, 65
Naringin, 62, 73
Natural
  pigments, 64
  process, 219
  products, 108
  source, 49, 83, 87
Neuroblastoma, 134
Neurodegenerative diseases, 50
Neuroprotective effect, 90
Neutrallipids, 208
Nitriles, 132
Nitrogen nutrients, 180
Nobiletin, 65, 73, 75
Nocardia, 205
  erythropolis, 211
Non-aqueous phase, 203, 214
Nonenzymatic
  antioxidants, 64
  lipid peroxidation, 135
Non-essential compounds, 63
Non-flavonoid phenolics, 133
Non-starch polysaccharides (NSP), 142, 148, 150
Nucleotides, 108
Nutraceutical
  diet, 74
  food, 116
  therapy, 49
Nutraceuticals, 42, 168, 190
Nutritional
  composition, 175
    mango seed, 235, 238
  properties, 33, 136, 170, 172, 234, 245

O

Ohmic heating, 82, 108
Oleic, 62, 177, 234, 240
Oligomeric cruciferin structure, 127
Oligomers, 5, 117
Oligosaccharides, 4–7, 9, 21, 129
Olive pomace, 149
O-methylated flavone, 75
Ophiostoma piliferum, 149
Orange
  bagasse composition, 71
  peel waste (OPW), 71
Organic
  acids, 34, 43, 45, 62, 117, 146, 159, 160, 176, 182, 185, 191
  loading rate, 179
  supplement, 177
Organoleptic
  modifications, 136
  properties, 82, 85
Ornithine, 208, 211
  lipids, 211
Oxazolidinethiones, 132
Oxidation, 42, 49, 63, 65, 69, 91, 131, 135, 254
Oxidative
  damage, 50, 91
  stress, 42, 44, 64, 85, 88, 89, 92, 135
Oxidizing effects, 89
Oxygenases, 12
Oxygenated compounds, 65

P

P.aeroginosa, 205
Packed-bed bioreactor, 149
Palmitic, 62, 234, 240
Paramithiotis, 136
Parenchymatous tissue, 132
Parkinson's disease, 50
Parthenium hysterophorus, 111
Pasteurization, 171
Pectin, 4, 6, 62, 66, 71, 74, 76, 172, 228
Pectinase, 6, 71, 142, 148
Pectinesterases, 6
Pectinolytic enzymes, 71
Pectins, 61, 66, 234
Pediococcus acidilactici, 183
Pelargonidin 3-glucoside, 119
Pelletierin, 116
Penicillium, 19, 147
  oxalicum, 254
Penta-o-galloyl glucose (PGG), 34, 39–41, 44–46, 53, 54, 244
Pentoses, 66
Peptidases, 144

Percolation, 68, 82
Peroxidase activity, 148
Petro hydrocarbons, 202
Petrochemical contaminants, 201
Petrohydrocarbon products, 212, 215
Petro-hydrocarbons, 201–203, 208, 215
Petroleum
  depletion, 175
  derivatives, 175
Pharmaceutical
  biological effects, 74
  industry, 81, 111, 244
  products, 66
Pharmaceuticals, 42, 63, 71, 109, 212
Pharmacological properties, 43, 86, 87, 107, 112
Phenolic
  acids, 40, 53, 62, 75, 84, 85, 87, 88, 90, 91, 107, 110, 126, 132, 137, 245
  antioxidants inhibit oxidation reactions, 49
  components, 51
  compounds, 10, 16, 17, 19, 34, 36, 37, 43–45, 51–54, 65, 75, 81–94, 109, 110, 125, 126, 132–135, 138, 171, 230, 235, 242–245
    ataulfo mango seed, 243
  extracts, 52
  fraction, 126
  glycosides, 10
  origin, 107
  rings, 64
Phenolics, 2, 10, 16, 72, 75, 91, 125, 126, 132, 136, 138, 145
Phenyl benzopyrone structure, 73
Phenylalanine, 131, 242
Phenylpropanoids, 65, 110
Phloroglucionol, 62
Phosphatases, 142, 144
Phospholipid, 208
Phosphorus, 117, 132, 144, 234, 239
Photobacterium phosphoreum, 209
Photo-oxidative damage, 258
Physicochemical
  approach, 216
  stability, 147
Phytase, 148
  activity, 148
  production, 147
Phytases, 144, 147, 148
  producers, 147
Phytic acid, 132, 144
Phytoalexins, 134, 135
Phytochemical
  compounds, 37, 40, 82
  constituents, 45
  in mango seed, 37
  origin, 67, 72
  profiles, 34
Phytochemicals, 33–37, 51, 53, 65, 69, 87, 109, 125, 130, 133, 134, 136, 138, 245
  cruciferous vegetables, 130
Phytocompounds, 69
Phytoremediation, 216
Pichia pastoris, 19, 148
Pinene, 65, 76
Piper carpunya, 111
Piperidine, 111
Plant cell
  components, 3
  membranes, 69
  vacuoles, 9
Plant
  matrix, 52, 67, 70
  phytochemicals, 136
  processing, 14
  secondary compounds (PSC), 145
  waste, 4
Plasma, 45, 74, 75, 91
Pollutants, 2, 8, 162, 201–203, 212, 214, 219, 220
Polycyclic aromatic hydrocarbons (PAHs), 202, 213
Polyethylene glycol, 15, 22
Polygalacturonase, 6
Polygalacturonatelyase, 6
Polyhydroxyalkanoates (PHA), 179–182
  degradation, 182
  production conditions, 182
  productivity, 180
Polymer purification, 182
Polymeric biosurfactants, 208
Polymerization, 4
Polymethoxyflavones, 75
Polymethoxylated
  flavones (PMFs), 73
  polyphenols, 65

Polymyxa, 208
Polypeptide chains, 127, 129
Polyphenol, 18, 37, 41, 52, 91, 121, 236, 241
Polyphenolic, 15, 34, 43, 49, 53, 64, 73, 81–83, 108, 110, 117, 241
compounds, 237
molecules, 65
Polyphenols, 10, 13, 40, 43, 50, 61, 62, 64, 65, 72, 73, 76, 81, 84–95, 107, 108, 110, 117, 121, 130, 132, 135, 137, 144, 171, 233, 234, 237, 241
Polyphenolxanthone, 42
Polysaccharides, 3–7, 9, 15, 21, 50, 62, 66, 85, 108, 129, 179, 208, 228
Polyterpenes, 109
Polyunsaturated fatty acids, 120
Pomegranate (*Punica granatum*), 10, 11, 13, 14, 81–83, 92, 107–109, 111–121, 241
juice, 116, 119, 120
peel, 83, 117–119
plants, 114
production, 113, 115
residues, 11, 14
seeds and husk composition, 118
waste, 119
Post-harvest processing, 125, 137
Potassium, 21, 117, 130, 160, 179, 234, 239
Potato starchy wastes, 149
Prebiotic properties, 165
Prebiotics, 6, 7, 21, 81, 82, 86, 92, 95, 150
Pressurized
liquid extraction (PLE), 52
low polarity water extraction, 52
Proanthocyanidins, 3, 65, 83, 92
Probiotic bacteria, 82, 92, 94, 167, 171
Problematics, 63
Process
ex-situ, 212
in-situ processes, 212
Production of cruciferous, 126, 127
Proinflammatory mediator genes, 135
Propionibacterium, 46
freudenreichii, 187
Protease chymosin, 160
Proteases, 3, 8, 45, 142
Protein, 18, 127, 150, 160, 161, 168–170, 174, 211, 238
composition, 161
concentrate, 160, 162, 167, 170, 185
denature, 163
extraction/purification, 15
fraction, 127, 129
infant food, 164
solubility, 174
supplement bars, 170
Protocathetic, 84
Pseudomonas, 47, 48, 181, 203, 211, 213–215
aeruginosa, 47, 48, 211, 215
chlororaphis, 211
fluorescens, 211
hydrogenovora, 181
Pseudopeletierine, 116
Pulsed electric field (PEF) extraction, 52
Punic pommel, 112
Punica granatum, 82, 83, 107, 111, 112, 241
Punicalagin, 117–121
Punicalin, 117
Purification factors, 17
Purine, 111
Pyrrolizidine, 111

Q

Quercetin, 20, 37, 39, 42, 83, 84, 86, 88, 90, 93, 110, 133
Quinoline, 111
Quinolizidine, 111
Quinone
reductase, 134
structures, 64

R

Radish, 126, 127, 130
Raphanus sativus, 130
Raw material, 2, 66, 82, 149, 175, 220, 235
Residues, 1, 2, 4, 6, 8, 10–17, 22, 33, 51, 61–63, 66, 69, 70, 83, 95, 141, 145, 146, 148–150, 191, 215, 226–230, 234
Reverse osmosis (RO), 164
Rhamnetin, 34, 39, 42, 53
Rhamnogalacturonan I, 66
Rhamnogalacturonans, 6
Rhamnolipid, 205
Rhamnose, 5, 131, 205

Rheumatoid, arthritis, 50
Rhizopus, 147
Rhodotorula glutinis, 211, 253–255, 261
Riboflavin, 62, 160, 190, 254
Ricinus communis, 17
Ricotta cheese, 174, 177
Rutaceae, 62
Rutinosides, 88

S

Saccharification, 2
Saccharomyces
  cerevisiae, 175
  lactis, 173
Salad dressing, 164
Salmonella
  typhi, 47
  typhimurium, 47
Saponins, 110, 111, 145
Sedimentation, 129, 172
Seed kernel, 235, 236, 238, 245
Serratia
  marcescens, 184, 211
  rubidea, 211
Sesquiterpenes, 65, 109
Shell, 8, 9, 13, 66, 82, 83, 107, 108, 116–118, 120, 121, 141, 226, 237
Shigella
  dysenteriae, 48
  flexnerri, 47
Sinensetine, 65
Single
  cell protein (SCP), 3, 159, 185, 188, 189, 191
  electron transfer (SET), 43
Sinigrin, 134, 135
Soil
  matrix, 203, 214, 218
  porous matrix, 146
  state
    culture, 12, 13
    fermentation (SSF), 2, 14, 17, 19, 22, 52, 146–151, 180, 182, 226, 228–231
    fermentation process, 180
Solubilization, 162, 202
Soluble coffee wastes, 150
Solvent consumption, 69, 70
Sophorolipid, 206, 211, 215
Sorghum, 84, 85
Soups, 164
Sources of biosurfactants, 210
Soursop (*Annona muricata*), 86
Soxhlet, 51, 52, 67, 68, 82
  extraction, 51
  method, 52
Spectrophotometric
  analysis, 37
  technique, 36, 43, 242
Sporotrichum thermophile, 148
Spray drying process, 164
Staphylococcus aureus, 46, 47
Starmerella bombicola, 206
Statistical analysis, 237, 255
Stearic, 62, 234, 240
Stilbenes, 91
Stomach infections, 116
Strainor species-dependent, 126
Streptococcus, 172, 180, 184
  thermophillus, 172, 180, 187
Submerged
  cultures, 13
  fermentation, 151
Substrate consumption, 255–258
Succinic acid, 62
Sugarcane bagasse, 14, 146, 148, 150
Sulfur amino acids, 127, 129
Supercritical fluid extraction (SFE), 52
Superoxide dismutases, 64
Surfactin, 206, 211
Sustainability, 146, 177, 207, 245
Syzygium aromaticum, 241

T

Talaromyces leycettanus, 150
Tannin type, 87
Tannins, 10, 40, 41, 87–89, 91, 110, 117, 119, 120, 137, 143–145, 241, 244, 245
Taraxacum officinale, 111
Tarbush (*Cassia fistula*), 89
Taurine, 208
Technological
  aspects, 34–36, 51
  process, 125
Terpenes, 65, 109, 110
Terpenoids, 65, 109, 145

Terpinene, 65, 76
Theobromine, 148
Thermal
gradients, 69
inactivation, 137
treatments, 170
Thermolabile compounds, 68, 108
Thermophiles, 150, 184
Thermostability, 150
Thermus thermophilus, 181
Thiobacillus thioxidans, 208, 211
Thiocyanates, 132
Thioglucosidase enzyme myrosinase, 132
Thioglucoside, 135
Tinctures, 68
Tocopherol, 64, 129
Torulopsis bombicola, 211
Total petroleum hydrocarbons (TPH), 215
Toxicity, 44, 45, 49, 111, 177, 202, 209, 211, 234
Traditional processes, 163
Tranquilizers, 111
Transferase activity, 7
Transglycosylation reaction, 166
Trehalolipid, 205
Trichoderma, 71
longibrachiatum, 145
viride, 145
Trichosporon
cutaneum, 188
species, 185
Triterpenes, 109
Trolox, 53, 234, 237, 240, 242
equivalent antioxidant capacity (TEAC), 237, 241
Tropane, 111
Tryptophan, 131
Tumorigenesis, 133
Turnip, 127, 130, 132
Type 2 diabetes, 49, 135
Tyrosinases, 144
Tyrosine, 131

U

UDP-glucoronosyl transferases, 134
Ulcer healing, 244
Ultrafiltered, 168, 169
Ultrafiltration (UF), 164, 169
Ultrasonic waves, 69
Ultrasound, 51, 52, 67, 70, 82, 108
assisted extraction (UAE), 52
extraction, 69, 72
freeze-drying, 51
Unconjugated acids, 85
Unique properties of biosurfactants, 209
Uric acid, 161
Uronic acids, 66
Ursolic acid, 108, 118
UV radiation, 65

V

Vacuum evaporation, 168, 169
Valine, 131
Valoneic acid dilactone, 39, 244
Valorization, 14, 16, 17, 22, 36, 52, 61, 63, 136, 159, 175, 177, 180, 182, 191, 233, 234, 240, 245
Value-added, 16, 21, 22, 34, 72, 107, 141, 159, 160, 162, 163, 175, 180, 191, 245
Vascular plants, 111
Vasoconstriction, 110
Vegetable
fermentation, 136
sources, 73, 82, 108, 120
Vegetal cells, 242
Victivallis, 92
Vitamins, 34, 61, 62, 72, 75, 83, 86–88, 108, 117, 126, 129, 161, 163, 173, 206, 208, 211, 253, 254

W

Waste treatment, 189
management, 13
Water, 10, 12, 13, 15, 17, 38, 51, 70, 75, 76, 86, 109, 111, 117, 120, 129, 146, 162–164, 182, 189, 203, 208, 210, 214–217, 219, 225–227, 229, 255
absorption index (WAI), 118
contents, 170
quality, 162
retention capacity, 66
Waxes, 65
Whey, 141, 159–180, 182, 183, 185–191
beverages, 170, 172
buttermilk, 174

cheeses, 174, 175
permeate processing, 164
protein
concentrate (WPC), 162, 167–169
concentrate 35 (WPC 35), 168
concentrate 50 (WPC 50), 169
concentrate 80 (WPC 80), 169
isolate (WPI), 168, 169
retentate, 167
Whipping agents, 164
World Health Organization (WHO), 43, 49

X

Xanthan gum, 172
Xanthine oxidase, 135
Xanthona, 42
Xanthones, 37, 39, 40, 42, 111
Xenobiotics, 134
Xylan, 5, 20, 21, 62
Xylanase, 10, 13, 17–21, 142–144, 148, 149
Xylanolytic, 71
Xylansarabinoxylans, 5
Xylobiose, 20, 21
Xylogalacturonan, 66
Xyloglucans, 5
Xylose, 5, 21, 129
Xylotriose, 20, 21

Y

Yarrowia, 186, 203
lipolytica, 186
Yeast, 3, 4, 12, 136, 149, 159, 172, 173, 175–177, 183–188, 191, 203, 208, 210, 254, 255, 258
Yield measurement, 235, 237

Z

Zacatecas, 115
Zeaxanthin, 72
Zebda, 38